Mechanical Science: Models, Systems and Applications

Mechanical Science: Models, Systems and Applications

Editor: Rene Sava

NY RESEARCH
P R E S S

New York

Published by NY Research Press
118-35 Queens Blvd., Suite 400,
Forest Hills, NY 11375, USA
www.nyresearchpress.com

Mechanical Science: Models, Systems and Applications
Edited by Rene Sava

International Standard Book Number: 978-1-63238-634-2 (Hardback)

Cataloging-in-Publication Data

Mechanical science : models, systems and applications / edited by Rene Sava.
    p. cm.
Includes bibliographical references and index.
ISBN 978-1-63238-634-2
1. Mechanics, Applied. 2. Mechanical engineering. I. Sava, Rene.
TA350 .M43 2019
620.1--dc23

Contents

Preface

Mechanical science is the study of mechanical systems, their analysis, designing, structuring and manufacturing by integrating principles of physics, material science and engineering. Mechanical science has applications across various scientific fields. Its systems are used for purposes of designing structures of varying sizes, building automobiles, industrial equipment design, etc. With the rapid industrial growth in the last few decades, mechanical science as a scientific field of study has experienced a significant progress and has emerged as a multidisciplinary subject. Some of the diverse topics covered in this book address the varied branches that fall under this category. It elucidates the concepts and innovative models around prospective developments with respect to mechanical science. The extensive content of this book provides the readers with a thorough understanding of the subject.

This book is a result of research of several months to collate the most relevant data in the field.

When I was approached with the idea of this book and the proposal to edit it, I was overwhelmed. It gave me an opportunity to reach out to all those who share a common interest with me in this field. I had 3 main parameters for editing this text:

1. Accuracy – The data and information provided in this book should be up-to-date and valuable to the readers.

2. Structure – The data must be presented in a structured format for easy understanding and better grasping of the readers

3. Universal Approach – This book not only targets students but also experts and innovators in the field, thus my aim was to present topics which are of use to all

Thus, it took me a couple of months to finish the editing of this book.

I would like to make a special mention of my publisher who considered me worthy of this opportunity and also supported me throughout the editing process. I would also like to thank the editing team at the back-end who extended their help whenever required.

Editor

Design and analysis of a bio-inspired module-based robotic arm

Zirong Luo[1], **Jianzhong Shang**[1], **Guowu Wei**[2], **and Lei Ren**[3]

[1] School of Mechatronics Engineering and Automation, National University of Defence Technology, 410073 Changsha, China
[2] School of Computing, Science and Engineering, University of Salford, Salford, M5 4WT, UK
[3] School of Mechanical, Aerospace and Civil Engineering, University of Manchester, Manchester, M13 9PL, UK

Correspondence to: Zirong Luo (luozirong@nudt.edu.cn), Guowu Wei (g.wei@salford.ac.uk) and Lei Ren (lei.ren@manchester.ac.uk)

Abstract. This paper presents a novel bio-inspired modular robotic arm that is purely evolved and developed from a mechanical stem cell. Inspired by stem cell whilst different from the other robot "cell" or "molecule", a fundamental mechanical stem cell is proposed leading to the development of mechanical cells, bones and a Sarrus-linkage-based muscle. Using the proposed bones and muscles, a bio-inspired modular-based five-degrees-of-freedom robotic arm is developed. Then, kinematics of the robotic arm is investigated which is associated with an optimization-method-based numerical iterative algorithm leading to the inverse kinematic solutions through solving the non-linear transcendental equations. Subsequently, numerical example of the proposed robotic arm is provided with simulations illustrating the workspace and inverse kinematics of the arm. Further, a prototype of the robotic arm is developed which is integrated with low-level control systems, and initial motion and manipulation tests are implemented. The results indicate that this novel robotic arm functions appropriately and has the virtues of lower cost, larger workspace, and a simpler structure with more compact size.

1 Introduction

In the field of robotics, researchers and engineers have a dream of developing robotic arms with agile characteristics like the human arm to perform dynamic tasks under complicated and unconstructed environment. However, more than sixty years passed since the first robotic manipulator being patented by George G. Devol in 1954 (see Siciliano and Khatib, 2008), it is still too far to say that this dream has come true. Nevertheless, new ideas and concepts have been putting forward to explore the ways of developing more efficient, dexterous and feasible robotic arms.

In order to achieve desired flexibility, manipulability and reconfigurability, one of the efforts is devoted to the development of novel, high performance module-based robotic arms. These include, to mention but a few, the one degree of freedom T-type and I-type modules proposed by Guan et al. (2009); the three degrees of freedom tendon-integrated spherical modules designed by Guckert and Naish (2009); the three degree of freedom active module suitable for space hyper-redundant robot presented by Shammas et al. (2006) and the bi-articular muscles developed by Yoshida et al. (2009); and through the construction of link modules and joints modules, Acaccia et al. (2008) developed a modular robotic system for generic industrial applications. However, most of these modules proposed are serial and rigid, they can only change the position and attitude of the end-effector within the allowable range of the linkage member. The degrees of flexibility of the structures of such serial rigid robotic arms are limited with simple functions such that they cannot adapt to the change of working environment and various customer demands. To overcome the shortcoming of serial robotic arms, effort has also been made to develop robotic arms based on parallel mechanism. Liu et al. (2006) proposed a novel five degree of freedom reconfigurable hybrid robotic arm composed of a 2-DOF parallel mechanism and a 3-DOF serial kinematic chain. Jin and Gao (2002) studied a new 6-SPS parallel manipulator with an orthogonal configuration. Wurst and Peting (2002) presented PKM concept for designing modular-based reconfigurable machine tools, and using slide module and swing module, Xi et al. (2006) proposed reconfigurable parallel robot consisting of two base tripods.

In addition, bio-inspired robotic arms have also been designed and developed. One of the earliest bio-inspired robot is a new robotic system coined DRRS (Dynamically Reconfigurable Robotic System) proposed by Fukufda and Nakagawa (1988). This robotic system consists of a number of intelligent cells each of which has a fundamental mechanical function. Each cell can detach or connect itself to the other cells autonomously depending on a specified task, the robot can then be of a manipulator or a mobile robot, the system also has the function of self-repair. In 1990, Fukuda and Kawauchi (1990) further proposed a new kind of robotic system called CEBOT (The Cellular Robotic System), which is a distributed robotic system consisting of reconfigurable autonomous units called "cell"s, the robot can reconfigure itself to an optimal structure depending on the specified purpose and environment. Further, Kotay and Rus (1999) proposed a robotic module called "Molecule" to build self-reconfiguring robots, the Molecules support multiple modalities of locomotion and manipulation. In the same period, Rus and Vona (1999) discussed a robotic system composed of crystalline modules that can aggregate together to form distributed robot systems. The crystalline modules can also move relative to each other by expanding and contracting. This actuation mechanism permits automated shape metamorphosis and provides a crystalline robot system the capability of self-reconfiguration. More recently, Yim et al. (2000) presented a modular reconfigurable robot called Poly-Bot. PolyBot is a modular reconfigurable robot system composed of two types of modules serving as segments and nodes. A segment module has one degree of freedom and two connection ports, and a node module is rigid with no mobility and has six connection ports. It was reported that two PolyBot systems have been built, and a robot with up to 32 modules being bolted together have been tested.

In this paper, we propose a stem-cell-inspired mechanical robotic arm. Different from the robotic arms discussed above that each of their mechanical cells or molecules is not separated but a completed robotic module being integrated with control and software system, the mechanical stem cell proposed in this paper, inspired by the character of stem cell, at a more basic and fundamental level, is a type of pure mechanical unit. This unit can be used to build different robotic modules in according to the practical applications.

Using the unit, a bio-inspired mechanical muscle based on Sarrus linkage is proposed, leading to the development of a 5-DOF bio-inspired robotic arm. This novel design helps augment workspace and overcome some shortcomings of con-ventional serial robotic arms. According to the available literature, although Sarrus linkage (Lee, 1996) has many applications in the field of robotics, including the crawling robot by Ranjana et al. (2006), the road vehicle by Gavin and Luis (2010) and the micro six-legged segmented robot by Katie and Robert (2011), there is no report on robotic arm that is constructed with Sarrus-linkage-based structures.

In this paper, a robotic arm evolved from a fundamental mechanical stem cell is for the first proposed, and its associated kinematics, simulation, prototype and initial test results are presented.

2 Bio-inspired robotic modules

2.1 Bio-inspired mechanical cells

Stem cells have the remarkable potential to be developed into different cell types during their life cycles, and under certain physiological or experimental conditions, they can be induced to become tissue- or organ-specific cells with special functions. Inspired by the evolution of stem cell, this paper proposes the concept of mechanical stem cells that are immature and have the potentiality to evolve into various robotic modules, such as bone, joint and muscle etc. A typical example of such a concept is illustrated in Fig. 1 where a mechanical stem cell is designed as a solid plate without any connections, in this sense, the cell is immature and its structure and function are indeterminate. However, with further evolution, the stem cell can grow connections at its edges and develop itself into mature mechanical cells. The connections can be of different types and two of them are considered in this paper, they are namely the male connection denoted as M and the female connection denoted as F. With the connections, mechanical cells such as FNMN cell and FFMM cell can be developed as shown in Fig. 1, where N standing for no connection. In Fig. 1, the FNMN cell consists of one mechanical stem cell and two connections, i.e. one female connection and one male connection; the FFMM cell has four connections which are arranged as F, F, M and M in clockwise sequence.

2.2 Development of mechanical bones

The FNMN and FFMM mechanical cells can further evolve themselves into mechanical bones forming structure modules that can be used as links or rigid bodies for constructing bio-inspired robotic arms. For example, the FNMN and FFMM mechanical cells can interact and grow into different mechanical bones such as the cubic bone, cuboid bone and polyhedral bone that are indicated in Fig. 2.

The cubic bone is a stable structure consisting of six FFMM cells, and by adding four more FFMM cells to a cubic bone, a cuboid bone can be generated as shown in Fig. 2b, the length of the cuboid bone is doubled that of the cubic bone. In theory, more FFMM cells may be incorporated in order

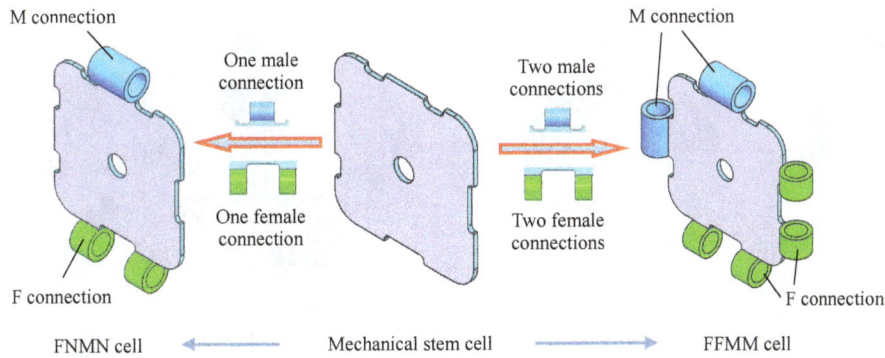

Figure 1. Bio-inspired mechanical cell.

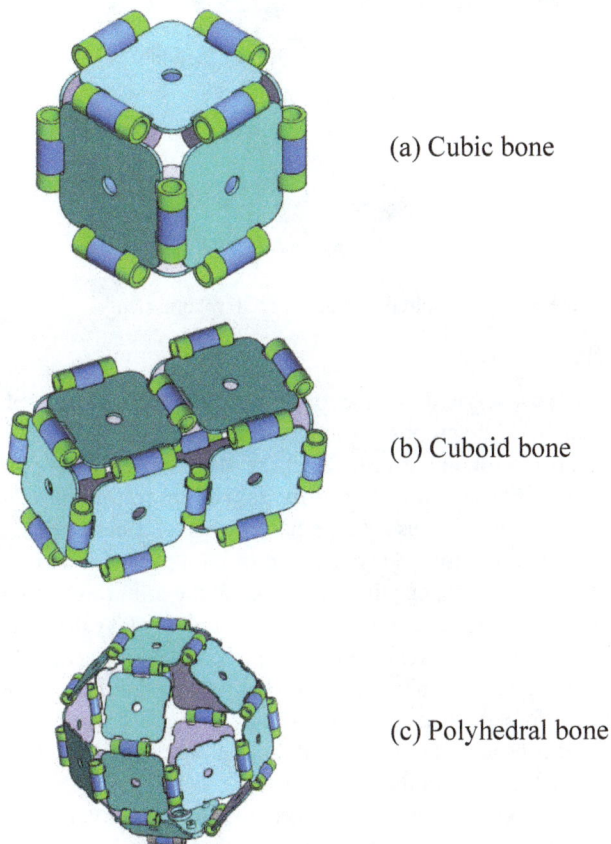

(a) Cubic bone

(b) Cuboid bone

(c) Polyhedral bone

Figure 2. Bio-inspired mechanical bones.

to extend the cuboid bones. Figure 2c shows a polyhedral bone consists of eight FFMM cells and twelve FNMN cells. Compared to the cuboid bone, a polyhedral bone has more inner space which can be used to install and store actuators, sensors and other member modules.

A bio-inspired mechanical bone developed through the mechanical cells have three advantages. First, its structure is simple, modular and apt-to-be fabricated. Second, its structure is reconfigurable and flexible. Third, the inner hollow space of these mechanical bones can be used to accommo-

date sub modules of robot, which is very useful for some special applications such as space technology.

2.3 Bio-inspired Sarrus-linkage-based mechanical muscle

Sarrus linkage (Sarrus, 1853) is a one degree of freedom spatial linkage that was invented by Pierre Frédéric Sarrus (1798–1861) in 1853, that can transfer rotary motion into linear motion, or inversely transfer the linear motion into standard rotary motion (Chen et al., 2013); sometimes it is also referred to as spatial crank mechanism. It provides a great variety of possible motions and may accomplish complex tasks beyond what planar linkages are capable of as indicated in He et al. (2014) and Li et al. (2013). The classic Sarrus linkage is a two-limb six-bar linkage of which each limb contains three parallel revolute joints, with the first joint of both limbs lying on the base plane and the third joint of both limbs lying on the platform plane.

Based on the Sarrus linkage, we can artificially nurture the FNMN and FFMM mechanical cells into a bio-inspired mechanical muscle that is composited of six mechanical cells connected by six revolute joints as indicated in Fig. 3a. Like the Sarrus linkage, the mechanical cells are jointed to form two sets of limbs, each of which being a four-link chain. The first chain is comprised of two FFMM cells as the bottom end labelled 1 and the top end labelled 4, and two FNMN cells serving as side links 2 and 3; the cells are connected by three parallel joints whose axes are denoted as A, B and C as indicated in Fig. 3a. Similarly, the other branch is composed of two FFMM cells serving as bottom end 1 and top end 4 which are of duplicates of the first branch, and two FNMN cells acting as side links 5 and 6; the links are joined by three parallel revolute joints whose axes are D, E and F, respectively.

To meet the requirement of practical engineering applications, a robuster Sarrus-linkage-based mechanical muscle is symmetrically developed by integrating four additional FNMN cells into the standard Sarrus structure associated with six extra joints, as illustrated in Fig. 3b. This design

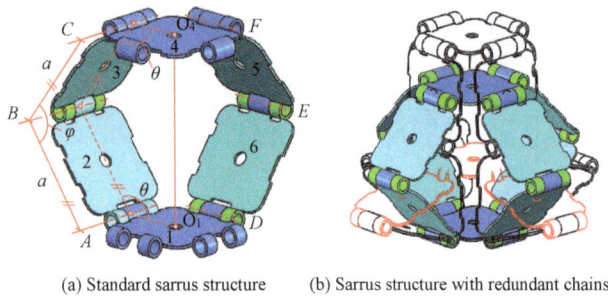

(a) Standard sarrus structure   (b) Sarrus structure with redundant chains

Figure 3. Bio-inspired Sarrus-linkage-based mechanical muscle.

does not alter the motion performance and output characteristics of a standard Sarrus structure but provides a more stable and reliable structure which is capable of carrying more payload. As illustrated in Fig. 3b, assuming that the bottom cell is fixed, the top cell can only has translational motion with respect to the bottom cell. The lower-side cells rotates about the bottom cell, and the upper-side cells rotate respectively about their adjacent lower-side cells and translate with the top cell. Obviously, the linear motion of the two parallel ends leads to the circular motion of the joints; and on the other hand, the circular motion of the joints results in the linear motion between the two parallel ends. Hence, this mechanical muscle can not only be used as a translational joint, but also acts as a supporting structure to expand further the workspace of a robotic arm.

Further, given the structure parameters of the Sarrus structure as illustrated in Fig. 3a, there exist $\varphi + 2\theta = 2\pi$ and $d_{O_1 O_4} = a\sqrt{2 - 2\cos\varphi}$ with $d_{O_1 O_4}$ standing for the distance between points O_1 and O_4.

3   Design and kinematics of the bio-inspired robotic arm

3.1   Design of a bio-inspired modular 5-DOF robotic arm

Based on the mechanical bones and muscle developed from the mechanical stem cells in the previous section, different type of robotic arms can be evolved and grown satisfying specific purpose and environment, and one bio-inspired modular-based 5-DOF robotic arm is developed in this paper as illustrated in Fig. 4. This bio-inspired robotic arm is composed of three bio-inspired bones, i.e. the shoulder bone, upper-arm bone and forearm bone, and two bio-inspired muscles as the upper-arm muscle and the forearm muscle, connected by three revolute joints, that is, the shoulder joint, the elbow joint and the wrist joint, and two rigid connections as the upper arm rigid connection and the forearm rigid connection. It can be seen that the robotic arm is purely developed from the mechanical stem cell presented in Sect. 2.1. Regarding the degrees of freedom of the robotic arm, there are three revolute joints, two 1-DOF Sarrus-linkage-based mus-

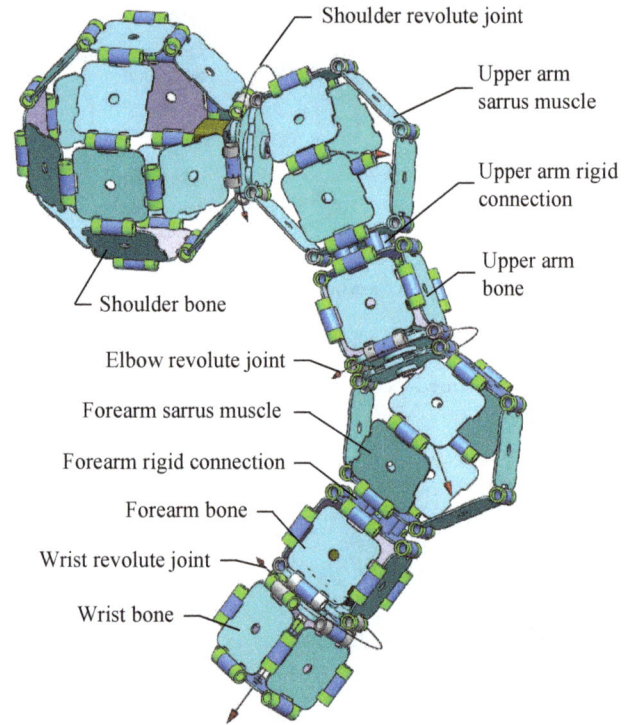

Figure 4. A bio-inspired modular 5-DOF robotic arm.

cles and two rigid connections, hence the total degrees-of-freedom of the proposed robotic arm is five.

For this robotic arm, through the shoulder revolute joint, it can realize an circular 360° rotation of upper arm; through the upper arm Sarrus-linkage-based muscle, it can realize a combined rotation and translation of the upper arm and below parts, and change the workspace of the arm; through the elbow revolute joint, it can realize an entire 360° rotation of forearm and the parts below; through forearm Sarrus muscle, it can realize a combined rotational and translational motion of the forearm and parts below, and change its workspace; further, through the wrist revolute joint, it can realize circular 360° rotation of the wrist bone.

As described above, the dexterous motion generated by the proposed robotic arm provides augmented workspace. Regarding its application, the shoulder can be fixed on the base or combined with the other robots, we can also directly connect the upper Sarrus arm to the base or the other robots with the shoulder revolute joint. By accurate controlling the motion of each joint, the robotic arm can execute specific mission satisfying various customer requirements.

3.2   Forward kinematics

As aforementioned, the bio-inspired 5-DOF robotic arm developed in this paper is evolved from the mechanical stem cell consisting of bone cells and Sarrus-linkage-based muscles. Kinematics of such a robotic arm is different from the conventional ones and is investigated in this section. In order to describe the position and orientation of the proposed robotic arm, a global coordinate system $\{O_0 - x_0\, y_0\, z_0\}$ is located at the shoulder joint, and four body-attached coordinate systems are established at the upper arm muscle, upper arm born, forearm muscle and forearm born, respectively, as shown in Fig. 5. As the robotic arm includes both serial and parallel kinematic chains, the traditional D–H method cannot be directly used to analyse its kinematics. In this paper, a more general approach, that is, homogeneous coordinate transformation method is used. By firstly solving the transformation matrices between the coordinate systems, and consequently deriving the total transformation matrix, we can obtain a matrix representing the posture of the wrist bone with respect to the global coordinate system leading to the investigation of the workspace and inverse kinematics of the proposed robotic arm.

Considering the motion sequence of the robotic arm, it involves both translational and rotational transformation. Herein, the homogeneous transformation matrix for translation is denoted as $\text{Trans}(a, b, c)$ with a, b and c being the coordinates of a displacement vector $a\,\boldsymbol{i} + b\,\boldsymbol{j} + c\,\boldsymbol{k}$, and the homogeneous transformation matrices for rotations about the x, y and z axes are denoted as $\text{Rot}(x, \theta)$, $\text{Rot}(y, \theta)$ and $\text{Rot}(z, \theta)$, respectively, as defined in Cai (2009). Further, for conciseness sake, the sine and cosine functions are denoted as s and c, respectively.

As discussed above, the bio-inspired robotic arm is built by the FFMM and FNMN mechanical cells evolved from the mechanical stem cell, thus all the unit modules used to construct the robotic arm have the same structure parameters and we assume that lengths of the side and thickness of a unit module are respectively a and b. According to the kinematics of Sarrus linkage, the angles θ_i and φ_i ($i = 2$ and 4) in the two Sarrus-linkage-based muscles in the robotic arm have the following relationship

$$2\theta_i + \varphi_i = 2\pi \quad (i = 2 \text{ and } 4). \tag{1}$$

Referring to Fig. 5, the position vector of the centroid point M in the wrist bone can be expressed in reference frame $\{O_4 - x_4\, y_4\, z_4\}$ in homogeneous coordinate form as

$$^4\boldsymbol{p} = [0\ 0\ a/2 + b\ 1]^T. \tag{2}$$

Coordinate system $\{O_4 - x_4\, y_4\, z_4\}$ is transformed from frame $\{O_3 - x_3\, y_3\, z_3\}$ by a rotation about the y_3 axis by an angle θ_5 followed by another rotation about the x_4 axis by $-\pi/2$, and subsequently a translation $\boldsymbol{q}_{34}$ along the z_4 axis. With respect to reference frame $\{O_3\}$, the translation $\boldsymbol{q}_{34}$ can be expressed as

$$^3\boldsymbol{p}_{34} = \overline{O_3 O_4} = \overline{O_3 D} + \overline{DE} + \overline{EF} + \overline{FO_4}$$
$$= [a/2\ 2a\sin\theta_4 + b\ 0\ 1]^T, \tag{3}$$

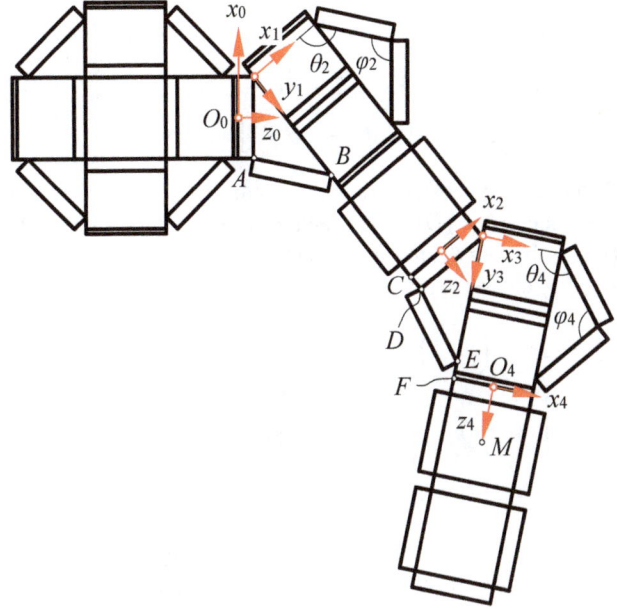

Figure 5. Geometry and coordinate systems of the robotic arm.

such that the homogeneous transformation matrix that defines reference frame $\{O_4\}$ relative to reference frame $\{O_3\}$ can be given as

$$\mathbf{T}_{34} = \text{Trans}\left(^3\boldsymbol{p}_{34}\right)\text{Rot}(y_3, \theta_5)\text{Rot}(x_4, -\pi/2)$$
$$= \begin{bmatrix} c\theta_5 & -s\theta_5 & 0 & a/2 \\ 0 & 0 & 1 & 2as\theta_4 + b \\ -s\theta_5 & -c\theta_5 & 0 & 0 \\ 0 & 0 & 0 & 1 \end{bmatrix}. \tag{4}$$

Similarly, according to Fig. 5, posture of coordinate system $\{O_3\}$ can be obtained from reference frame $\{O_2\}$ by a rotation about the z_2 axis by angle θ_3 with a following rotation about the x_2 axis by $\pi/2$ and a translation given by vector $\boldsymbol{q}_{23}$. The translational vector $\boldsymbol{q}_{23}$ can be presented in reference frame $\{O_2\}$ as

$$^2\boldsymbol{p}_{23} = [a/2\ b\ 0\ 1]^T, \tag{5}$$

and therefore transformation matrix that defines frame $\{O_3\}$ with respect to frame $\{O_2\}$ can be obtained as

$$\mathbf{T}_{23} = \text{Trans}\left(^2\boldsymbol{p}_{23}\right)\text{Rot}(z_2, \theta_3)\text{Rot}(y_2, \theta_4 - \pi)\text{Rot}(x_2, \pi/2)$$
$$= \begin{bmatrix} -c\theta_3\, c\theta_4 & -c\theta_3\, s\theta_4 & s\theta_3 & a/2 \\ -s\theta_3\, c\theta_4 & -s\theta_3\, s\theta_4 & -c\theta_3 & 0 \\ s\theta_4 & -c\theta_4 & 0 & b \\ 0 & 0 & 0 & 1 \end{bmatrix}. \tag{6}$$

Further, by defining the displacement vector of the origin of frame O_2 with respect to the origin of frame $\{O_1\}$ as

$$^1\boldsymbol{p}_{12} = \overline{O_1 O_2} = \overline{O_1 A} + \overline{AB} + \overline{BC} + \overline{C O_2}$$
$$= [a/2 \; a(1 + 2\sin\theta_2) + 3b \; 0 \; 1]^T. \tag{7}$$

Transformation of coordinate system $\{O_2\}$ with respect to reference frame $\{O_1\}$ that is achieved by a rotation about the x_1 axis by $-\pi/2$ and a translation with displacement vector $^1\boldsymbol{p}_{12}$ can be expressed as

$$\mathbf{T}_{12} = \mathrm{Trans}\left(^1\boldsymbol{p}_{12}\right)\mathrm{Rot}(x_1, -\pi/2)$$

$$= \begin{bmatrix} 1 & 0 & 0 & a/2 \\ 0 & 0 & 1 & a(1+2s\theta_2)+3b \\ 0 & -1 & 0 & 0 \\ 0 & 0 & 0 & 1 \end{bmatrix}. \tag{8}$$

From Fig. 5, the displacement of the origin of reference frame $\{O_1\}$ with respect to the origin of frame $\{O_0\}$ can be given as

$$^0\boldsymbol{p}_{01} = [a/2 \; b \; 0 \; 1]^T, \tag{9}$$

thus, the transformation from frame $\{O_1\}$ to frame $\{O_0\}$ which is performed by a rotation about the z_0 axis by angle θ_1, a following rotation about the x_0 axis by $\pi/2$, and a subsequent translation $\boldsymbol{p}_{01}$ can be derived as

$$\mathbf{T}_{01} = \mathrm{Trans}\left(^0\boldsymbol{p}_{01}\right)\mathrm{Rot}(z_0, \theta_1)\mathrm{Rot}(y_0, \theta_2 - \pi)\mathrm{Rot}(x_0, \pi/2)$$

$$= \begin{bmatrix} -c\theta_1 c\theta_2 & -c\theta_1 s\theta_2 & s\theta_1 & a/2 \\ -s\theta_1 c\theta_2 & -s\theta_1 s\theta_2 & -c\theta_1 & 0 \\ s\theta_2 & -c\theta_2 & 0 & b \\ 0 & 0 & 0 & 1 \end{bmatrix}. \tag{10}$$

Finally, multiplying all the four transformation matrices in Eqs. (4), (6), (8) and (10), we can have the final coordinate transformation matrix relating frame $\{O_4\}$ to frame $\{O_0\}$ as

$$\mathbf{T}_{04} = \mathbf{T}_{01}\mathbf{T}_{12}\mathbf{T}_{23}\mathbf{T}_{34}. \tag{11}$$

Furthermore, combining Eq. (2) and Eq. (11), the position vector of point M relative to be base coordination system $\{O_0\}$ can be expressed as

$$^0\boldsymbol{p} = \mathbf{T}_{04}{}^4\boldsymbol{p}. \tag{12}$$

According to Eqs. (2) and (12), it can be found that the wrist revolute joint has no effect on the position of the wrist bone centroid M, but it affects the orientation of wrist bone.

3.3 Inverse kinematics

Inverse kinematics provides background for position control of a robotic arm. Given the Cartesian coordinate of the end-effector in the transformation matrix $\mathbf{T}_{04}$, we attempt to derive the joint space formed by the five variables $\vartheta = \{\theta_1, \theta_2,$

$\theta_3, \theta_4, \theta_5\}$. Since the proposed robotic arm is formed in a hybrid configuration, the inverse kinematics is highly non-linear such that it is difficult to find closed-form solutions. Thus, in this section, a numerical approach is proposed for solving the inverse kinematics of this bio-inspired robotic arm.

Dividing matrix $\mathbf{T}_{04}$ into blocks, it has

$$\mathbf{T}_{04} = \begin{bmatrix} \mathbf{R}_{04} & \boldsymbol{q}_{04} \\ \mathbf{0} & 1 \end{bmatrix}, \tag{13}$$

where, $\mathbf{R}_{04} = [x_{04} \; y_{04} \; z_{04}]^T$ is the rotation matrix lying in $SO(3)$ defining orientation of the wrist bone with respect to the base frame, and $\boldsymbol{q}_{04} = [q_x \; q_y \; q_z]^T$ is a vector in $\mathbb{R}^3$ that locates the position of the centroid M of the wrist bone relative to the base frame.

Assuming that elements of the matrix in Eq. (13) are specified, equalizing Eqs. (11) and (13) leads to a set of non-linear equations containing 5 variables and 12 equations which are not all independent. In this way, the inverse kinematic problem of the proposed robotic arm is converted into finding solutions for the 12 dependent equations.

However, as aforementioned, due to the hybrid structure of the proposed bio-inspired robotic arm, there is no analytical solution for the set of non-linear transcendental equations, there exist only numerical solutions or approximate solutions. Further, unless in the case of special circumstances, there is no general direct method to get the numerical solution structure of non-linear equations, including the existence of solution and multiple solutions. At present, there are two approaches for obtaining the numerical solution, i.e. the indirect method based on fixed point theorem, and the optimization method based on variational principle, both of these methods are based on iterative numerical solution adopting the strategy of the "time for accuracy" termed in Zhou et al. (2008). Considering that the kinematics equations to be solved herein is a hyper-redundant non-linear transcendental one without direct solution available, we convert this problem into a global optimization problem based on optimization methods as follows.

Given elements to matrix $\mathbf{T}_{04}$ in Eq. (13) and equalizing it with Eq. (11), a set of dependent equation can be formed in terms of the five variables as

$$\begin{cases} f_1(\theta_1, \theta_2, \theta_3, \theta_4, \theta_5) = 0 \\ f_2(\theta_1, \theta_2, \theta_3, \theta_4, \theta_5) = 0 \\ \quad \cdots \\ f_{12}(\theta_1, \theta_2, \theta_3, \theta_4, \theta_5) = 0 \end{cases}, \tag{14}$$

where the joint space $\vartheta = \{\theta_1, \theta_2, \theta_3, \theta_4, \theta_5\}$ contains the five variables whose variation ranges are defined as

$$\pi/2 \le \theta_2, \; \theta_4 \le \pi \text{ and } -\pi \le \theta_1, \; \theta_2, \; \theta_3 \le \pi. \tag{15}$$

In order to convert the equation solving problem into an equivalent optimization problem, Eq. (14) is converted to construct an energy function as

$$P(\vartheta) = \sum_{i=1}^{12} |f_i(\theta_1, \theta_2, \theta_3, \theta_4, \theta_5)|. \tag{16}$$

Thus, the problem of solving non-linear transcendental equations is transformed into the problem of solving the energy function complying with a given constraint. That is, given a sufficiently small positive number ε, the aim is to search for a set of values $\vartheta^*(\theta_1, \theta_2, \theta_3, \theta_4, \theta_5)$ such that $P(\vartheta^*) < \varepsilon$ holds, and as a result, $\vartheta^*(\theta_1, \theta_2, \theta_3, \theta_4, \theta_5)$ provides as a set of approximate solutions for the above non-linear transcendental equations.

The numerical solutions of non-linear transcendental equations can be implemented in three steps as presented in Lu et al. (2008):

1. the existence of solutions: checking whether the equations have solutions;

2. solution isolation: dividing the solution interval into smaller sub-intervals, for each sub-interval, there may exist a solution or not, and if so, we can refer to any point in the sub-intervals as an approximation solution;

3. solution precise: efforts to improve the accuracy of the approximation solution, make it satisfy a certain accuracy requirement.

Considering the approach of solving non-linear equations based on global optimization method, a numerical iterative algorithm is proposed with its procedures shown in Fig. 6 and the specific process is interpreted as follows.

Firstly, define the maximum number of iteration N and the threshold error ε: if the calculation accuracy reaches the threshold error ε, stop the operation and the corresponding set of variable obtained may provide a set of inverse kinematic solutions; if the number of operations has reached the maximum number of iteration N, even if accuracy error is larger than the threshold, also stop operations. Obviously, if the number of operations has reached the maximum number of iteration, and the calculation error is still big, this means that the accuracy of the solutions is not high enough to meet the threshold, we need to increase the maximum number of iteration N so as to achieve a set of more precise solutions.

Secondly, allocate values of the five variables $\theta_1, \theta_2, \theta_3, \theta_4$ and θ_5 according to their individual intervals defined in Eq. (15), and substitute them into the energy function $P(\vartheta)$, then perform the iterative operation and select the minimum value of the energy function $P(\vartheta)$ searching for the optimized set of variable values $(\theta_{k1}, \theta_{k2}, \theta_{k3}, \theta_{k4}, \theta_{k5})$ that leads to minimum value of the energy function denoted as $P(\vartheta)_{\min}$.

Thirdly, continue allocating the five variables $\theta_1, \theta_2, \theta_3, \theta_4$ and θ_5 in a smaller interval near $(\theta_{k1}, \theta_{k2}, \theta_{k3}, \theta_{k4}, \theta_{k5})$, and

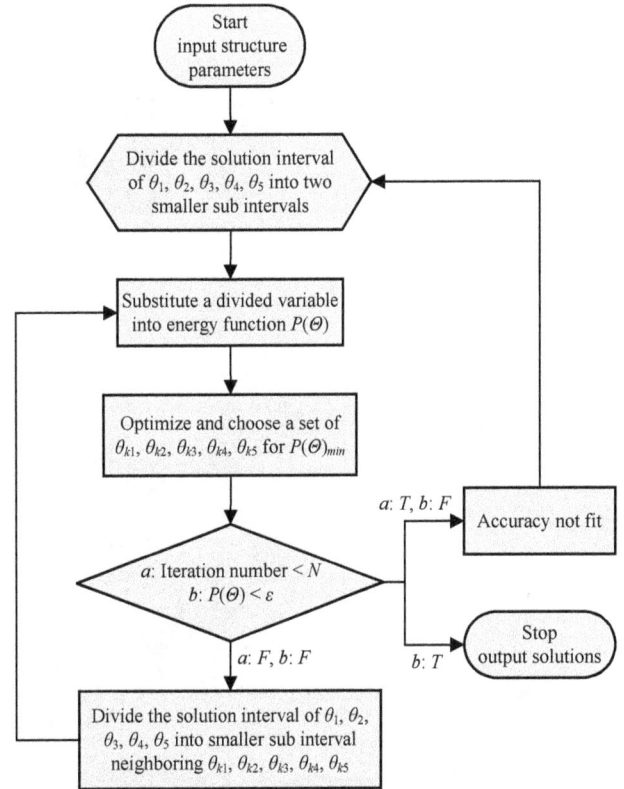

Figure 6. Flow chart for numerical solution of the proposed inverse kinematics.

running iterative operations. This will result in better minimum energy function value $P(\vartheta^*)$ and the refined solutions for the five variables.

Repeat the above steps until the value of energy function is less than the threshold error or the iteration times reached the maximum number, the final output of the five variables $\theta_1, \theta_2, \theta_3, \theta_4$ and θ_5 will be a set of optimal solutions for the non-linear equations, that is, the inverse kinematic solution of the proposed robotic arm.

Obviously, this algorithm can reliably obtain a set of solutions satisfying the equations through a series of continuous divide and iteration. To seek a more accurate solution, more divided intervals and iteration times are required, which naturally increases the operation time. It can be found that the above algorithm can satisfy the general application requirements without involving any kind of complex algorithm such as genetic algorithm or neural networks. In this paper, the results obtained indicate that this algorithm is sufficient to compute the inverse kinematics and provide results for position and orientation control, the results are indicated by numerical simulation and physical prototype validation.

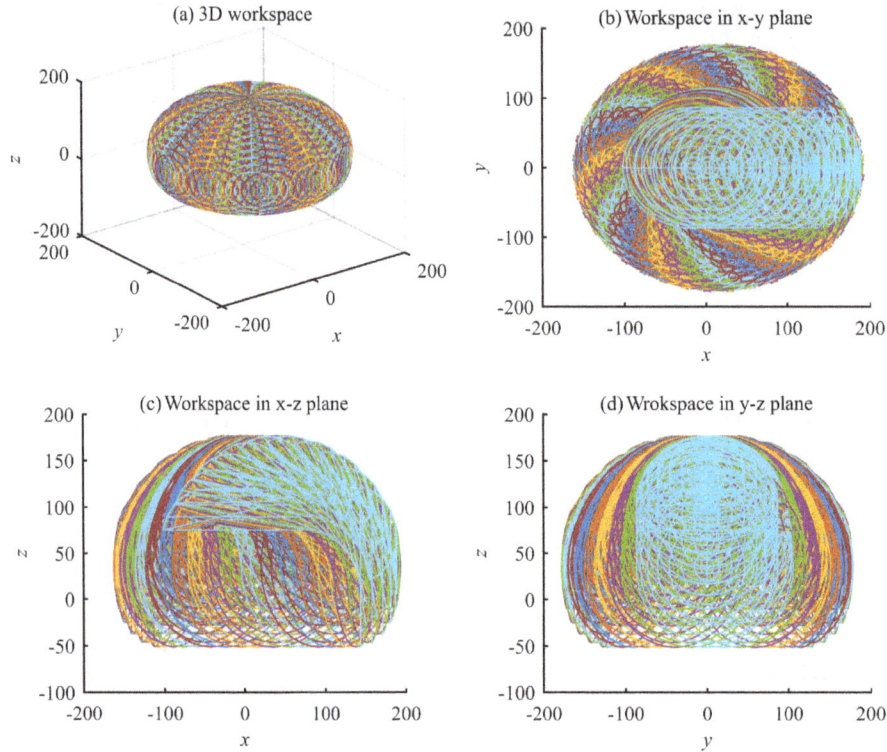

Figure 7. Workspace of the bio-inspired robotic arm.

4 Numerical simulation and prototype of the robotic arm

4.1 Workspace and the typical configurations

Based on the transformation matrix of the wrist bone centroid M given in Eq. (12), using Matlab package, we can plot the trajectory of point M relative to the fixed reference frame $\{O_0\}$, that is, workspace of the wrist bone centroid M relative to the universal coordinate system, as shown in Fig. 7. In this example, the size of each stem mechanical cell is $a \times a \times b = 30\,\text{mm} \times 30\,\text{mm} \times 6\,\text{mm}$, in addition, the geometric collisions among member modules during spatial motion are not considered.

Figure 7a and b show that the projection of the workspace of the robotic arm in any plane that is perpendicular to the z axis is symmetric, the arm can make a whole circular motion around the z axis without any interference. Figure 7b and c show that the wrist can perform both forward and backward motions in a certain range in the negative direction of the z axis, which provides potential capability of mimicking the motion of a human arm.

The workspace of the robotic arm in Fig. 7 also demonstrates that, without considering the size of the shoulder, when all the Sarrus-linkage-base muscles are fully retracted, the entire arm length in the z axis direction is 108 mm, while the wrist centroid M can reach in a the range from -71 to 200 mm along z axis, as illustrated in Fig. 7c and d. Like-

wise, when retracted, the position of wrist centroid M in the x- and y-direction can be zero, and the actual workspace is in the range from -175 to 175 mm along the x axis and the y axis, respectively.

In addition, Fig. 8 demonstrates the typical configurations and workspace augmentation principle of the proposed robotic arm. Figure 8a shows one extreme phase of the robotic arm with two Sarrus-linkage-based muscles being fully retracted, compared with the fully expanded phase indicated in Fig. 8b and the intermediate phase in Fig. 8c, the robotic arm has the smallest volume of workspace at this phase, and in this configuration, the lengths of three equivalent links l_1, l_2 and l_3 are $l_1 = \sqrt{a^2 + (a + 4b)^2}$, $l_2 = \sqrt{a^2 + 4b^2}$ and $l_3 = a/2 + b$, respectively; where, a and b are respectively the side length and thickness of the mechanical stem cells. Figure 8b shows another extreme phase of the robotic arm with the two Sarrus-linkage-based muscles being fully expanded, compared with Fig. 8(a), it can be found that the lengths between O_0 and O_2, and between O_2 and O_4 increase, such that the lengths of the three equivalent links become $l_1 = \sqrt{(a/2 + b)^2 + (5a/2 + 3b)^2}$, $l_2 = \sqrt{(a/2 + b)^2 + (3a/2 + b)^2}$ and $l_3 = a/2 + b$. Figure 8c shows an intermediate phase in which configuration the lengths of the three equivalent links are $l_1 = \sqrt{(a/2(1 - c\alpha) + bs\alpha)^2 + (a(1 + 3s\alpha/2) + b(3 - c\alpha))^2}$, $l_2 = \sqrt{(a/2(1 - c\eta) + bs\eta)^2 + (3a/2s\eta - bc\eta)^2}$ and $l_3 = a/2 + b$.

(a) Fully retracted configuration

(b) Fully expanded configuration

(c) Intermediate phase

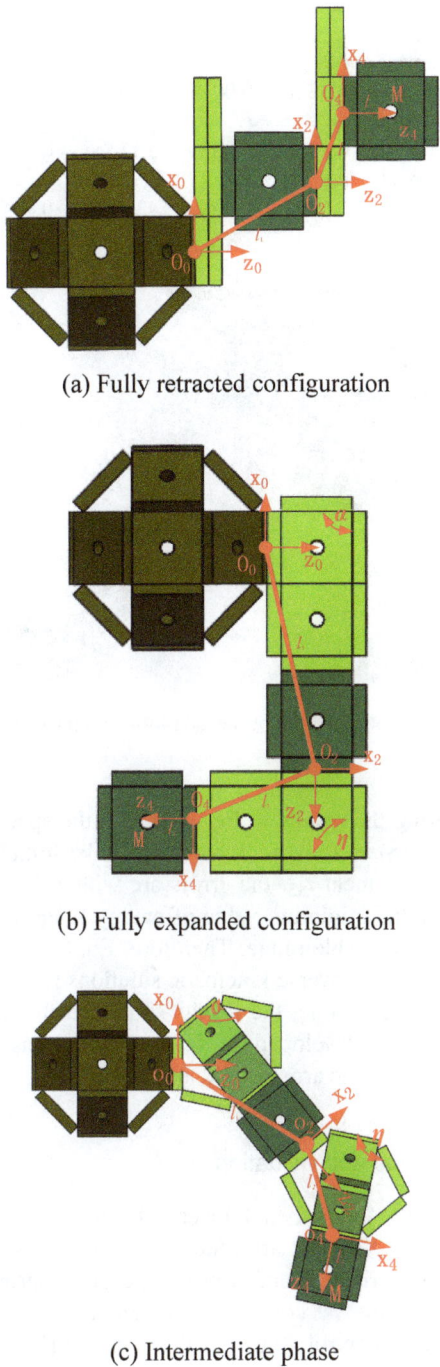

Figure 8. Geometric configurations of the bio-inspired robotic arm.

Based on the changes of the lengths of the three equivalent links l_1, l_2 and l_3, the workspace augmentation function of the two Sarrus-linkage-based muscles can be formulated and illustrated in Fig. 9.

Figure 9a shows the relationship between length of the equivalent link l_1 and the configuration angle α of the upper arm Sarrus-linkage-based muscle. It can be found that as the upper arm muscle transform from the fully retracted

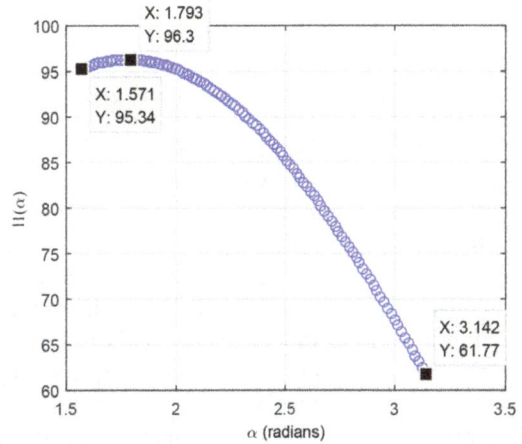

(a) Length change of upper arm muscle

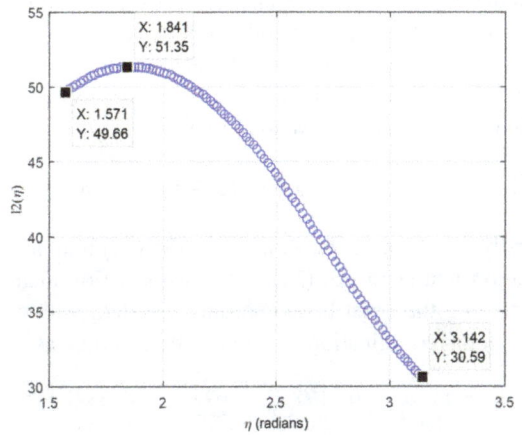

(b) Length change of forarm muscle

Figure 9. Workspace augmentation function of the two Sarrus-linkage-based muscles.

phase to the fully expanded phase, the actual length of link l_1 increases from 61.77 to 95.34 mm with the maximum length of 96.33 mm; which provides an extra 34 mm flexible length to the upper arm to achieve a linear motion along the z axis.

Similarly, as shown in Fig. 9b, as the forearm Sarrus-linkage-based muscle transforms from the fully retracted phase to the fully expanded phase, the actual length of the equivalent link l_2 increase from 30.59 to 49.66 mm, and the maximum length is 51.35 mm, which implies that an extra 20 mm flexible length is provided to the forearm for the linear motion along the z_4 axis.

Hence, with the two Sarrus-linkage-based muscles integrated in this novel bio-inspired robotic arm, it can break through the limitation of conventional robotic structure and enlarge its workspace, which makes the robotic arm particularly suitable for performing large-scale motion in a compact space, not only as a general industrial robotic arm, but also as a device carried by mobile robot to perform investigation, detection, rescue, transportation and other tasks.

4.2 Numerical solution for the inverse kinematics

Further, in order to validate the feasibility and accuracy of the proposed algorithm for inverse kinematic analysis, one numerical example is provided in this section. The validation method and process is: first, give an initial value for the five variables θ_1, θ_2, θ_3, θ_4 and θ_5, the two structure parameters a and b of the mechanical stem cell, and substitute them into the forward kinematics matrix in Eq. (11) to achieve the transformation matrix; second, by equating the specified transformation matrix to Eq. (13), construct a set of non-linear transcendental equations containing 12 equations; third, by using the algorithm presented in Sect. 3.3, solve the equations and check the group of optimal solutions and minimum energy function value, this leads to the appropriate inverse kinematic solutions of the robotic arm.

Here, we assume that the mechanical stem cell has structure parameters $a = 30$ mm, $b = 6$ mm, and a set of values within the ranges of the five joint variables is given as

$$\vartheta = (\theta_1, \theta_2, \theta_3, \theta_4, \theta_5) = (\pi/1.5, \pi/1.2, -\pi/2, -\pi/6, \pi/3). \quad (17)$$

Firstly, substituting the variables in Eq. (17) into the transformation matrix in Eq. (11), the transformation matrix for representing the wrist bone coordinate system with respect to the global coordination system can be specified as

$$\mathbf{T}_{04} = \begin{bmatrix} 0.6250 & 0.3460 & -0.6998 & -43.3841 \\ -0.6495 & -0.2667 & -0.7120 & -96.1147 \\ -0.4330 & 0.8995 & 0.0580 & 100.5428 \\ 0 & 0 & 0 & 1 \end{bmatrix}. \quad (18)$$

Then, by equating matrices Eqs. (18) and (13), a set of non-linear transcendental equation containing 5 variables and 12 equations can be constructed.

Subsequently, substituting the above matrix into the algorithm developed in Matlab^{TM} program, and setting subdivision number as $n = 48$, the threshold error $\varepsilon = 0.5$ and the maximum number of iterations $N = 20$, we get a set of inverse kinematic solutions as

$$\vartheta = (\theta_{k1}, \theta_{k2}, \theta_{k3}, \theta_{k4}, \theta_{k5})$$
$$= (2.0982, 2.6129, -1.6797, -0.3984, 1.0433), \quad (19)$$

where, the number of iterations is 16, the minimum energy function value $P(\vartheta)_{\min} = 0.2250$, such that $P(\vartheta)_{\min} < \varepsilon$ and therefore the operation terminates.

Finally, by substituting the results in Eq. (19) into the transformation matrix in Eq. (11), we can get a optimized numerical transformation matrix as

$$\mathbf{T}'_{04} = \begin{bmatrix} 0.6645 & 0.3459 & -0.6624 & -43.3883 \\ -0.6073 & -0.2667 & -0.7484 & -96.1120 \\ -0.4356 & 0.8996 & 0.0328 & 100.5100 \\ 0 & 0 & 0 & 1 \end{bmatrix}. \quad (20)$$

Table 1. Errors of the inverse kinematic solutions.

Element	t_{11}	t_{12}	t_{13}	t_{14}
Error	6.32 %	0.00 %	5.34 %	0.00 %
Element	t_{21}	t_{22}	t_{23}	t_{24}
Error	6.50 %	0.00 %	5.11 %	0.00 %
Element	t_{31}	t_{32}	t_{33}	t_{34}
Error	0.60 %	0.00 %	43.45 %	0.00 %

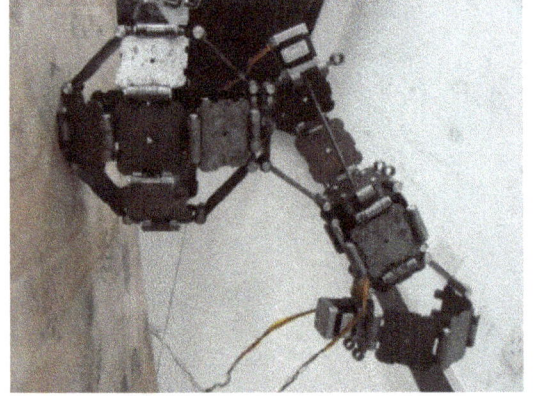

Figure 10. Prototype of the proposed bio-inspired robotic arm.

Comparing the results in Eq. (20) with the specified values in Eq. (18) as show in Table 1, it can be found that except for the element t_{33}, the errors are within the tolerance ranges, and the minimum value of energy function $P(\vartheta)$ is within the acceptable range. Therefore, Eq. (20) can be accepted as a set of inverse kinematic solutions for the robotic arm. At the same time, the results show that the numerical iteration method developed for the non-linear transcendental equations of robotic arm is valid and feasible.

4.3 Prototype and validation

Based on the forward and inverse kinematic analysis of the bio-inspired robotic arm, the mechanical cells were designed and fabricated, and a prototype of the proposed 5-DOF robotic arm was constructed, assembled and integrated with low-lever control system as illustrated in Fig. 10. It can be seen that, the mechanical system was completely built by using only the FFMM and FNMN cells that are developed from the mechanical stem cell, the three revolute joints at the shoulder, elbow and wrist are driven by rotary stepper motors, and the Sarrus-linkage-based muscles are driven by linear motors.

To validate the forward and inverse kinematic models of the robotic arm, a virtual obstacles is designed to construct obstacle avoidance motion test of the robotic arm. As shown in Fig. 11a, the robotic arm is in the rear of a cuboid obstacle, the purpose of obstacle avoidance motion test is to locate the robotic arm in front of the obstacle without any interference.

(a) Initial configuration of the robot

(b) Robotic arm shrinking

(c) Robotic arm swinging

(d) Robotic arm expanding

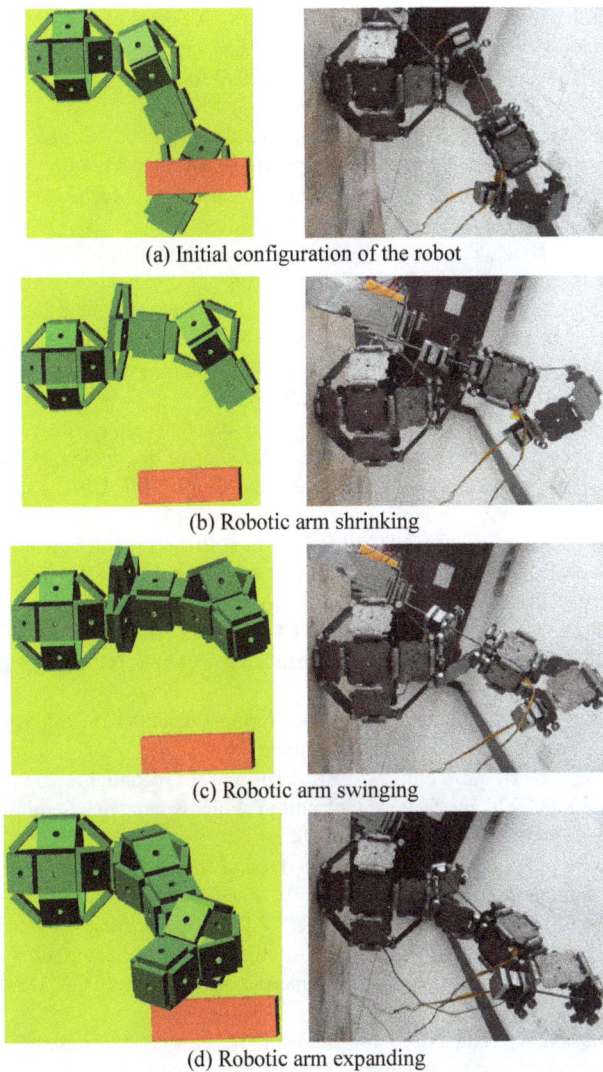

Figure 11. Obstacle avoidance motion test of the robotic arm.

It can be seen that if the shoulder joint rotates directly, the whole arm will swing toward the obstacle, which may lead to an interference between the arm and the obstacle. Therefore, in order to avoid potential collision with the obstacle, a simple but efficient motion control strategy is used based on the property of the Sarrus-linkage-based muscles. Figure 11a shows the initial configuration of the robot in simulation corresponding with the physical test. By only driving the upper arm Sarrus-linkage-based muscle, the upper arm shrinks and the arm arrives at a position over the obstacle, as shown in Fig. 11b. Following a rotation of the shoulder's revolute joint, the robotic arm swings ahead and points in the front direction of the obstacle, as shown in Fig. 11c. Finally, by releasing the upper arm muscle, the upper arm expands to make the robotic arm reach the desired position avoiding the obstacle, as illustrated in Fig. 11d.

Initial tests indicate that the proposed robotic arm can not only perform the functions desired in the design but also overcome obstacles through the shrinking motion of the upper arm and forearm Sarrus-linkage-based muscles, which greatly simplifies control strategy and reduces the financial cost for establishing complex control system.

5 Conclusions

In this paper, a novel bio-inspired robotic arm was for the first time proposed and presented. This robotic arm was designed and developed based on a single type of mechanical stem cells.

Inspired by the function and characteristics of the stem cell but different from the other robot "cell" or "molecule", the mechanical stem cell presented in this paper is simple but capable of evolving into different functional cells, bones and muscles. Using the bones and a mechanical muscle developed based on the Sarruse linkage, a 5-DOF bio-inspired robotic arm was designed and its associated kinematics was investigated. In order to solve the inverse kinematics of the proposed robotic arm, an optimization-method-based numerical iterative algorithm was proposed and verified with a numerical example and computer simulations. Further, a physical prototype of the proposed 5-DOF robotic arm was developed and initial tests were carried out to validate the correctness of forward kinematics and the applicability of inverse kinematics solving algorithm.

Overall, the paper has indicated that the stem-cell inspired pure mechanical stem cell has parallels in biology and provides a flexible modular way to build mechanical bones and muscles for robotic arm development. Advantages of the proposed bio-inspired robotic arm can be summarized in three aspects: first, its structure is simple, modular and apt to be fabricated; second, its structure is reconfigurable and flexible; and third, the inner hollow space of these robot bones can be used to settle sub modules of robot, which is very useful for some special application such as space technology.

Author contributions. Designed and developed the robotic arm: Zirong Luo. Kinematics Analysis: Zirong Luo, Jianzhong Shang, Guowu Wei and Lei Ren. Simulation ad prototype: Zirong Luo, Jianzhong Shang. Wrote the paper: Zirong Luo, Jianzhong Shang, Guowu Wei and Lei Ren.

Acknowledgements. The author wishes to thank Ernest Appleton, Bo Liao, Yunkai Yang and Jun Zhang for their valuable contributions in developing the prototype.

Edited by: K. Mianowski

References

Acaccia, G., Bruzzone, L., and Razzoli, R.: A modular robotic system for industrial applications, Assembly Automation, 28, 151–162, 2008.

Cai, Z.: Robotics, 2nd Edn., Tsinghua University Press, Beijing, 2009.

Chen, G., Zhang, S., and Li, G.: Multi-stable behaviors of compliant Sarrus mechanisms, J. Mech. Robot., 5, 021005, doi:10.1115/1.4023557, 2013.

Fukuda, T. and Kawauchi, Y.: Robotic system (CEBOT) as one of the realization of self-organizing intelligent universal manipulator, in: Proceedings IEEE Conference on Robotics and Automation, Cincinnati, Ohio, 662–667, 1990.

Fukufda, T. and Nakagawa, S.: Dynamically reconfigurable robotic system, in: Proceedings IEEE Conference on Robotics Automation, Franklin Plaza Hotel, Philadelphia, Pennsylvania, 1581–1586, 1988.

Gavin, D. and Luis, R.: Biologically inspired telescopingactive suspension arm vehicle: Preliminary results, in: IEEE/ASME International Conference on Advanced Intelligent Mechatronics, Montreal, Canada, 1380–1384, 2010.

Guan, Y., Jiang, L., Zhang, K. X., Qiu, J., and Zhou, X.: 1-DOF joint modules and their applications in new robotic systems, in: IEEE International Conference on Robotics and Biomimetics, Bangkok, 1905–1910, 2009.

Guckert, M. and Naish, M. D.: Design of a novel 3 degree of freedom robotic joint, in: IEEE/RSJ International Conference on Intelligent Robots and Systems, NJ, USA, 5146–5152, 2009.

He, X., Kong, X., Chablat, D., Caro, S., and Hao, G.: Kinematic analysis of a single-loop reconfigurable 7R mechanism with multiple operation modes, Robotica, 32, 1171–1188, 2014.

Jin, Z. and Gao, F.: Novel 6-SPS parallel3-dimensional platform manipulator and its force/motion transmission analysis, Chinese J. Mech. Eng., 15, 298–302, 2002.

Katie, L. and Robert, J.: Myriapod-like ambulation of a segmented micro robot, Autonomous Robots, 31, 103–114, 2011.

Kotay, K. and Rus, D.: Locomotion versatility through self-reconfiguration, Robot. Autonom. Syst., 2, 17–23, 1999.

Lee, C.-C.: Kinematic analysis and dimensional synthesis of general-type Sarrus mechanism, JSME Int. J. Ser. C, 39, 790–799, 1996.

Li, J., Zhang, G., Muller, A., and Wang, S.: A family of remote centerof motion mechanisms basedon intersecting motion planes, T. ASME J. Mech. Design, 135, 091009, doi:10.1115/1.4024848, 2013.

Liu, H., Huang, T., Mei, J., Zhao, X., Chetwynd, D. G., Li, M., and Hu, S.: Kinematic design of 5-DOF hybrid robot with largeworkspace/limb-stroke ratio, T. ASME J. Mech. Design, 129, 530–537, 2006.

Lu, T., Kang, Z., and Fang, X.: Numerical calculation method, Tsinghua University Press, Beijing, 2008.

Ranjana, S., Srinath, A., and Richard, G.: Towards a 3g crawling robot through the integration of micro robot technologies, in: IEEE International Conference on Robotics and Automation, Orlando, Florida, USA, 296–302, 2006.

Rus, D. and Vona, M.: Self-reconfiguration planning with compressible unit module, in: IEEE Proceeding of International Conference on Robotics and Automation, Detroit, MI, USA, 2513–2520, 1999.

Sarrus: Note sur la transformation des mouvements rectilignes alternatifsen mouvements circulaireset re ciproquement, Acad. Sci. C. R. Hebd. Seances Acad. Sci., 36, 1036–1038, 1853.

Shammas, E., Wolf, A., and Choset, H.: Three degrees-of-freedomjoint for spatial hyper-redundant robots, Mech. Mach. Theory, 41, 170–190, 2006.

Siciliano, B. and Khatib, O. (Eds.): Springer handbook of robotics, Springer-Verlag, Berlin, Heidelberg, 2008.

Wurst, K. H. and Peting, U.: PKM concept for reconfigurable machine tools, in: The 3rd Chemnitz Parallel Kinematics Seminar, Zwickau, Germany, 63–66, 2002.

Xi, F., Xu, Y., and Xiong, G.: Design and analysis of a reconfigurable parallel robot, Mech. Mach. Theory, 41, 2006.

Yim, M., Duff, D., and Roufas, K.: PolyBot: a modular reconfigurable robot, in: Proceedings of the 2000 IEEE International Conference on Robots & Automation, San Francisco, CA, USA, 514–520, 2000.

Yoshida, K., Hata, N., Oh, S., and Hori, Y.: Extended manipulability measure and application for robot arm equipped with bi-articular driving mechanism, in: Proc. 35th Annu. Conf. IEEE Ind. Electron. Soc., Porto, 3083–3088, 2009.

Zhou, G., Song, B., and Xie, J.: Numerical analysis, Higher Education Press, Beijing, 2008.

Design and analysis of a 3-DOF planar micromanipulation stage with large rotational displacement for micromanipulation system

Bingxiao Ding[1], Yangmin Li[1,2,3], Xiao Xiao[1], Yirui Tang[1], and Bin Li[3]

[1]Department of Electromechanical Engineering, University of Macau, Taipa, Macao SAR, China

[2]Department of Industrial and Systems Engineering, The Hong Kong Polytechnic University, Hung Hom, Hong Kong SAR, China

[3]Tianjin Key Laboratory for Advanced Mechatronic System Design and Intelligent Control, Tianjin University of Technology, Tianjin, China

Correspondence to: Yangmin Li (yangmin.li@polyu.edu.hk)

Abstract. Flexure-based mechanisms have been widely used for scanning tunneling microscopy, nanoimprint lithography, fast servo tool system and micro/nano manipulation. In this paper, a novel planar micromanipulation stage with large rotational displacement is proposed. The designed monolithic manipulator has three degrees of freedom (DOF), i.e. two translations along the X and Y axes and one rotation around Z axis. In order to get a large workspace, the lever mechanism is adopted to magnify the stroke of the piezoelectric actuators and also the leaf beam flexure is utilized due to its large rotational scope. Different from conventional pre-tightening mechanism, a modified pre-tightening mechanism, which is less harmful to the stacked actuators, is proposed in this paper. Taking the circular flexure hinges and leaf beam flexures hinges as revolute joints, the forward kinematics and inverse kinematics models of this stage are derived. The workspace of the micromanipulator is finally obtained, which is based on the derived kinematic models.

1 Introduction

Recently, flexure based mechanism with ultrahigh precision plays an increasing important role in many kinds of fields where high resolution and high repeatability accuracy are demanded, such as optical fiber alignment (Culpepper and Anderson, 2004), bioengineering (Li and Xu, 2006), scanning tunnel microscopy (STM) (Schitter et al., 2008). In precision applications such as following issues have attracted much attention: (1) resolution; (2) number of degrees of freedom(DOF); (3) workspace. The resolution refers to the capability of the system that can distinguish and detect the smallest change of the variable, in micromanipulation system this variable refers to the displacement of the end-effector other than the actuator, because of the amplification mechanism will degrade the system resolution (Ku et al., 2000). For planar mechanism, the maximum number of the DOF is three, i.e. translations along X/Y axis and rotation around Z axis,

it can meet most of planar applications. The workspace of a manipulator is often defined as the set of points that can be reached by its end-effector, namely, it is the space in which the manipulator can work in either a 3-D space or a 2-D surface. The motion range of the micro manipulation stage is typically within several microns, for the reason of limited rotational scope of flexure hinges and limited stroke of piezoelectric actuators which are widely utilized in all kinds of flexure-based stages (Polit and Dong, 2011). For the purpose of getting better performance of the micromanipulation stage, more and more optimized flexure hinges with outstanding characteristics, such as more accurate in position and large motion scope, are designed by researchers (Acer and Sabanovic, 2013; Yi et al., 2003; Qin et al., 2013; Hao and Li, 2014).

The stage should meet the design requirements when it is used in this specific application, this paper proposes a compliant parallel manipulator which adopts lever amplification mechanism to magnify the displacement of the piezoelectric actuators. Based on the structure of the flexure-based stage and the distribution of the actuators, manipulators can also be classified into two categories, i.e. serial structure and parallel structure. The serial structure of the stage enables a simple control strategy, however it endows with high inertia, low natural frequency and cumulative errors. To overcome such drawbacks, parallel kinematic structure is designed because it can provide high load capacity, low inertia, high accuracy and high stiffness. However, the disadvantages are limited workspace and complicated control strategies. Although the compliant parallel mechanism possesses such drawbacks, it has attracted much attention and become an hot issue.

During the literature review, the current research about flexure-based micromanipulation stages for planar application can be classified into 1-DOF, 2-DOF and 3-DOF manipulation stage. Although 1-DOF micromanipulator possesses the advantages of easy to control and no parasitic motion, it has limited applications. So the 2-DOF parallel symmetric micromanipulation stages with decoupled motion are proposed for many planar applications. However, to fully describe the movement of an object needs two translations and one rotation in planar applications, this paper presents a 3-DOF flexure-based parallel micromanipulation stage with large rotational scope for micromanipulation system. In micromanipulation application field, the operator often needs to adjust the position and orientation of biological objects under microscope to do cutting and filtering operation. As shown in Fig. 1, the stage is placed under the microscope, the operator can adjust the position of the object with the 3-DOF micro stage, it can make the manipulation more dexterous by operators.

During the literature review, a myriad of 3-DOF flexure-based compliant parallel mechanism has been fabricated by previous researchers. Generally speaking, the flexure-based micromanipulation stage has limited workspace, to overcome this drawback, many different kinds of amplification mechanisms are adopted to amplify the stroke of piezoelectric actuators. Lu et al. designed compliant parallel micro motion stage with two translations (along X and Y direction) and one rotation (around Z axis), also adopted piezoelectric actuators, but without amplification mechanism, the drawback of this design is the very small workspace (Lu et al., 2004); Bhagat et al. also designed a planar 3-DOF micromanipulator, this mechanism adopted three piezoelectric actuators to achieve required displacements in X, Y and θ and utilized lever amplifier to enhance the displacement of the mechanism (Bhagat et al., 2014); Hao fabricated a flexure-based spatial 3-DOF compliant parallel mechanism with three translational motions along X, Y and Z axis, respectively, moreover this monolithic CPM can be used as positioning stage, acceleration sensor and energy harvesting device (Hao, 2013); Wang and Zhang proposed a 3-DOF nanopositioning stage with two-level lever amplifier with over two hundreds of microns translational displace-

ment along x/y, however the rotational scope and natural frequency are relatively small in Wang and Zhang (2016). In addition, other researchers have also designed many kinds of 3-DOF compliant parallel stages with different characteristics (Dong et al., 2016; Tian and Shirinzadeh, 2009).

The proposed flexure-based positioning stage is featured with two kinds of flexure hinges, right circular hinge and leaf beam flexure hinge in different places and an optimized lever amplification mechanism is adopted to compensate the limited stroke of piezoelectric actuators. The remainder of this paper is organized as follows: The flexure-based parallel positioning stage with optimized structure is proposed in Sect. 2, and also the rotational stiffness is compared between right circular hinge and leaf spring beam flexure hinge; In Sect. 3, the flexure-based mechanism is modeled based on PRB method and the pivot drifting analysis is also conducted; The kinematic analysis of this manipulation stage and the numerical simulation is conducted in Sects. 4 and 5 respectively; The dynamic characteristics and performance of this stage are evaluated in Sect. 6; Finally, the whole works of this research are concluded in Sect. 7 with future works indicated.

2 Three-DOF micromanipulation-stage design

For a planar 3-DOF parallel kinematic micromanipulation stage, the end-effector can translate along X/Y axis and rotate around Z axis. During the literature review, there are many kinds of 3-DOF planar mechanism configurations (Gao et al., 2002). Several typical planar parallel 3-DOF stages are shown in Fig. 2. For the (a) mechanism structural configuration, each chain has only revolute joint; the mechanisms (b) and (c) have prismatic joints and revolute joints. In this study, the (a) structure configuration is adopted to design the micromanipulation stage due to its easy to design and fabricate. When designing a flexure based micromanipulation stage, the structure, flexure hinges and actuators are needed to take into consideration, particularly, the goal of this study is to design a micromanipulation stage with large workspace and high resolution. Compared with shape memory alloy (SMA) and giant magnetostrictive actuator (GMA), the piezoelectric actuator is much cheaper and easily obtained, moreover the PEAs can achieve sub-nano level resolution and has fast response characteristics (Hubbard et al., 2006). Although the voice coil motor(VCM) and electromagnetic actuator (EMA) possess the advantages of large output stroke and driven force, the disadvantages are low resolution and slowly response. Considering all aspects, this design adopts PEA to drive the stage integrated with amplification mechanism to amplify the stroke of the PEAs.

2.1 Flexure hinges

Flexure hinges are basically designing elements in the flexure-based micromanipulation stages, which have been

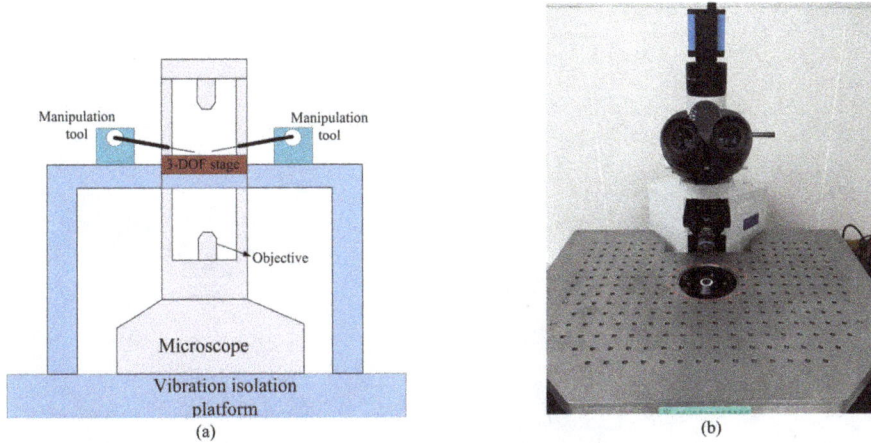

Figure 1. Micromanipulation system. **(a)** Schematic illustration. **(b)** Microscope picture.

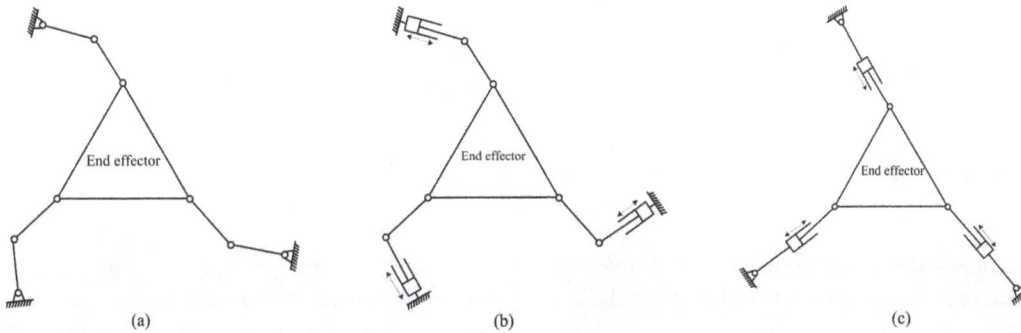

Figure 2. The typical planar parallel 3-DOF kinematic stages.

widely used in these applications where ultrahigh precision motion is needed such as aerospace field, high precision machine and bioengineering field. Furthermore, micromanipulation technology has become an important technology along with the appearance of using flexures. Compared with conventional joints, flexure hinges possess the advantages of no needing for lubrication, no hysteresis, no clearance and no wear. Therefore, flexure-based micromanipulation stages are capable of achieving highly precise positioning accuracy. The goal of this study is to design a compliant parallel micromanipulation stage with large workspace. Taking fabrication process and motion precision into consideration, this design chooses two types of flexures: leaf spring flexure and right circular cut flexure. The rotational stiffness of each type of flexure hinge can be calculated by the following equations, respectively (Smith, 2000):

$$K_{\theta_z M_z} = \frac{EI}{2a_x} \tag{1}$$

$$K_{\theta_z M_z} = \frac{2Ebt^{5/2}}{9\pi a_x^{1/2}} \tag{2}$$

Here, E represents the Young's modulus of aluminum alloy material, I denotes the second moment of area about

the neutral axis and b is depth of the flexure hinge. What's more, the relationship between rotational stiffness of different notch types of flexure hinges and t, a_x are illustrated in Fig. 3a, b respectively, here $E = 71.7\,\text{GPa}$, $b = 8\,\text{mm}$. As Fig. 3 depicted, the leaf spring flexure has lower stiffness than right circular cut flexure with the same t and a_x, therefore the leaf spring flexure hinge has a larger rotation than right circular cut flexure with the same driven force.

2.2 Lever amplification mechanism

The lever amplification mechanism is adopted to compensate the stoke of piezoelectric actuators for its advantage of simple mechanism and easy fabrication comparing with Scott-Russell (SR) Mechanism and Bridge Type Mechanism (BTM). As shown in Fig. 4, three types of lever amplification with different geometric forms have the same amplification ratio and the black dot line denotes the centroid of lever amplification mechanism. Moreover, the geometric form of the lever amplification mechanism will affect the acceleration of the end-effector. The inertia moment of each amplifier mechanism can be calculated by the following equation:

$$J_z = \sum m_i r_i^2 \tag{3}$$

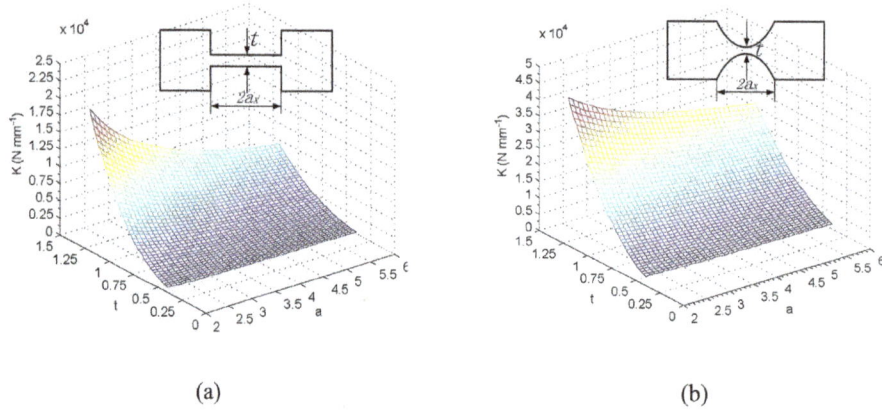

Figure 3. Different types of single DOF flexure hinge and its rotational stiffness with t and a_x. **(a)** leaf spring flexure and its rotational stiffness; **(b)** right circular cut flexure and its rotational stiffness.

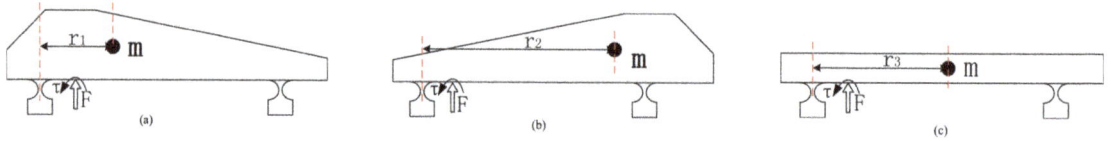

Figure 4. The lever amplification mechanism with different geometric forms.

where the m_i denotes mass distribution and r_i denotes the distance between each mass and rotational axis. So the following equation can be derived:

$$\begin{cases} J_1 = mr_1^2 \\ J_2 = mr_2^2 \\ J_3 = mr_3^2 \end{cases} \quad (4)$$

For the reason of $r_1 < r_3 < r_2$, the following relationship can be obtained:

$$J_1 < J_3 < J_2 \quad (5)$$

With the same driving force F, inducing the same moment τ, the acceleration of the end-effector with different geometries can be written as:

$$\alpha_i = \frac{\tau}{J_i} \quad (6)$$

where $i = 1, 2, 3$. From above analysis, inducing $\alpha_1 > \alpha_3 > \alpha_2$, it means that Fig. 4a is the reasonable geometry of lever amplification mechanism, (b) and (c) will degrade the response time of the manipulation stage.

2.3 Preload mechanism

It is well known that the piezoelectric actuator possesses the advantages of high resolution and fast response, however, it can not bear the lateral force since the lateral force or moment may cause the damage to the piezoelectric actuators, so shear stress and tensile stress must be avoided during the

actuation. Also during the positioning process, the piezoelectric actuator needs to maintain a continuous connection state with the mobile platform. During the literature review, the typical pre-tightening mechanism uses the bolt to generate the thrust force and the bolt contact with piezoelectric directly, as shown in Fig. 5a, however the lateral force and moment can not be avoided and pre-load force can not be measured. A new pre-tightening mechanism is proposed in this study, as shown in Fig. 5b, the piezoelectric actuator and mobile platform maintain a line-face contact. The interaction of the semi-cylinder and the adjustable pre-load force block preclude the generation of the lateral force and bending moment.

2.4 Structural configuration

For a 3-RRR (R refers to revolute joint) parallel mechanism, it has two interesting configurations of the structure. As depicted in Fig. 6a and b, the same symbol denotes the same meaning, it means that the two schematic diagrams have the same geometric architecture configuration. The 3-RRR diagram (a) and (b) represent the structure with minimal rotation and maximal rotation, respectively. With the equal actuation force, the torque can be calculated by following equation:

$$M_1 = F_1 \cdot (DB_1 + C_1 O) + F_2 \cdot (DB_2 + C_2 O) + F_3 \quad (7)$$
$$\cdot (DB_3 + C_3 O)$$

$$M_2 = F_1 \cdot DB_1 + F_2 \cdot DB_2 + F_3 \cdot DB_3 \quad (8)$$

Obviously, $M_1 > M_2$, it means that with the equal driving force, the (b) will generate a large rotational angle. So the

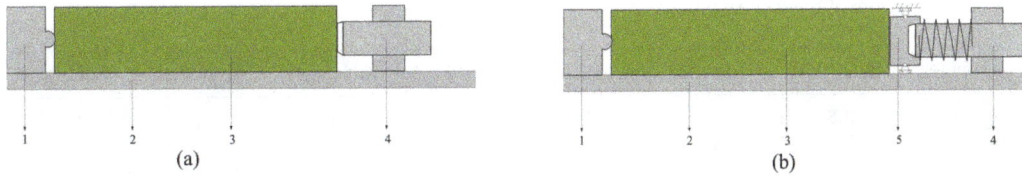

Figure 5. Pre-tightening mechanism. **(a)** traditional pre-tightening mechanism; **(b)** proposed pre-tightening mechanism; 1, mobile platform; 2, fixed platform; 3, piezoelectric actuator; 4, bolt; 5, adjustable pre-load force block.

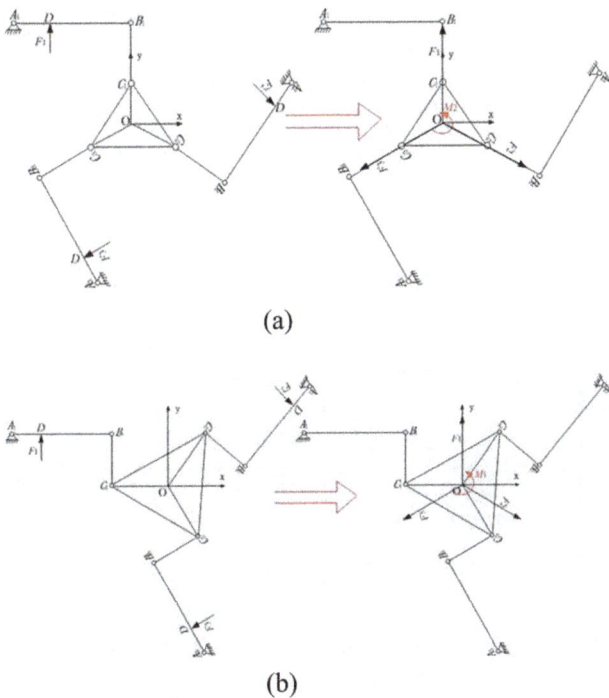

(a)

(b)

Figure 6. Two different 3-RRR structural configurations.

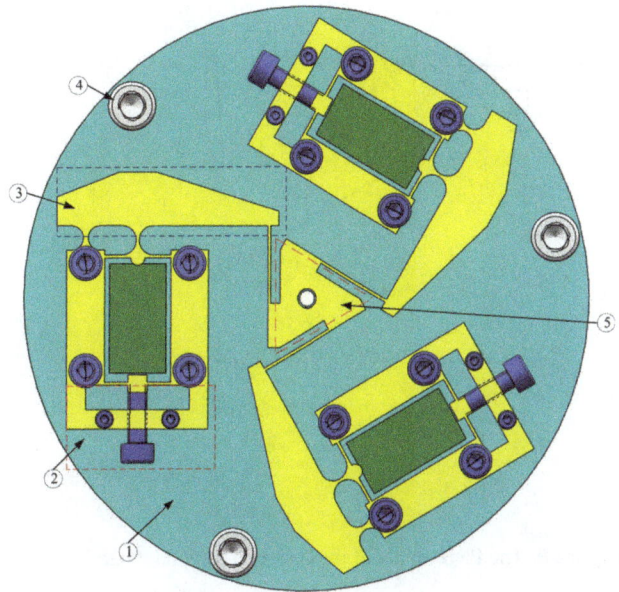

Figure 7. The whole assembled diagram of this three-DOF microstage: (1) The base platform; (2) The pre-tightening mechanism; (3) The lever amplifier; (4) The grounding bolt; (5) The central motion platform.

structural configuration (b) is adopted to design the 3-DOF micromanipulation stage for a large rotational scope.

Based on the above analysis, the designed monolithic 3-DOF compliant parallel mechanism integrated with lever amplification mechanism is presented in Fig. 7. The proposed compliant parallel mechanism is designed to be used as a high precision planar positioning stage to manipulate objects under the microscopy and the dimension scale of this stage is approximately $120 \times 120 \times 8 \, \text{mm}^3$. In order to validate the correction of the designed mechanism, the Workbench software is adopted to simulate this parallel mechanism. As shown in Fig. 8, it does not exist any physical interferences under the condition of maximum input displacement, which indicates that the structural design is reasonable. The main design criteria of this compliant mechanism is to get a large workspace of the end-effector, to achieve this goal the beam flexures are exploited to connect the motion platform and lever amplification mechanism, meanwhile right circular hinges are adopted as revolute joints. The ex-

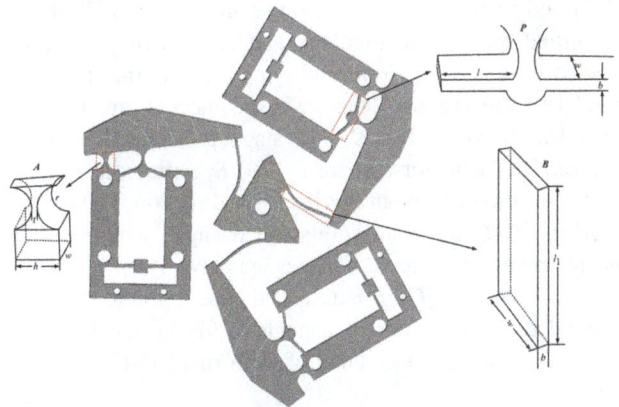

Figure 8. The motion simulation of manipulation stage with the condition of maximum input displacement.

act architectural parameters of flexure hinges are presented in Table 1.

Table 1. Architectural parameters of the flexure hinges.

parameter	value(mm)	parameter	value(mm)	parameter	value(mm)	parameter	value(mm)
t	0.8	h	8	l	8	l_1	16
r	2.5	w	8	b	0.6		

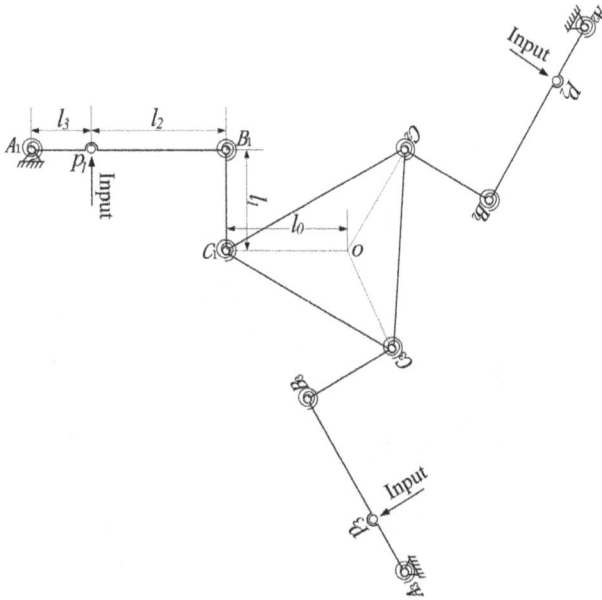

Figure 9. The PRB model of the designed compliant parallel stage.

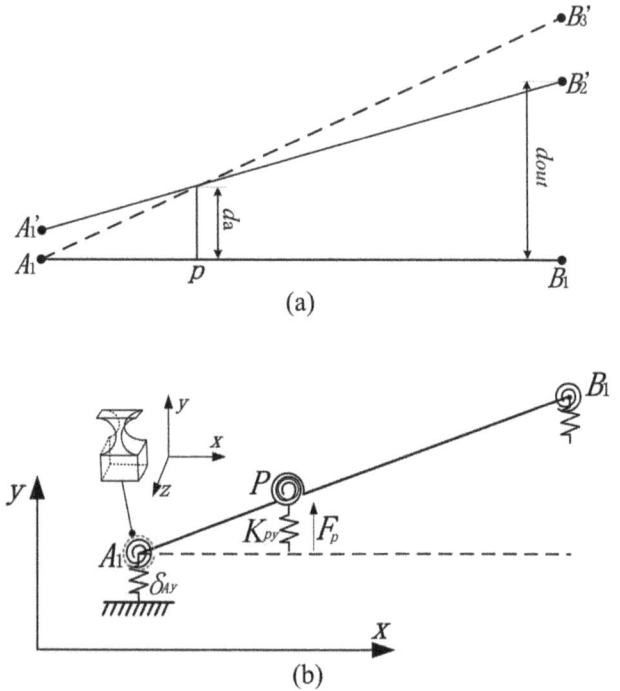

Figure 10. Lever amplifier mechanism. **(a)** Bending schematic diagram; **(b)** analysis model.

3   The micromanipulation stage modeling

Compliant mechanisms involving with nonlinearity characteristics make the analysis and modeling very difficult. To facilitate the analysis of the manipulation stage, the pseudo-rigid-body (PRB) model is adopted in this study. With the identified mechanism topology and each flexure hinge replaced by a revolute joint and a torsional spring, the PRB model of the proposed flexure-based mechanism is developed as shown in Fig. 9. The stage is actuated by three stacked PZT actuators, where P_1, P_2, P_3 represents each actuator location and input driving force direction. The mobile platform $C_1C_2C_3$ is an equilateral triangle, which is connected by three identical linkages, namely, $A_1B_1 = A_2B_2 = A_3B_3$, $B_1C_1 = B_2C_2 = B_3C_3$. The global frame XOY is grounded at the center of triangle $A_1A_2A_3$ and the local frame is attached at the center of platform $C_1C_2C_3$.

3.1   Pivot drifting analysis

The workspace of the manipulation stage depends on the actual input displacement of the piezoelectric actuator and the actual amplification ratio of the lever mechanism, not the theoretical amplification ratio. From the experience, the lever mechanism will produce bending deformation and pivot

drifting when a loading is applied, the bending deformation can be ignored in this research for the geometric form of the lever amplifier mechanism. As shown in Fig. 10a, the theoretical magnification ratio of the mechanism is A_1B_1/A_1p, and the output displacement is B_1B_3'. However, the actual output displacement of the lever amplifier mechanism is B_1B_2', when considering the displacement loss caused by the pivot drifting. Obviously, the actual amplification ratio $A_{amp} = d_{out}/d_a$ is smaller than the theoretical ratio due to the displacement loss of the lever mechanism. The displacement loss needs to be taken into consideration in order to get a more accurate relationship between the input stroke and output displacement of the mobile platform.

As shown in Fig. 10b, when the driving force F_p is applied on the mechanism, a pivot drift displacement δ_{Ay} will be produced along the y direction of the circular flexure hinge for its compliance in y axis. And δ_{Ay} can be solved by the following equation:

$$\delta_{Ay} = \frac{F_p}{K_{Ay}} \tag{9}$$

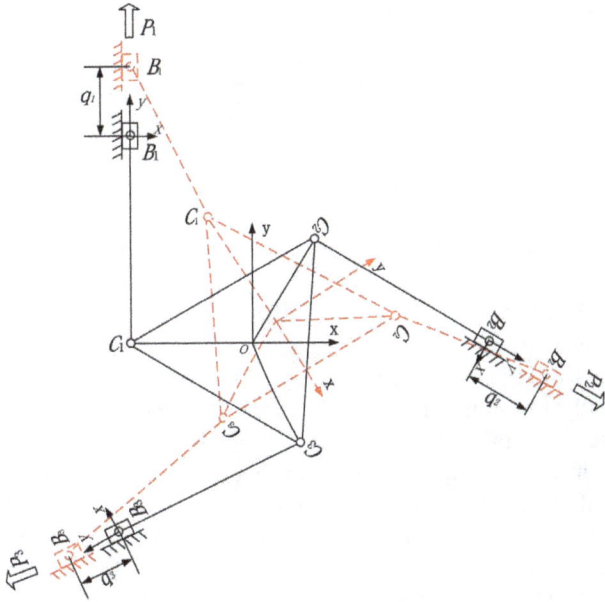

Figure 11. The kinematic schematic of 3-PRR planar mechanism.

where K_{Ay} is the stiffness of the circular hinge along the Y axis, and it can be calculated by the following equation: Wu and Zhou (2002):

$$K_{Ay} = Ew / \left\{ \frac{2(2r/t+1)}{\sqrt{4r/t+1}} \arctan\sqrt{4r/t+1} - \frac{\pi}{2} \right\} \quad (10)$$

$$F_p = K_{py} \cdot d_a \quad (11)$$

where $K_{py} = K_{pc} + K_{pb}$ denotes the total stiffness of the position p along the driven direction, $K_{pc} = K_{Ay}$ is the stiffness of the circular hinge along the y axis and $K_{pb} = \frac{Eb^3 w}{4l^3}$ represents the stiffness of the beam flexure hinge; d_a is the actual output displacement of the piezoelectric actuator; and the output displacement is $d_{out} = (d_a - \delta_{Ay})(1 + \frac{l_2}{l_3}) + \delta_{Ay}$. So the relationship between the input displacement and the output displacement of the lever mechanism A_{amp} can be derived as follows:

$$\frac{d_{out}}{d_a} = 1 + \frac{l_2}{l_3} \cdot \frac{b^3 \cdot \left\{ \frac{2(2r/t+1)}{\sqrt{4r/t+1}} \arctan\sqrt{4r/t+1} - \frac{\pi}{2} \right\}}{4l^3} \quad (12)$$

Here, E denotes the Young's modulus of the aluminum alloy material, and the parameters w, t, r and l are illustrated in Fig. 8. The amplification ratio is 3.2780 when taking the pivot drift into consideration, it exists about 8.16 % deviation compared with nominal amplification ratio (3.5455); and exists about 3.68 % deviation error compared with simulation result (3.1573).

4 Kinematic analysis

The kinematic diagram of the designed mechanism is presented in Fig. 11, q_i, $(i = 1, 2, 3)$ denotes the output dis-

placement of the lever mechanism, it means that $q_i = d_{out}$. The workspace of mobile platform is defined by a vector $\boldsymbol{\mu} = (x, y, \theta)^T$, which represents the position and orientation of the reference point. And the input stroke of each stacked actuator can be denoted by $\boldsymbol{d}_i$, $(i = 1, 2, 3)$; the Jacobian matrix which connects the input stroke and output displacement of end-effector is denoted by J, so we can get the following equation:

$$\begin{bmatrix} x \\ y \\ \theta \end{bmatrix} = J \begin{bmatrix} d_1 \\ d_2 \\ d_3 \end{bmatrix} \quad (13)$$

Here, we use l_0, l_1 to denote the length of $C_i O$ and $B_i C_i$, $(i = 1, 2, 3)$, respectively, homogeneous transformation matrix from the global frame XOY to the coordinates XO_iY can be described by this equation:

$$^O T_i = \begin{bmatrix} c_i & -s_i & l_0 \\ s_i & c_i & -l_1 \\ 0 & 0 & 1 \end{bmatrix} \quad (14)$$

where $c_i = \cos\alpha_i$, $s_i = \sin\alpha_i$, $\alpha_1 = 0$, $\alpha_2 = \frac{2\pi}{3}$, $\alpha_3 = \frac{4\pi}{3}$, $i = 1, 2, 3$.

Also, the motion of the mobile platform can be described in the global frame XOY.

$$^O T_O = \begin{bmatrix} c & -s & x \\ s & c & y \\ 0 & 0 & 1 \end{bmatrix} \quad (15)$$

where, $c = \cos\theta$, $s = \sin\theta$. Here, we adopt $^O T_i$ to represent the homogeneous transformation matrix from the mobile platform to each coordinate XO_iY. Combining the Eqs. (13) and (14) together, we can get the following equation:

$$^O T_i = \begin{bmatrix} c_i & -s_i & l_0 \\ s_i & c_i & -l_1 \\ 0 & 0 & 1 \end{bmatrix} \begin{bmatrix} c & -s & x \\ s & c & y \\ 0 & 0 & 1 \end{bmatrix} \quad (16)$$

Here, we define the $s = \theta$, $c = 1$, because the rotational angle of the platform is very tiny. After a further processing of Eq. (15), we can obtain the relatively simple $^O T_i$ as follows:

$$^O T_i = \begin{bmatrix} c_i - s_i\theta & -c_i\theta - s_i & xc_i - ys_i + l_0 \\ s_i + C_i\theta & c_i - s_i\theta & xs_i + yc_i - l_1 \\ 0 & 0 & 1 \end{bmatrix} \quad (17)$$

In the frame of XOY, the coordinates value of C_i, $(i = 1, 2, 3)$ can be expressed as $^O C_1 = (-l_0, 0)$, $^O C_2 = \left(\frac{l_0}{2}, \frac{\sqrt{3}l_0}{2}\right)$, $^O C_3 = \left(\frac{l_0}{2}, \frac{-\sqrt{3}l_0}{2}\right)$.

$$\begin{cases} ^i C_i x = (c_i - s_i\theta)^O C_i x - (c_i\theta + s)^O C_i y + xc_i - ys_i + l_0 \\ ^i C_i y = (s_i + c_i\theta)^O C_i x - (s_i\theta cs_i)^O C_i y + xs_i - yc_i - l_1 \end{cases} \quad (18)$$

where $i = 1, 2, 3$. For the length of $B_i C_i$ is constant constraint, we can obtain the following equation:

$$(q_i + {}^i C_i y)^2 + {}^i C_i x^2 = l_1^2 \qquad (19)$$

So the q_i can be calculated by the following equation:

$$q_i = l_1 \sqrt{1 - \left(\frac{{}^i x}{l_1}\right)^2} - {}^i C_i y \qquad (20)$$

The above Eq. (19) is a nonlinear equation, after further analysis, we can get ${}^i C_i x / l_1 \approx 0$, subsequently, we can get the following linearized equation:

$$q_i = l_1 - {}^i C_i y \qquad (21)$$

We can obtain the following equations after substituting (17) into (20).

$$\begin{cases} q_1 = l_0 \theta + y \\ q_2 = l_0 \theta - \frac{\sqrt{3}}{2} x - \frac{1}{2} y \\ q_3 = l_0 \theta + \frac{\sqrt{3}}{2} x - \frac{1}{2} y \end{cases} \qquad (22)$$

So the relationship between input stroke of piezoelectric actuators and the position and orientation of the central platform can be written as follows:

$$\begin{bmatrix} x \\ y \\ \theta \end{bmatrix} = A_{\mathrm{amp}} \begin{bmatrix} 0 & -\frac{\sqrt{3}}{3} & -\frac{\sqrt{3}}{3} \\ \frac{2}{3} & -\frac{1}{3} & -\frac{1}{3} \\ \frac{1}{3l_0} & \frac{1}{3l_0} & \frac{1}{3l_0} \end{bmatrix} \begin{bmatrix} d_1 \\ d_2 \\ d_3 \end{bmatrix} \qquad (23)$$

where the A_{amp} is the actual amplification ratio of the lever amplifier.

5 Workspace analysis

The workspace of the 3-DOF manipulator can be calculated by the above aforementioned kinematic analysis. Actually, the reachable workspace of the manipulator not only depends on the above Eq. (22), but also has the relationship with material property. The following list is the constraint condition of the 3-DOF micromanipulator:

$$\begin{cases} 0 \le d_i \le d_{\max}, \quad i = 1, 2, 3 \\ \sigma < \sigma_m \end{cases} \qquad (24)$$

where d_i denotes the output displacement of the piezoelectric actuators; and σ is the stress of the material.

Here, the piezoelectric actuators $P - 820.20$ are adopted to drive the stage, so the first constraint condition is the input stroke of each piezoelectric actuator:

$$0 \le d_i \le 20 \qquad (25)$$

where $i = 1, 2, 3$.

Table 2. Length parameters of microstage.

parameters	values(mm)
$A_1 B_1$	39
$B_1 C_1$	16
$C_1 C_2$	12
$A_1 P_1$	11

Based on above analysis, the workspace of the micromanipulator can be obtained by calculating the outputs of the reference point. The 3-D workspace is graphically illustrated in Fig. 12a. From the simulation results, the reachable workspace of the end-effector is a hexahedron, and the projection area on XY plane is varying with the θ value. The maximum projection on the XY plane is shown in Fig. 12b. From the simulation results, the reachable workspace along X/Y axis and around Z is $[-42.31, 42.31]\,\mu\mathrm{m}$, $[-48.56, 48.56]\,\mu\mathrm{m}$, $[0, 10.28]\,\mathrm{mrad}$, respectively.

6 Dynamic characteristics

The dynamic response has to be modeled in order to fully describe the free vibrations of the 3-DOF micro manipulation stage and ensure the flexure-based mechanism operate properly in the dynamic range. Referring to the PRB method, the flexure hinges generated rotational motion around Z axis are treated as an ideal revolute joint with torsional spring. The output variables of the end-effector are defined as $\overrightarrow{\boldsymbol{O}} = [x_o, y_o, \theta_o]^T$. During the simulation analysis, the motion type of the lever $A_i B_i (i = 1, 2, 3)$ is mainly rotational motion, and beam links $B_i C_i (i = 1, 2, 3)$ will be generated bending deformation when the input is applied. So the total kinematic energy of the manipulation stage is composed by the translational kinematic energy T_t of the end-effector and rotational kinematic energy T_r of the lever $A_i B_i (i = 1, 2, 3)$ and the end-effector.

$$T_r = \frac{1}{2} \sum_{i=1}^{3} J_{A_i B_i} \omega_{A_i B_i}^2 + \frac{1}{2} J_o \dot{\theta}_o^2$$

$$T_t = \frac{1}{2} m_o \left(\dot{x}_o^2 + \dot{y}_o^2 \right) \qquad (26)$$

where $J_{A_i B_i}$ denotes the moment of inertia of each lever; $\omega_{A_i B_i}$ represents the angular velocity of each lever. As illustrated in Fig. 8, the potential energy mainly concentrates on beam type flexure hinges at position $P_i (i = 1, 2, 3)$ and beam links $B_i C_i (i = 1, 2, 3)$ when input force is applied. So the total potential energy V of the manipulation stage can be calculated by the following equation:

$$V = \frac{1}{2} K_x x_o^2 + \frac{1}{2} K_y y_o^2 + \frac{1}{2} K_\theta \theta_o^2 + \frac{3}{2} K_{pb} \left(\frac{d_{\mathrm{out}}}{A_{\mathrm{amp}}} \right)^2 \qquad (27)$$

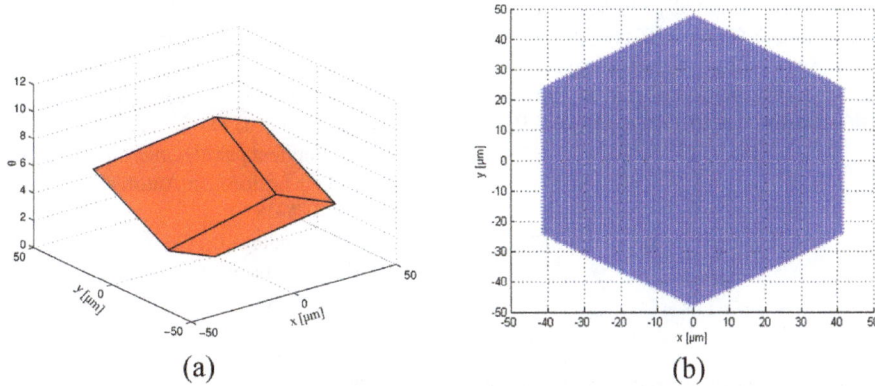

Figure 12. Reachable workspace of the stage. **(a)** 3-D workspace; **(b)** maximum projection on XY plane.

where the K_x, K_y, K_θ denotes the equivalent stiffness of the micro manipulation stage along X/Y axis and around Z axis, respectively. The numerical solution of K_x, K_y, and K_θ can be obtained by some experimental techniques. Substituting the potential energy and kinematic energy into the Lagrange's equation can yield to the following equation:

$$M[\ddot{x}_o \quad \ddot{y}_o \quad \ddot{\theta}_o]^T + K[x_o \quad y_o \quad \theta_o]^T = [F_1 \quad F_2 \quad F_3]^T \qquad (28)$$

where $\boldsymbol{F}_i (i = 1, 2, 3)$ is the ith input force generated by piezoelectric actuator. Then the following mass-stiffness matrix characteristic equation can be obtained:

$$|K - M\lambda_i^2| = 0 \qquad (29)$$

So the natural frequencies can be calculated by $f_i = \lambda_i / 2\pi$.

In this section, the dynamic characteristics of the manipulation stage are obtained through the modal analysis by Workbench. The first four modal shapes of the micromanipulation stage without piezoelectric actuator assembled are represented in Fig. 13. The first four natural frequencies are 1280.3, 1308.4, 1618.6 and 1918.2 Hz. It should be noticed that dynamic performance of micro-manipulation stage will be mainly determined by some experimental procedures, which are planned in our future work and the aforementioned analysis and simulation results will be validated.

7　Conclusions

A flexure-based monolithic micro manipulation stage with large workspace is designed and analyzed in this paper. The piezoelectric actuators are adopted to drive the manipulation stage because they can provide with the fast response and large output driven force. The optimized lever amplifier is integrated into the mechanism in order to compensate the stroke of the piezoelectric actuators. For the purpose of getting larger workspace, different kinds of flexure hinges, different structural configurations and lever amplifiers with different geometric forms are compared in this paper. To validate the design, the kinematics model and reach-

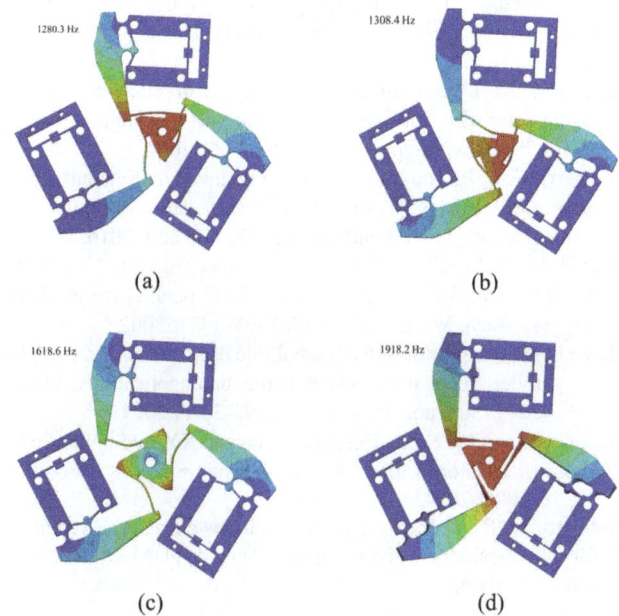

Figure 13. The lever amplification mechanism with different geometric forms.

able workspace of the manipulation stage are analytically obtained. The working range of the manipulation stage along each axis is ± 42.31 μm, ± 48.56 μm, 0–10.28 mrad, respectively. Then the dynamic characteristics and performance evaluation of the designed mechanism are conducted via the Workbench software. The prototype fabrication and experimental validations, and control of the micro manipulation stage will be performed in our future work.

Competing interests. The authors declare that they have no conflict of interest.

Acknowledgements. This work is partially supported by National Natural Science Foundation of China (51575544,51275353), Macao Science and Technology Development Fund (110/2013/A3,108/2012/A3), Research Committee of University of Macau (MYRG2015-00194-FST, MYRG203(Y1-L4)-FST11-LYM).

Edited by: G. Hao

References

Acer, M. and Sabanovic, A.: Micro position control of a designed 3-PRR compliant mechanism using experimental models, 9th Asian Control Conference, 1–6, 2013.

Bhagat, U., Shirinzadeh, B., Clark, L., Chea, P., Qin, Y. D., Tian, Y. L., and Zhang, D. W.: Design and analysis of a novel flexure-based 3-DOF mechanism, Mech. Mach. Theory, 7, 173–187, 2014.

Culpepper, M. L. and Anderson, G.: Design of a low-cost nanomanipulator which utilizes a monolithic, spatial compliant mechanism, Precision Engineering, 28, 469–482, 2004.

Dong, Y., Gao, F., and Yue, Y.: Modeling and experimental study of a novel 3-RPR parallel micro-manipulator, Robotics and Computer-IntegratedManufacturing, 37, 115–124, 2016.

Gao, F., Li, W. M., Zhao, X. C., Jin, Z. L., and Zhao, H.: New kinematic structures for 2-, 3-, 4-, and 5-DOF parallel manipulator designs, Mech. Mach. Theory, 37, 1395–1411, 2002.

Hao, G.: Towards the design of monolithic decoupled XYZ compliant parallel mechanisms for multi-function applications, Mech. Sci., 4, 291–302, doi:10.5194/ms-4-291-2013, 2013.

Hao, G. B. and Li, H. Y.: Design of 3-legged XYZ compliant parallel manipulators with minimised parasitic rotations, Robotica, 33, 787–806, 2014.

Hubbard, N. B., Culpepper, M. L., and Howell, L. L.: Actuators for micropositioners and nanopositioners, Appl. Mech. Rev., 59, 324–334, 2006.

Ku, S. S., Pinsopon, U., Cetinkunt, S., and Nakajima, S. I.: Design, fabrication, and real-time neural network control of a three-degrees-of-freedom nanopositioer, IEEE/ASME Transactions on Mechatronics, 5, 273–280, 2000.

Li, Y. M. and Xu, Q. S.: A novel design and analysis of a 2-DOF compliant parallel micromanipulator for nanomanipulation, IEEE Transactions on Automation Science and Engineering, 3, 248–254, 2006.

Lu, T. F., Handley, D. C., Yong, Y. K., and Eales, C.: A three-DOF compliant micromotion stage with flexure hinges, Industrial Robot: An International Journal, 31, 355–361, 2004.

Polit, S. and Dong, J.: Development of a high-bandwidth XY nanopositioning stage for high-rate micro-/nanomanufacturing, IEEE/ASME Transactions on Mechatronics, 16, 724–733, 2011.

Qin, Y. D., Shirinzadeh, B., Zhang, D. W., and Tian, Y. L.: Design and kinematics modeling of a novel 3-dof monolithic manipulator featuring improved Scott-Russell mechanisms, Journal of Mechanical Design, 135, 1–9, 2013.

Schitter, G., Thurner, P. J., and Hansma, P. K.: Design and input-shaping control of a novel scanner for high-speed atomic force microscopy, Mechatronics, 18, 282–288, 2008.

Smith, S. T.: Flexures: elements of elastic mechanisms, CRC Press, 2000.

Tian, Y. L. and Shirinzadeh, B.: Development of a flexure-based 3-RRR parallel mechanism for nano-manipulation, IEEE/ASME International Conference on Advanced Intelligent Mechatronics, 1324–1329, 2009.

Wang, R. Z. and Zhang, X. M.: A planar 3-DOF nanopositioning platform with large magnification, Precision Engineering, 46, 221–231, 2016.

Wu, Y. F. and Zhou, Z. Y.: Design of flexure hinges, Eng. Mech., 19, 136–140, 2002.

Yi, B., Chung, G., Na, H., Kim, W., and Suh, I.: Design and experiment of a 3-DOF parallel micromechanism utilizing flexure hinges, IEEE Transactions on Robotics and Automation, 19, 604–612, 2003.

3

An Algebraic Formulation for the Configuration Transformation of a Class of Reconfigurable Cube Mechanisms

Chin-Hsing Kuo, Jyun-Wei Su, and Lin-Chi Wu

Department of Mechanical Engineering, National Taiwan University of Science and Technology, Taipei 106, Taiwan

Correspondence to: Chin-Hsing Kuo (chkuo717@mail.ntust.edu.tw)

Abstract. This paper presents an algebraic strategy for formulating the configuration transformation of a special class of reconfigurable cube mechanism (RCM) made by 2^3 cyclically connected sub-cubes. The RCM studied here is kinematically equivalent to a spatial eight-bar linkage having eight transformable configurations. In this paper, the reconfiguration characteristics of the RCM are figured out first. Then, the initial configuration of the RCM is described by a joint-screw matrix, from which all the consecutive joint-screw matrices that represent the configuration transformation of the RCM can be derived. An illustrative example is provided to determine the eight joint-screw matrices of an RCM at an initial configuration. This reconfiguration formulation is further applied to enumerate all feasible topological configurations of such a special reconfigurable mechanism. The results show that, for such a special kind of reconfigurable cube mechanisms, there is only one feasible initial topological configuration for the RCM to perform a complete cycle of reconfiguration.

1 Introduction

There is a special class of reconfigurable cube mechanism (RCM) equivalent to an eight-bar closed-loop spatial linkage with cyclic reconfiguration, Fig. 1. The mechanism can demonstrate eight different operation configurations, i.e., configuration A to H in Fig. 1, where nine different figures are exposed on the outer surface of the RCM. An earlier study (Kuo and Su, 2017) to this special mechanism has shown that the manipulation of this elegant artifact can be interpreted by using the mechanism theories in terms of variable mobility and isomorphism identification.

When investigating the topological properties of reconfigurable mechanisms, configuration transformation is one interesting topic to be explored. In literature, a couple of studies have made significant contributions to the configuration transformation analysis for some specific reconfigurable mechanisms. For example, Wohlhart (1996) introduced a special linkage, namely "kinematotropic linkage," that could permanently change its mobility by reconfigure the linkage into different working configurations. Dai and Rees

Jones (1999a, b, 2005) studied a special foldable/erectable mechanism and developed an EU-elementary matrix operation for formulating its topology transformation (Dai and Rees Jones, 2005). Zhang and Dai (2008, 2009) proposed an evolutionary reconfiguration algorithm of general spatial metamorphic mechanisms. Yan and Kang (2009a) studied the configuration transformation of variable topology mechanisms based on the concept of mapping function, leading to a general methodology for configuration synthesis of variable topology mechanisms (Yan and Kang, 2009b). Yan and Kuo (2007, 2009) put forward a systematic approach for configuration analysis and synthesis for general variable topology mechanisms where the topological reconfiguration can be described by graph (Yan and Kuo, 2006a), finite state machine (Yan and Kuo, 2006b), and screw matrix (Kuo and Yan, 2007). Ghrist and Peterson (2007) realized the reconfiguration of reconfigurable systems in robotics and biology by using state complex technique.

In addition to the general theories of reconfiguration, some tailor-made approaches for formulating the configuration transformation were available for several specific applica-

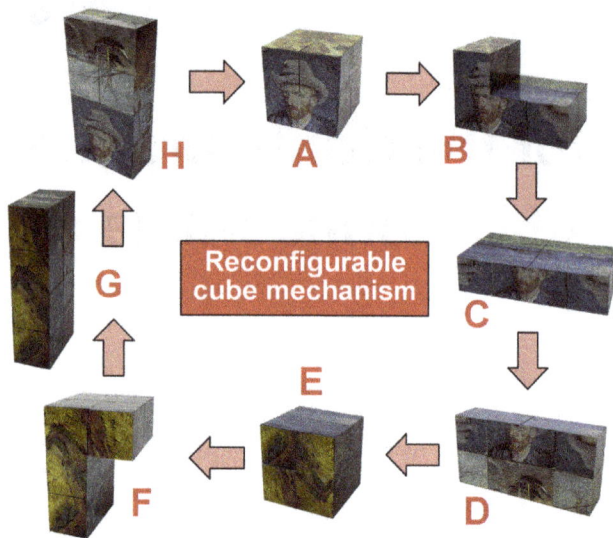

Figure 1. The reconfigurable cube mechanism (RCM).

tions. For instance, Liu and Dai (2002) investigated the folding process of packaging cartons, leading to a reconfiguration methodology and algorithm for the reconfigurable carton folding. Ding and Yang (2012) evaluated the geometry and reconfiguration principles of a special Mandala-type artifact and discussed its application for aerospace engineering. Wei et al. (2010, 2011) analyzed the configuration singularity and reconfiguration properties of a Hoberman switch-pitch ball. Ding et al. (2013) designed a deployable polyhedral linkage for which the changeable configuration was presented by joint screws. Ding and Lu (2013) analyzed the motion sequence and isomorphism of a chain-type cube mechanism. Gan et al. (2009, 2010, 2013a, b) proposed a metamorphic parallel mechanism whose configuration was changed via a special reconfigurable joint. Zhang et al. (2010a, 2012) presented other fantastic metamorphic parallel mechanisms via the concepts of variable-axis joints, origami folding (Zhang et al., 2010b), and kirigami (Zhang and Dai, 2014).

On the other hand, topology synthesis of reconfigurable mechanisms is also an interesting and challenging problem. In the past decades, topology synthesis of reconfigurable mechanisms has been attempted for several special mechanisms, e.g., the kinematotropic linkages (Galletti and Fanghella, 2001), metamorphic mechanisms (Zhang et al., 2008; Zhang and Dai, 2009), variable topology mechanisms (Yan and Kuo, 2009; Kuo and Chang, 2014; Shieh et al., 2011), and reconfigurable mechanisms (Kuo et al., 2009; Huang et al., 2010). However, those studies were mostly focused on the deployable or folding mechanisms. For the reconfigurable mechanisms formed by connected sub-cubes that we study here, its topology synthesis and enumeration tasks are still an open problem.

Therefore, the aim of this work is to develop an algebraic formulation for describing the configuration transformation

of the RCM made by 2^3 sub-cubes. Then, based on the developed formulation, all topologically non-isomorphic configurations of the RCMs are enumerated. In what follows, Sect. 2 firstly studies the reconfiguration characteristics of the RCM. Accordingly, Sect. 3 puts forward an algebraic computational procedure for representing the configuration transformation of the studied RCM. Based on this computational procedure, Sect. 4 briefs the enumeration of all possible topologically non-isomorphic configurations for the RCM with 2^3 sub-cubes. Finally, Sect. 5 concludes the works and contribution of this paper.

2 Configuration Characteristics

By observing the RCM in Fig. 1, the topological and reconfiguration characteristics of the RCM can be concluded as follows.

2.1 Topological Characteristics

The topological characteristics of the RCM include:

1. The RCM is always a single-loop closed kinematic chain during reconfiguration.

2. All the sub-cubes are topologically similar class. In the RCM, the links are all binary links, the joints are all revolute joints, and the linkage is a single-loop chain. Therefore, each link is topologically identical to each other, i.e., the links are similar class (Harary, 1964).

3. The orientation of each joint is not changed after the joint is reconfigured. Referring Fig. 4 in (Kuo and Su, 2017), for example, when the blue sub-cube is grounded, the RCM can verify a series of configuration changes as shown from Fig. 4a to h. Then, it can be verified that the orientation of each joint will remain the same in all configurations, even if the joint axis is displaced from some configuration to another one. For example, joint c in the figure always points at the z-direction in the eight configurations.

4. Each configuration must have joints pointing at the x-, y-, and z-directions, respectively. Subject to point 3), since the RCM has x-, y-, and z-direction joints at any initial configuration, each configuration will possess joints corresponding to the three directions, respectively. This fact can be verified from Fig. 4 in (Kuo and Su, 2017).

5. Since all the joints are incident to the edges of the sub-cubes, all the joints form an orthogonal pattern in each configuration, i.e., they are either parallel or orthogonal to one another in each configuration.

6. In each configuration, there must have exactly two, two, and four joints pointing at the three axial directions,

respectively. For example, there may have two joints pointing at the z-direction, two joints at the x-direction, and two joints at the y-direction in some configuration.

2.2 Reconfiguration Characteristics

The reconfiguration characteristics of the RCM include:

1. When any two joints are being coaxial, they will become a pair of workable joints, i.e., the degree of freedom of motion of the joint is not restrained by the configurations or the link shape.

2. The workable joints must appear in pairwise—there is no single workable joint on an axis.

3. When there exists a pair of workable joints in the configuration, this configuration is able to transform into the next one. For example, in Fig. 4h of (Kuo and Su, 2017), joints (b, f) and (d, h) are the only two pairs of workable joints. So joints (b, f) or (d, h) may be actuated to transfer configuration H into G or A, respectively.

4. The configuration of the RCM can be classified into "operation status" or "transition status." When the RCM is at an operation status, it has two or more pairs of workable joints that are pointing at two different orientations. On the other hand, when the RCM is at a transition status, it has only one pair of workable joints. Therefore, referring Fig. 4 in (Kuo and Su, 2017), configurations A, C, D, E, G, and H are at operation status, whereas configurations B and F are at transition status.

5. The transformation from configuration X to Y (X, Y = A, B, ..., H) is called "forward" if it follows the transferring direction in the previous one. Otherwise, it is called a "backward" transformation. For example, Fig. 2 shows the reconfiguration sequence of the RCM of Fig. 4 in (Kuo and Su, 2017). The circles denote the configurations A to H and two trivial configurations B′ and F′ (that are equivalent to configurations B and F, respectively). Notation (α, β) between two circles indicates the working joints under which the configurations can be transformed. For example, configuration A can be transformed to B by operating joints (a, g) and vice versa. Therefore, as shown on the figure, configuration A has three pairs of workable joints, i.e., (a, g), (d, h), and (c, e), and it can be transformed into different configurations as different workable joints are operated. So, if the RCM is reconfigured via a sequence of A → B → C → D, the transformations from B to C and from C to D are both forward. If the RCM is reconfigured, for example, via a sequence of A → B → C → D → C, the transformation from D to C is backward. As shown in the figure, it is noticed that configuration A can be transformed to C via sequence A → B → C or A → B′ → C. For sequence A → B → C

the working joints are (a, g) and (c, e) in turn, whereas for sequence A → B′ → C the working joints are (c, e) and (a, g) in turn. Obviously, the sets of the working joints for these two sequences are identical but have different orders of actuation. It is further observed that the two transition configurations between A and C, i.e. B and B′, are configurationally isomorphic (Kuo and Su, 2017) in essential. Therefore, configuration B′ is trivial as it plays the same role of configuration B. This situation also happens between configurations E and G where the trivial configuration F′ plays the same function of configuration F.

6. The operation of a working joint is either a forward or a backward operation. A configuration is said under a forward operation when its working joints are different from the ones in the previous two transformations[1]. On the other hand, a configuration is said under a backward operation when its working joints are as same as the ones in any of the previous two configurations. For example, if the RCM is reconfigured at a sequence of A → B → C → D, then the operation of joints (d, h) between C and D is a forward operation since it is different from the two previous operations (c, e) and (a, g) in this reconfiguring sequence. On the other hand, if the RCM is reconfigured by B → A → B′ → C, then the operation of joints (a, g) between B′ and C is a backward operation since it is identical to the second previous operation between configurations B and A.

7. A backward operation of the working joints will induce a repeated transformation or reverse the direction of transformation of the reconfiguration. For instance, in the previous example, the operation of joints (a, g) between B′ and C is a backward operation for transformation B → A → B′ → C. This operation leads the RCM into configuration C, where the RCM has three options of the workable joints: (c, e), (d, h), and (a, g). If joints (c, e) are selected as the working joints for configuration C, the configuration will be changed from C to B, where either the transformation B → A will be repeated or the reverse transformation C → B will happen. Alternatively, if either joint (d, h) or (a, g) are selected as the working joints for configuration C, then a reverse transformation (i.e., the transformation in the main loop will be reversed from counterclockwise to clockwise or the transformation will be reversed from B′ → C to C → B′) will be resulted.

8. Each configuration has only one forward operation for forming a non-repeated, cyclic transformation. This property can be verified from Fig. 2.

---

[1]The two coaxial working joints of a configuration can point at either x-, y-, or z-directions only. So, the working joints merely need to compare with those in the previous two transformations.

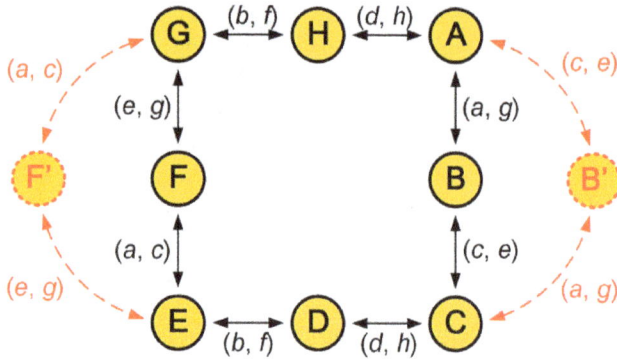

Figure 2. The reconfiguration sequence of the RCM.

3   Configuration Transformation

The configuration transformation of the RCM is an interesting problem In our previous study (Kuo and Su, 2017), we have shown that the configurations of the RCM can be represented by a matrix of joint screws. In what follows, a computational procedure is presented for deriving all the configurations of the RCM based on a given screw matrix of the initial configuration.

3.1   A Computational Procedure for Formulating the Configuration Transformation Process

The flowchart of the computational procedure for formulating the configuration transformation process is given in Fig. 3. The detailed procedure is introduced as follows.

3.1.1   Defining the referencing coordinate system

When an initial configuration of the RCM is given, a Cartesian coordinate system is attached to some link as a referencing coordinate system for the reconfiguration. For convenience, the origin of the system is set to some corner of the sub-cube, and the three coordinate axes are to point along the edges of the cube. For example, Fig. 4a is a given initial configuration of an RCM (same as Fig. 5a in Kuo and Su, 2017) and a coordinate system is attached onto link 2 as shown. Note that the arrangement of the coordinate system is independent of the derived results, i.e., it can be arbitrarily set to any links and any corner.

3.1.2   Determining the initial joint-screw matrix

Now, the joint-screw matrix Π_{initial} (Kuo and Su, 2017) can be written with respect to the defined coordinate system at the initial configuration. For example, the joint-screw matrix

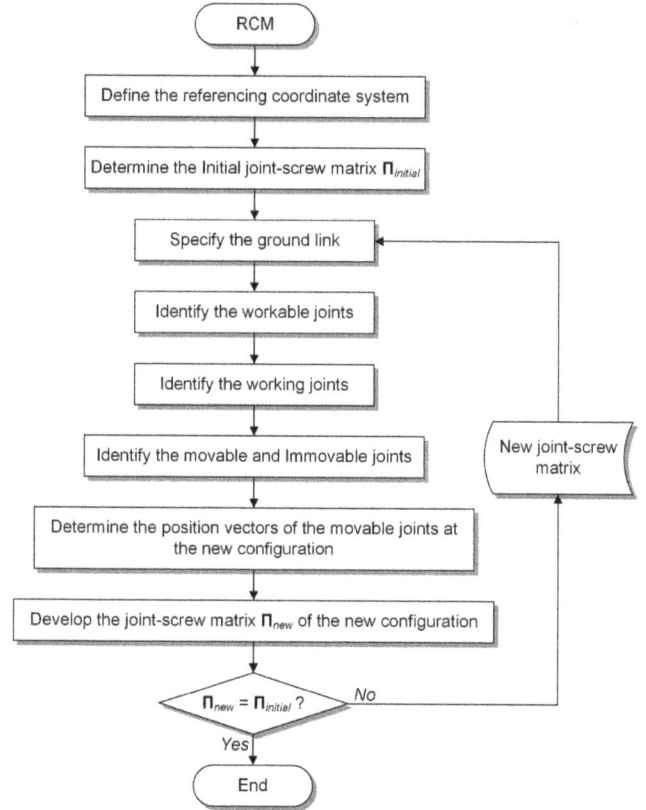

Figure 3. A computational procedure for determining all feasible configurations of the RCM.

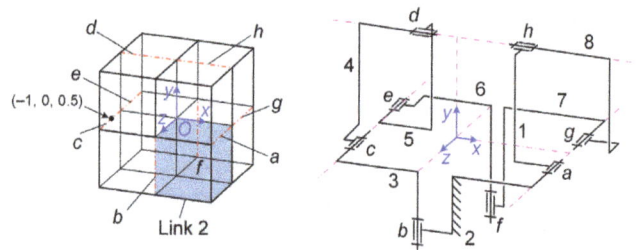

Figure 4. The RCM at the initial configuration. **(a)** The RCM; **(b)** equivalent linkage.

for the configuration in Fig. 4a is:

$$\Pi_{\text{initial}} = \begin{bmatrix} S_a^T \\ S_b^T \\ S_c^T \\ S_d^T \\ S_e^T \\ S_f^T \\ S_g^T \\ S_h^T \end{bmatrix} = \begin{bmatrix} 0 & 0 & 1 & 0 & -1 & 0 \\ 0 & 1 & 0 & -1 & 0 & 0 \\ 0 & 0 & 1 & 0 & 1 & 0 \\ 1 & 0 & 0 & 0 & 0 & -1 \\ 0 & 0 & 1 & 0 & 1 & 0 \\ 0 & 1 & 0 & 1 & 0 & 0 \\ 0 & 0 & 1 & 0 & -1 & 0 \\ 1 & 0 & 0 & 0 & 0 & -1 \end{bmatrix} \quad (1)$$

Where S_a^T represents the transpose of the screw vector of joint a and so does to other joints. Note that, as explained in

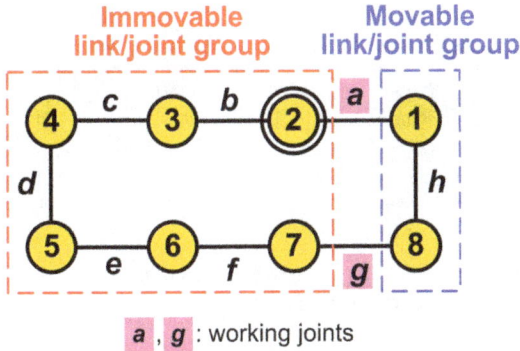

a, g : working joints

Figure 5. The movable and immovable link/joint groups.

Kuo and Su (2017), each sub-cube is defined as a unit cube and all the unit vectors **s** for the joint screws should point at the positive x, y, and z directions, respectively.

3.1.3   Specifying the ground link

Since the RCM linkage has no specific ground link, one link should be grounded to be the reference of the relative motions of the other sub-cubes. Any link of the RCM can be selected as the ground. For example, for the RCM in Fig. 4a, link 2 is selected as the ground link. Accordingly, the RCM becomes a linkage mechanism as depicted in Fig. 4b.

3.1.4   Identifying the workable joints

As stated in Sect. 2, a workable joint is a joint whose degree of freedom of motion is not restrained by the configuration or the link shapes. According to the reconfiguration characteristics described in Sect. 2, when two joints become coaxial, they will together form a pair of workable joints. So our next step is to identify the group of the workable joints from the joint-screw matrix. For example, it can be easily identified from Eq. (1) that vectors (S_a, S_g), (S_c, S_e), and (S_d, S_h) form three groups with identical vectors within it. Therefore, it concludes that there are three groups of workable joints, (a, g), (c, e), and (d, h), at the shown configuration.

3.1.5   Identifying the working joints

Now one group of the workable joints is chosen as the group of working joints for actuating the RCM at the configuration. In order to derive all follow-up feasible configurations of the RCM, i.e., a cyclic reconfiguration without repeated configurations, only the group of working joints that is at forward operation as introduced in Sect. 2 can be chosen. To do this, the selected working joint group should be compared with the working joint groups in the previous two configurations. If the selected working joint group is not as same as that in the previous two configurations, it will be a feasible working joint group. This comparison should be continued until the

working joint group has been identified. Since each configuration has only one forward operation, the selection result of the working joint group will be unique. If there are no previous configurations to be compared, e.g., the current configuration is the first or second configuration, this check can be ignored. For example, for the configuration in Fig. 4, joint group (a, g) is selected as the working joints.

3.1.6   Identifying the movable and immovable joints

After the working joint group is identified, all the remaining joints can be divided into two kinds, the movable joints and immovable joints. A movable joint means that its joint axis will be displaced as the working joints are actuated. Reversely, the location of the joint axis of an immovable joint will be unchanged when the working joints are functioning. The identification of movable and immovable joints is illustrated via the topological graph in Fig. 5. In this graph, the vertices represent the links with their numbering and the edges are the joints with their labeling. The ground link, link 2, is labeled with a concentric circle. In the previous step, joint group (a, g) was identified as the working joints, so the cyclic connection of the vertices and edges in the graph can be divided by joint (a, g) so as to form two groups, i.e., the two dashed-line blocks. Then, the group that contains the ground link is the immovable joint/link group and the other one is the movable joint/link group. So, in this configuration, the movable joints are joint h only, and the immovable joints are joints a, b, c, d, e, f, and g. The above identification of the movable joints can be manipulated by the joint-screw matrix by using the following algorithm:

Given: the m-th and n-th (m < n) row vectors represent the two selected working joints
Set   joint group A = the collection of the 1st to the (m−1)-th and the (n+1)-th to the last row vectors
*   joint group B = the collection of the (m+1)-th to (n−1)-th row vectors*
If   any joint in group A is incident to the ground link
*   joint group A = immovable joint group*
*   joint group B = movable joint group*
*   else*
*   joint group A = movable joint group*
*   joint group B = immovable joint group*
end

3.1.7   Deriving the position vectors of the movable joints at the new configuration

When a pair of joints (J_{w1}, J_{w2}) is selected as the working joints at certain configuration, a couple of joints $(J_{m1}, J_{m2}, \ldots, J_{mi})$ will become the movable joints of a forward transformation for the configuration. The next step is to determine the new screw vectors of these movable joints in the next configuration. Since reconfiguring a joint will not change its orientation (as analyzed in Sect. 2), only the position vector of the joint screw will be changed after reconfiguration. In what follows, the derivation of the new position vectors of the movable joints at the next configuration will be illustrated.

Table 1. Orientation and position vectors of the RCM at the new configuration.

Joint	Orientation vector	Position vector
a	$[0, 0, 1]^T$	$[1, 0, 0.5]^T$
b	$[0, 1, 0]^T$	$[0, -0.5, 1]^T$
c	$[0, 0, 1]^T$	$[-1, 0, 0.5]^T$
d	$[1, 0, 0]^T$	$[-0.5, 1, 0]^T$
e	$[0, 0, 1]^T$	$[-1, 0, -0.5]^T$
f	$[0, 1, 0]^T$	$[0, -0.5, -1]^T$
g	$[0, 0, 1]^T$	$[1, 0, -0.5]^T$
h	$[1, 0, 0]^T$	$[1.5, -1, 0]^T$

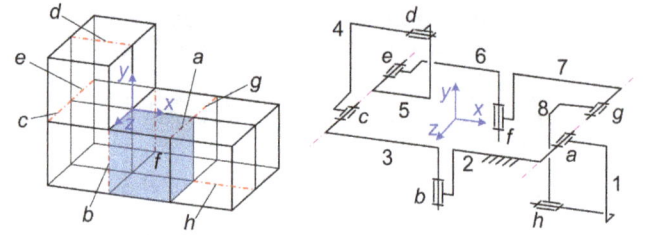

Figure 6. The RCM at the new configuration. **(a)** The RCM; **(b)** equivalent linkage (Kuo and Su, 2017).

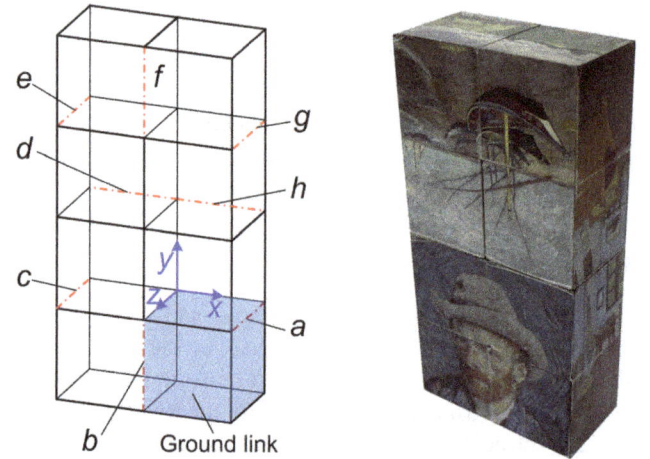

Figure 7. Initial configuration of the illustrative RCM. **(a)** Mechanism drawing (Kuo and Su, 2017); **(b)** physical model.

The position vector of a joint in the unit RCM is measured from the origin of the reference system to the middle point of the joint. For example, the position vector of joint c in Fig. 4 is $[-1, 0, 0.5]^T$. Assume that a working joint J_w has a position vector $[p_x, p_y, p_z]^T$ and some movable joint J_m has a position vector $[q_x, q_y, q_z]^T$ at the current configuration. If a working joint J_w points at the α-direction ($\alpha = x$, y, or z), the new position vector, $[\bar{q}_x, \bar{q}_y, \bar{q}_z]^T$, of the movable joint J_m at the next configuration can be calculated by

$$\bar{q}_\alpha = q_\alpha \tag{2a}$$

$$\bar{q}_\beta = 2p_\beta - q_\beta \tag{2b}$$

where β represents the two axial directions other than α. For example, it is known that joint a is the working joint and joint h the movable joint at the configuration in Fig. 5. So the position vector for joint a can be written as $[p_x^a, p_y^a, p_z^a]^T = [1, 0, 0.5]^T$, and the position vector for joint h as $[q_x^h, q_y^h, q_z^h]^T = [0.5, 1, 0]^T$. Since the working joint points at the z-direction ($\alpha = z$), the new position vector for joint h after reconfiguration can be derived by Eq. (2) as

$$\begin{bmatrix} \bar{q}_x^h \\ \bar{q}_y^h \\ \bar{q}_z^h \end{bmatrix} = \begin{bmatrix} 2p_x^a - q_x^h \\ 2p_y^a - q_y^h \\ q_z^a \end{bmatrix} = \begin{bmatrix} 2 \times 1 - 0.5 \\ 2 \times 0 - 1 \\ 0 \end{bmatrix} = \begin{bmatrix} 1.5 \\ -1 \\ 0 \end{bmatrix} \tag{3}$$

By following this logic, the new position vectors of all the joints can be derived as summarized in Table 1.

3.1.8 Developing the joint-screw matrix of the new configuration

After obtaining the new position vectors of all movable joints, the joint-screw matrix of the new configuration can be constructed. First, the joint screws of the immovable joints will be the same in the new configuration. Second, for the moveable joints, the joint orientations will be invariant in all configurations and the position vectors of the joints have been derived in the previous step. So the joint-screw matrix Π_{new} can be written accordingly. For example, the joint-

screw matrix of the new configuration in Fig. 4 is

$$\Pi_{\text{new}} = \begin{bmatrix} S_a^T \\ S_b^T \\ S_c^T \\ S_d^T \\ S_e^T \\ S_f^T \\ S_g^T \\ S_h^T \end{bmatrix} = \begin{bmatrix} 0 & 0 & 1 & 0 & -1 & 0 \\ 0 & 1 & 0 & -1 & 0 & 0 \\ 0 & 0 & 1 & 0 & 1 & 0 \\ 1 & 0 & 0 & 0 & 0 & -1 \\ 0 & 0 & 1 & 0 & 1 & 0 \\ 0 & 1 & 0 & 1 & 0 & 0 \\ 0 & 0 & 1 & 0 & -1 & 0 \\ 1 & 0 & 0 & 0 & 0 & 1 \end{bmatrix} \tag{4}$$

It is verified that this joint-screw matrix corresponds to the next configuration of the RCM, Fig. 6.

3.1.9 Distinguishing the derived matrix with the one for the initial configuration

After the joint-screw matrix Π_{new} of the new configuration is determined, it should be compared with the joint-screw matrix, Π_{initial}, of the initial configuration. If they are not identical, then this matrix can be used to compute the screw matrix for its next configuration and so on. Oppositely, if the resulted matrix and the one from the initial configuration are identical, it means that the computation has verified all joint-

No.	Configuration	Joint-screw matrix	Working joints (W) Movable joints (M) Immovable joints (I)	Configuration in transition
1	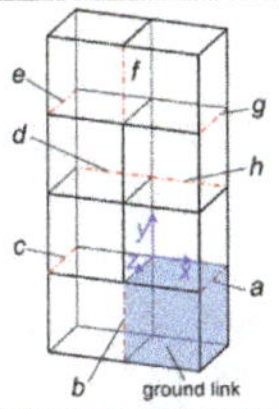	$\begin{bmatrix} 0 & 0 & 1 & 0 & -1 & 0 \\ 0 & 1 & 0 & -1 & 0 & 0 \\ 0 & 0 & 1 & 0 & 1 & 0 \\ 1 & 0 & 0 & 0 & 0 & -1 \\ 0 & 0 & 1 & 2 & 1 & 0 \\ 0 & 1 & 0 & -1 & 0 & 0 \\ 0 & 0 & 1 & 2 & -1 & 0 \\ 1 & 0 & 0 & 0 & 0 & -1 \end{bmatrix}$	W: d, h M: e, f, g I: a, b, c	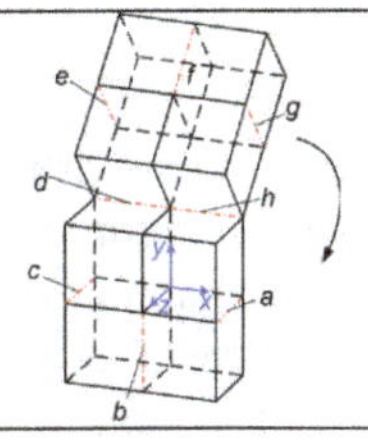
2	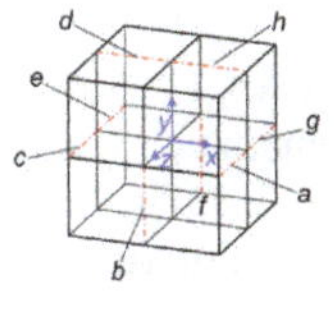	$\begin{bmatrix} 0 & 0 & 1 & 0 & -1 & 0 \\ 0 & 1 & 0 & -1 & 0 & 0 \\ 0 & 0 & 1 & 0 & 1 & 0 \\ 1 & 0 & 0 & 0 & 0 & -1 \\ 0 & 0 & 1 & 0 & 1 & 0 \\ 0 & 1 & 0 & 1 & 0 & 0 \\ 0 & 0 & 1 & 0 & -1 & 0 \\ 1 & 0 & 0 & 0 & 0 & -1 \end{bmatrix}$	W: a, g M: h I: b, c, d, e, f	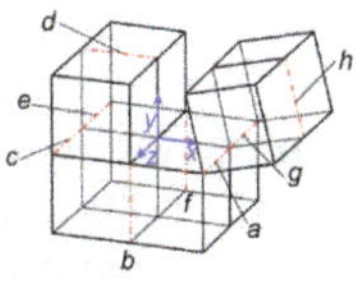
3	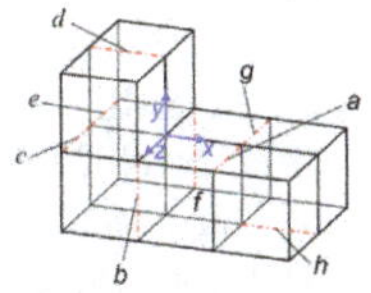	$\begin{bmatrix} 0 & 0 & 1 & 0 & -1 & 0 \\ 0 & 1 & 0 & -1 & 0 & 0 \\ 0 & 0 & 1 & 0 & 1 & 0 \\ 1 & 0 & 0 & 0 & 0 & -1 \\ 0 & 0 & 1 & 0 & 1 & 0 \\ 0 & 1 & 0 & 1 & 0 & 0 \\ 0 & 0 & 1 & 0 & -1 & 0 \\ 1 & 0 & 0 & 0 & 0 & 1 \end{bmatrix}$	W: c, e M: d I: a, b, f, g, h	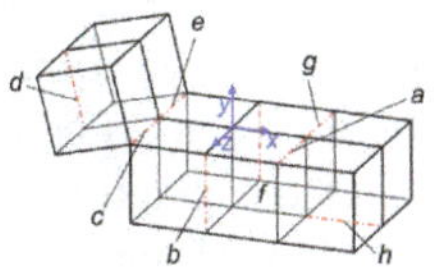
4	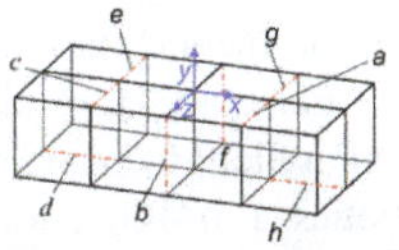	$\begin{bmatrix} 0 & 0 & 1 & 0 & -1 & 0 \\ 0 & 1 & 0 & -1 & 0 & 0 \\ 0 & 0 & 1 & 0 & 1 & 0 \\ 1 & 0 & 0 & 0 & 0 & 1 \\ 0 & 0 & 1 & 0 & 1 & 0 \\ 0 & 1 & 0 & 1 & 0 & 0 \\ 0 & 0 & 1 & 0 & -1 & 0 \\ 1 & 0 & 0 & 0 & 0 & 1 \end{bmatrix}$	W: d, h M: e, f, g I: a, b, c	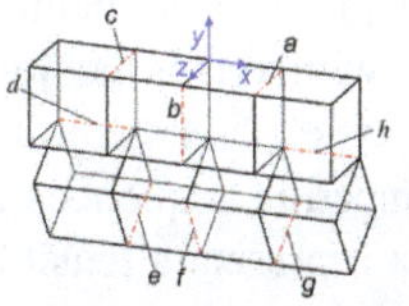
5	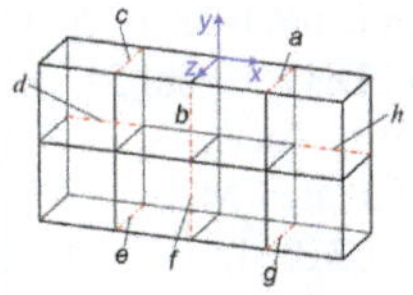	$\begin{bmatrix} 0 & 0 & 1 & 0 & -1 & 0 \\ 0 & 1 & 0 & -1 & 0 & 0 \\ 0 & 0 & 1 & 0 & 1 & 0 \\ 1 & 0 & 0 & 0 & 0 & 1 \\ 0 & 0 & 1 & -2 & 1 & 0 \\ 0 & 1 & 0 & -1 & 0 & 0 \\ 0 & 0 & 1 & -2 & -1 & 0 \\ 1 & 0 & 0 & 0 & 0 & 1 \end{bmatrix}$	W: b, f M: c, d, e I: a, g, h	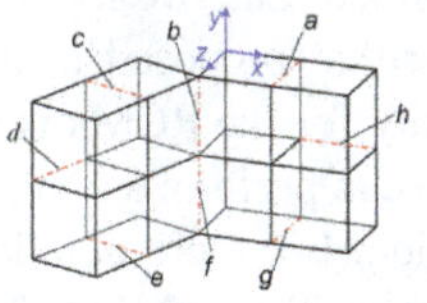
6	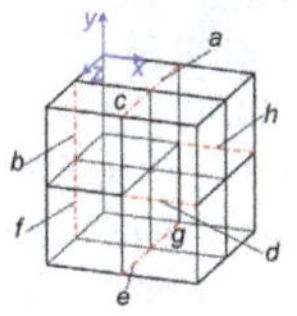	$\begin{bmatrix} 0 & 0 & 1 & 0 & -1 & 0 \\ 0 & 1 & 0 & -1 & 0 & 0 \\ 0 & 0 & 1 & 0 & -1 & 0 \\ 1 & 0 & 0 & 0 & 2 & 1 \\ 0 & 0 & 1 & -2 & -1 & 0 \\ 0 & 1 & 0 & -1 & 0 & 0 \\ 0 & 0 & 1 & -2 & -1 & 0 \\ 1 & 0 & 0 & 0 & 0 & 1 \end{bmatrix}$	W: a, c M: d, e, f, g, h I: b	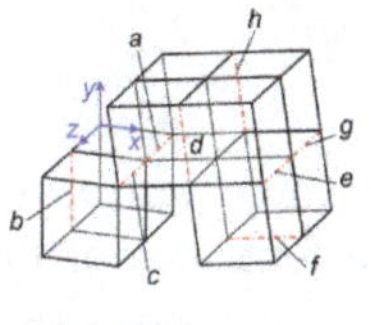
7	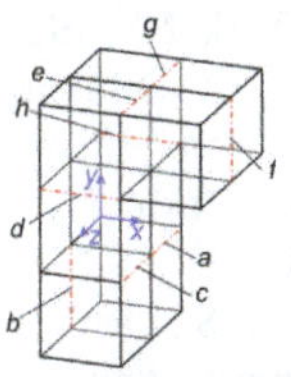	$\begin{bmatrix} 0 & 0 & 1 & 0 & -1 & 0 \\ 0 & 1 & 0 & -1 & 0 & 0 \\ 0 & 0 & 1 & 0 & -1 & 0 \\ 1 & 0 & 0 & 0 & 2 & -1 \\ 0 & 0 & 1 & 2 & -1 & 0 \\ 0 & 1 & 0 & -1 & 0 & 2 \\ 0 & 0 & 1 & 2 & -1 & 0 \\ 1 & 0 & 0 & 0 & 0 & -1 \end{bmatrix}$	W: e, g M: f I: a, b, c, d, h	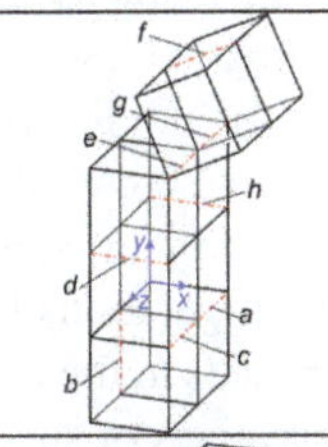
8	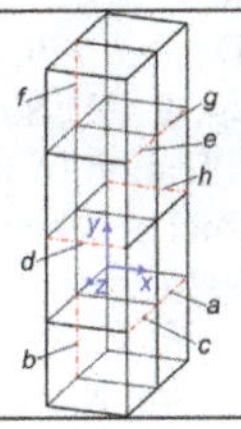	$\begin{bmatrix} 0 & 0 & 1 & 0 & -1 & 0 \\ 0 & 1 & 0 & -1 & 0 & 0 \\ 0 & 0 & 1 & 0 & -1 & 0 \\ 1 & 0 & 0 & 0 & 2 & -1 \\ 0 & 0 & 1 & 2 & -1 & 0 \\ 0 & 1 & 0 & -1 & 0 & 0 \\ 0 & 0 & 1 & 2 & -1 & 0 \\ 1 & 0 & 0 & 0 & 0 & -1 \end{bmatrix}$	W: b, f M: c, d, e I: a, g, h	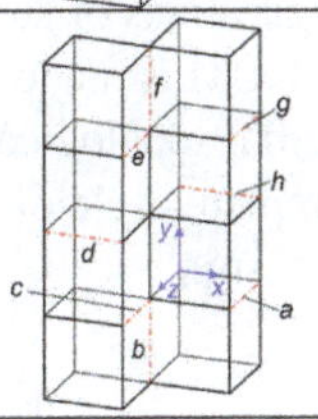

Figure 8. Computation results for the configuration transformation of the illustrative RCM.

screw matrices for a cycle of the reconfiguration. In this case, the computation procedure can be terminated.

3.2 Example

Here the presented computational procedure is illustrated by taking Fig. 7 as an example. Figure 7 shows the initial configuration of an RCM to be manipulated. For illustration, the lower-right sub-cube is selected as the ground link in each following configuration. The computation results are summarized in Fig. 8.

4 Topological Enumeration

Based on the configuration transformation algorithm proposed above, all possible topological configurations of the RCMs with 2^3 sub-cubes can be enumerated. The synthesized RCMs should have the same cubic geometry as the existing design and should be able to complete cyclic configuration transformation. A computational enumeration procedure has been introduced in our another work (Wu and Kuo, 2015). The major steps for the enumeration is summarized as follows. First, all possible connecting sequences among the eight sub-cubes with rotational joints are enumerated. Next, according to some observed joint arrangement rules, the synthesized connecting sequences are examined to see whether or not it can represent a feasible mechanism. Then, the configurational isomorphism of the resulted configurations is further examined. Last, by applying the configuration transformation algorithm proposed in this paper, the cyclic reconfiguration ability for the RCMs is verified. It turns out that only one design can perform a complete cycle of configuration transformation. For more detailed enumeration procedure, one can refer to (Wu and Kuo, 2015).

5 Conclusions

A computational procedure was presented for formulating the configuration transformation of the RCM with 2^3 sub-cubes. The reconfiguration characteristics of the RCM were first investigated. According to these characteristics, the configuration transformation of the RCM can be described by the manipulation on its joint-screw matrix. That is, given by one initial configuration, the joint-screw matrices of all its following transformed configurations in the RCM can be derived. The result further leads to the enumeration of all possible topological configurations of such an RCM. The result showed that when the RCM is made by 2^3 sub-cubes, there is only one feasible initial topological configuration for the RCM to perform a complete cycle of reconfiguration, which is exactly the existing design.

Competing interests. The authors declare that they have no conflict of interest.

Edited by: X. Ding

References

Dai, J. S. and Rees Jones, J.: Configuration Transformations in Metamorphic Mechanisms of Foldable/Erectable Kinds, Proceedings of the 10th World Congress on the Theory of Machines and Mechanisms, Oulu, Finland, 20–24 June, 1999a.

Dai, J. S. and Rees Jones, J.: Mobility in Metamorphic Mechanisms of Foldable/Erectable Kinds, ASME Journal of Mechanical Design, 121, 375–382, 1999b.

Dai, J. S. and Rees Jones, J.: Matrix Representation of Topological Changes in Metamorphic Mechanisms, ASME Journal of Mechanical Design, 127, 837–840, 2005.

Ding, X. and Lu, S.: Fundamental Reconfiguration Theory of Chain-Type Modular Reconfigurable Mechanisms, Mech. Mach. Theory, 70, 487–507, 2013.

Ding, X. and Yang, Y.: Reconfiguration Theory of Mechanism From a Traditional Artifact, ASME Journal of Mechanical Design, 132, 114501, doi:10.1115/1.4002692, 2012.

Ding, X., Yang, Y., and Dai, J. S.: Design and Kinematic Analysis of a Novel Prism Deployable Mechanism, Mech. Mach. Theory, 63, 35–49, 2013.

Galletti, C. and Fanghella, P.: Single-Loop Kinematotropic Mechanisms, Mech. Mach. Theory, 36, 743–761, 2001.

Gan, D., Dai, J. S., and Liao, Q.: Mobility Change in Two Types of Metamorphic Parallel Mechanisms, ASME Journal of Mechanisms and Robotics, 1, 041007, doi:10.1115/1.3211023, 2009.

Gan, D., Dai, J. S., and Liao, Q.: Constraint Analysis on Mobility Change of a Novel Metamorphic Parallel Mechanism, Mech. Mach. Theory, 45, 1864–1876, 2010.

Gan, D., Dai, J. S., Dias, J., and Seneviratne, L.: Unified Kinematics and Singularity Analysis of a Metamorphic Parallel Mechanism With Bifurcated Motion, ASME Journal of Mechanisms and Robotics, 5, 031004, doi:10.1115/1.4024292, 2013a.

Gan, D., Dai, J. S., Dias, J., and Seneviratne, L.: Reconfigurability and Unified Kinematics Modeling of a 3rTPS Metamorphic Parallel Mechanism with Perpendicular Constraint Screws, Robot. Cim.-Int. Manuf., 29, 121–128, 2013b.

Ghrist, R. and Peterson, V.: The Geometry and Topology of Reconfiguration, Adv. Appl. Math., 38, 302–323, 2007.

Harary, F.: Combinatorial Problems in Graphical Enumeration, in: Applied Combinatorial Mathematics, edited by: Beckenbach, E. F., J. Wiley, New York, NY, 1964.

Huang, H., Li, B., Liu, R., and Deng, Z.: Type Synthesis of Deployable/Foldable Articulated Mechanisms, 2010 International Conference on Mechatronics and Automation (ICMA), Xi'an, China, 4–7 August, 2010.

Kuo, C.-H. and Yan, H.-S.: On the Mobility and Configuration Singularity in Mechanisms with Variable Topologies, ASME Journal of Mechanical Design, 129, 617–624, 2007.

Kuo, C.-H. and Chang, L.-Y.: Structure Decomposition and Homomorphism Identification of Planar Variable Topology Mechanisms, ASME Journal of Mechanisms and Robotics, 6, 021002, doi:10.1115/1.4026336, 2014.

Kuo, C.-H. and Su, J.-W.: Configuration Analysis of a Reconfigurable Cube Mechanism: Mobility and Configurational Isomorphism, Mech. Mach. Theory, 107, 369–383, 2017.

Kuo, C.-H., Dai, J. S., and Yan, H.-S.: Reconfiguration Principles and Strategies for Reconfigurable Mechanisms, ASME/IFToMM International Conference on Reconfigurable Mechanisms and Robots (ReMAR 2009), London, United Kingdom, 2009,

Liu, H. and Dai, J. S.: Carton Manipulation Analysis Using Configuration Transformation, Proceedings of the Institution of Mechanical Engineers, Part C, J. Mech. Eng. Sci., 216, 543–555, 2002.

Shieh, W.-B., Sun, F., and Chen, D.-Z.: On the Operation Space and Motion Compatibility of Variable Topology Mechanisms, ASME Journal of Mechanisms and Robotics, 3, 021007, doi:10.1115/1.4003579, 2011.

Wei, G., Ding, X., and Dai, J. S.: Mobility and Geometric Analysis of the Hoberman Switch-Pitch Ball and Its Variant, ASME Journal of Mechanisms and Robotics, 2, 031010, doi:10.1115/1.4001730, 2010.

Wei, G., Ding, X., and Dai, J. S.: Geometric Constraint of an Evolved Deployable Ball Mechanism, Journal of Advanced Mechanical Design, Systems and Manufacturing, 5, 302–314, 2011.

Wohlhart, K.: Kinematotropic Linkages, in: Recent Advances in Robot Kinematics, edited by: Lenarčič, J. and Parenti-Castelli, V., Kluwer Academic Publishers, Dordrecht, Netherlands, 359–368, 1996.

Wu, L.-C. and Kuo, C.-H.: Enumerating the Topological Configurations of the Reconfigurable Cube Mechanism with Eight Subcubes, The 3rd ASME/IFToMM International Conference on Reconfigurable Mechanisms and Robots (ReMAR 2015), Beijing, China, 20–22 July, 2015.

Yan, H.-S. and Kuo, C.-H.: Topological Representations and Characteristics of Variable Kinematic Joints, ASME Journal of Mechanical Design, 128, 384–391, 2006a.

Yan, H.-S. and Kuo, C.-H.: Representations and Identifications of Structural and Motion State Characteristics of Mechanisms with Variable Topologies, T. Can. Soc. Mech. Eng., 30, 19–40, 2006b.

Yan, H.-S. and Kang, C.-H.: Configuration Transformations of Mechanisms with Variable Topologies, J. Chin. Soc. Mech. Eng., 30, 311–321, 2009a.

Yan, H.-S. and Kang, C.-H.: Configuration Synthesis of Mechanisms with Variable Topologies, Mech. Mach. Theory, 44, 896–911, 2009b.

Yan, H.-S. and Kuo, C.-H.: Structural Analysis and Configuration Synthesis of Mechanisms with Variable Topologies, ASME/IFToMM International Conference on Reconfigurable Mechanisms and Robots, London, United Kingdom, 2009,

Zhang, K. and Dai, J. S.: A Kirigami-Inspired 8R Linkage and Its Evolved Overconstrained 6R Linkages With the Rotational Symmetry of Order Two, ASME Journal of Mechanisms and Robotics, 6, 021007, doi:10.1115/1.4026337, 2014.

Zhang, K., Dai, J. S., and Fang, Y.: Topology and Constraint Analysis of Phase Change in the Metamorphic Chain and Its Evolved Mechanism, ASME Journal of Mechanical Design, 132, 121001, doi:10.1115/1.4002691, 2010a.

Zhang, K., Fang, Y., Fang, H., and Dai, J. S.: Geometry and Constraint Analysis of the Three-Spherical Kinematic Chain Based Parallel Mechanism, ASME Journal of Mechanisms and Robotics, 2, 031014, doi:10.1115/1.4001783, 2010b.

Zhang, K., Dai, J. S., and Fang, Y.: Geometric Constraint and Mobility Variation of Two $3S_vPS_v$ Metamorphic Parallel Mechanisms, ASME Journal of Mechanical Design, 135, 011001, doi:10.1115/1.4007920, 2012.

Zhang, L. and Dai, J. S.: Genome Reconfiguration of Metamorphic Manipulators Based on Lie Group Theory, ASME 2008 International Design Engineering Technical Conferences (IDETC 2008), Brooklyn, New York, USA, 2008.

Zhang, L. and Dai, J. S.: Reconfiguration of Spatial Metamorphic Mechanisms, ASME Journal of Mechanisms and Robotics, 1, 011012, doi:10.1115/1.2963025, 2009.

Zhang, L., Wang, D., and Dai, J. S.: Biological Modeling and Evolution Based Synthesis of Metamorphic Mechanisms, ASME Journal of Mechanical Design, 130, 072303, doi:10.1115/1.2900719, 2008.

4

Design and Modelling of a Cable-Driven Parallel-Series Hybrid Variable Stiffness Joint Mechanism for Robotics

Cihat Bora Yigit and Pinar Boyraz

Department of Mechanical Engineering, Istanbul Technical University, Inonu Cd. No:65, 34437, Beyoglu, Istanbul, Turkey

Correspondence to: Cihat Bora Yigit (yigitci@itu.edu.tr) and Pinar Boyraz (pboyraz@itu.edu.tr)

Abstract. The robotics, particularly the humanoid research field, needs new mechanisms to meet the criteria enforced by compliance, workspace requirements, motion profile characteristics and variable stiffness using lightweight but robust designs. The mechanism proposed herein is a solution to this problem by a parallel-series hybrid mechanism. The parallel term comes from two cable-driven plates supported by a compression spring in between. Furthermore, there is a two-part concentric shaft, passing through both plates connected by a universal joint. Because of the kinematic constraints of the universal joint, the mechanism can be considered as a serial chain. The mechanism has 4 degrees of freedom (DOF) which are pitch, roll, yaw motions and translational movement in z axis for stiffness adjustment. The kinematic model is obtained to define the workspace. The helical spring is analysed by using Castigliano's Theorem and the behaviour of bending and compression characteristics are presented which are validated by using finite element analysis (FEA). Hence, the dynamic model of the mechanism is derived depending on the spring reaction forces and moments. The motion experiments are performed to validate both kinematic and dynamic models. As a result, the proposed mechanism has a potential use in robotics especially in humanoid robot joints, considering the requirements of this robotic field.

1 Introduction

In recent years, there has been an emergent need in robotics to develop new mechanisms that go beyond the conventional structures, focusing on compliant, lightweight and energy efficient designs. Following this need, increasing number of studies related with non-conventional robot mechanism design are reported (Grioli et al., 2015). In addition, large workspace, smooth motion profiles, and new mechanical structures with certain redundancies to ease the control applications can be considered as desired properties of such mechanisms. Mizuuchi et al. (2002), Yang et al. (2005), Ham et al. (2009) and Vanderborght et al. (2013) also emphasizes that such joint designs are needed in robotics instead of conventional structures.

Most of non-conventional mechanisms are studied in continuum and hyper-redundant robotics and used especially in minimally invasive surgery (MIS) robotics. These structures combine the compliancy and lightweight compact design requirements, although they may fall short of controllability

and introduce more complexity in modelling. For instance, in Gravagne et al. (2003) a planar continuum robot is introduced accounting the large deflection dynamics, examining the dynamics of a planar backbone section. Another approach for designing compliant and lightweight design can be seen in Wendlandt and Sastry (1994), employing a parallel kinematic mechanism (PKM) together with a spring and a spherical joint in the middle. Instead of including PKM or spring, a typical approach is to propose multi-sections as given in Jones and Walker (2006) in order to increase the controllability of the mechanical structure. Additionally, a PKM mechanism, called cable-driven universal joint (CPUJ), is presented as a module (Lim et al., 2009) and multisection properties are also examined in Lim et al. (2012). These mechanisms are used widely in continuum robotics, although satisfying most of the requirements that come with the disadvantage of complex dynamics. One way to overcome the problem of complex and highly-nonlinear dynamics, spatial models considering the large deformation are used (Tunay,

2013). On the other hand, an extra difficulty comes from the compactness requirements. Since the continuum backbone structures have often small diameter in nature, there is very limited space to fit the actuation units within the module. Therefore, a cable-driven remote manipulation is preferred with often an antagonist arrangement (Potkonjak et al., 2011). Although, it can be used for a different purpose, a very similar humanoid neck mechanism is proposed in Gao et al. (2012b) and is based on a compression spring and two parallel plates, driven by four cables. The same research group also studied inverse kinematics of the structure in Gao et al. (2012a); however, it differs from the mechanism proposed in this work since in their mechanism, there is no serial link to restrict the highly complex spring movement. A similar mechanism is introduced in Nori et al. (2007) and a partial kinematic model with control strategy is presented. In order to handle the nonlinear dynamics of continuum robots, the elastic rod dynamic behavior can be taken as a model using Cosserat Theory in Cao and Tucker (2008). The continuum robotics literature is diverse in terms of pointing the new directions in mechanism design and a detailed review can be found in Walker (2013). However, there are still alternatives to elastic back-bone and continuum structures that may lead to more feasible structures. For example novel 3 DOF fully parallel manipulators with rotational capabilities which is given in Liu et al. (2005) can also be considered as good candidates for especially humanoid neck and joint design. The alternative multi-section designs can be structures such as given in Woehrmann et al. (2013) with effective magnetic actuation and interleaved continuum-rigid manipulators as presented in Conrad et al. (2013). Moreover, a non-compliant but similar structure to the mechanism proposed in this paper, known as 3-SPS mechanism is studied in Alici and Shirinzadeh (2004) and Kim et al. (2015).

In addition to given studies in continuum robotics and cable-driven robotic mechanisms, robotic joint with varying stiffness/compliance is also required. In the last decade, many inspiring developments occured in this area. In Ham et al. (2009) and Vanderborght et al. (2013), the most important ones are summarized and classified. Among this classification, the structure-controlled stiffness in Ham et al. (2009) uses the natural characteristics of the elastic element providing the compliance. The "Jack Spring" mechanism in Hollander et al. (2005) controls the number of active coils by using a screw mechanism to adjust the stiffness without using any additional elements in the mechanism ,which simplifies the design. The proposed mechanism in this paper can be included in this class since it uses natural mechanical behaviour of the helical spring under bending and compression effects.

In this study, a cable-driven, compression spring-supported hybrid mechanism is proposed. The advantages of the proposed mechanism are the unique combination of multiple traits of variable compliancy, hybrid parallel-serial structure for better controllability and lightweight design.

The mechanism can be used in robotic joints especially in humanoid design due to its smooth motion profile, its potential in design of a multi-section robot as a module or section which is individually controllable. The main contribution of this paper is to present a new joint mechanism design which combines advantages of two different joint design approaches. The first approach, which is given in Nori et al. (2007) and Gao et al. (2012b), is to use compression spring and cable-driven actuation. The most important disadvantage of these structures is modelling and control difficulties. These mechanisms do not have accurate mathematical models as serial manipulators have. In the proposed mechanism, the additional shaft and the universal joint constrains the motion, which in turn allows treating it as a serial mechanism and facilitates calculations of the kinematic and dynamic modelling which are also presented. Furthermore, the shaft allows transferring yaw motion directly. In the second approach, Yang et al. (2005), Lim et al. (2009, 2012), Alici and Shirinzadeh (2004) and Kim et al. (2015) use the two part shaft and the universal joint however the absence of compression spring results in a stiff structure. On the other hand, proposed mechanism shows a compliant behaviour as a result of the additional translation motion and the helical compression spring. Linear helical compression spring provides nonlinear stiffness characteristics under combined bending and compression effects. Besides, nonlinear stiffness characteristics is essential for a variable stiffness actuator design and most of the designs make use of complex nonlinear spring mechanisms in Ham et al. (2009) and Vanderborght et al. (2013). Another contribution of this study to design literature on variable stiffness actuators is the simplification of nonlinear stiffness mechanism. Therefore, the helical spring can be assumed as nonlinear stiffness mechanism under compression and bending effects. Although it is a commonly used machine element, there exist few studies analysing combined effect of bending and compression on a helical spring. In Leech (1994), shape memory alloy wires are used as actuators and two different spring loading scenarios are analysed. First a single-sided load is applied, second a pure bending is analysed. Both solutions are based on Castigliano's Theorem. In this study, the same theorem is used and the solution is improved numerically with the combination of bending and compression scenario.

This paper first presents the main ideas in the concept of cable-driven parallel-series hybrid mechanism (CDPS) in Sect. 2. Then, the kinematic modelling is presented in Sect. 3. Next, in Sect. 4, the full dynamic modelling of this hybrid mechanism is reported. In Sect. 5, both experimental structure and results are detailed. Finally, in Sect. 6, the conclusions are drawn and further work on the mechanism design are proposed.

2 Description of Cable-Driven Parallel-Series (CDPS) Hybrid Mechanism

The proposed mechanism includes a lower plate and an upper plate which are used to hold the compression spring in concentric position as shown in Fig. 1. The structure is driven by three cables pulled or set free by three servo motors located underneath the CDPS mechanism. The upper plate is able to perform rotation in two axes providing roll and pitch angles. This configuration is further supported by a concentric shaft with a universal joint in the middle, which passes through the mechanism restricting and defining the bending motion of the compression spring. The concentric shaft is actuated by a fourth motor to provide the movement about z axis – yaw angle. These four motors are encapsulated in a separate case and can be installed at a far location from the mechanism which makes the mechanism lightweight and remotely-actuated. Despite the fact that the mechanism can be classified as a parallel mechanism because of its main construction properties, it is modelled as a serial mechanism due to serial kinematics imposed by the middle shaft and the universal joint. Therefore, the mechanism is called as CDPS hybrid mechanism.

The mechanism has 4 DOF in total, 3 of them are related with motion and the other one is a translational motion along z axis for adjusting the stiffness. Two rotations of upper plate (roll and pitch) are actuated in a cable-driven way, however the motion is constrained by the shaft inside the spring and the universal joint. The restriction in the motion simplifies the kinematic and dynamic calculations. The last DOF of the mechanism is the translational motion of the upper plate along z axis and is designed to adjust the stiffness value of the spring which determines the combined stiffness of the mechanism in pitch and roll axes. Within this study, the control is achieved over stiffness via a structure-dependent way without using additional screw mechanism as in Hollander et al. (2005). The yaw motion herein is not inherently compliant, therefore modelling and experiments sections do not include yaw motion in this work and examine other 3 DOF (pitch, roll and translation) . However, when required, the compliant behavior can be induced using a torsional spring at the lower plate.

3 Kinematic Modelling

The upper and lower plates of the CDPS mechanism are connected by a shaft with a universal joint. Therefore, excluding the yaw motion, the neck mechanism is essentially considered as a serial manipulator with rotation around two different axes at the center of the shaft such as pitch and roll and a translation at the end of the shaft, summing up to 3 DOF. The shaft having the universal joint in the middle of the structure transmits the torque to change the yaw angle while providing a geometric constraint for resolving the spring forces because it determines the bending point. Therefore, in this mechanism

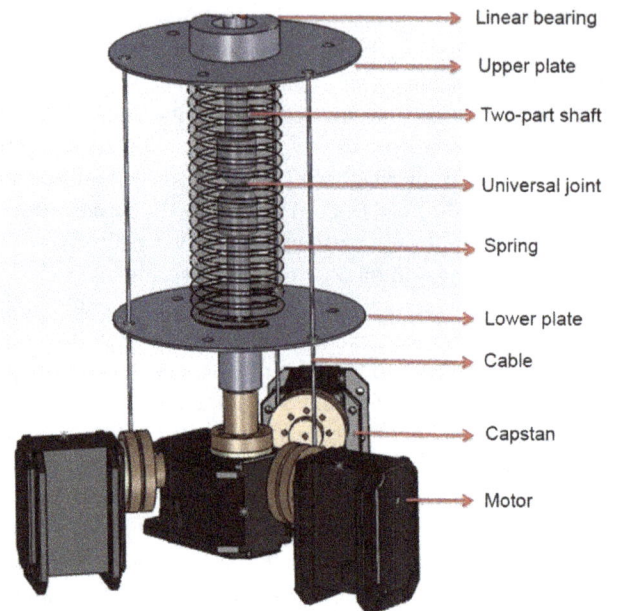

Figure 1. CDPS mechanism and its components, with motors, bearings and capstans.

the amount of the compression can be taken into account and be described by distance between the center of the universal joint and the upper plate. The yaw motion of shaft does not affect the upper plate, because it is directly transmitted to the robotic head platform. The pitch and roll (θ, ϕ) angles of the upper plate are determined by the cable lengths. Therefore, the relative position of the upper plate with reference to the lower plate can be defined by using three generalised coordinates (θ, ϕ, d_2). They are included in a vector which is denoted as q, noting that the vectors are shown with bold italic characters. The variable d_2 represents the distance between the center of the universal joint and the upper plate. The geometric variables used in deriving the kinematic model are given schematically in Fig. 2.

It is clearly seen in Fig. 3 that the neck mechanism is an RRP structure with 3 DOF, having the variables roll, pitch and translation (θ, ϕ, d_2). The local frames (X_1, Y_1, Z_1) and (X_2, Y_2, Z_2) intersect and one of them is rotated by angle $\pi/2$ with respect to the other one. The distance d_1 is constant since it is structurally fixed.

Having defined the variables of the kinematic model, now one can formulate and solve the forward and inverse kinematics. In forward kinematics, the cable lengths (l_1, l_2, l_3) or the motor shaft angles are inputs and the roll, pitch angles and translation of the upper plate (θ, ϕ, d_2) are the outputs. On the other hand, in inverse kinematics, (θ, ϕ, d_2) are inputs and the cable lengths or necessary motor shaft angles are outputs. In order to derive the kinematics, we can use transformation between the upper plate and the lower plate. Since the points where the cables are attached on the upper plate (P_1, P_2, P_3) can define the upper plane and the points

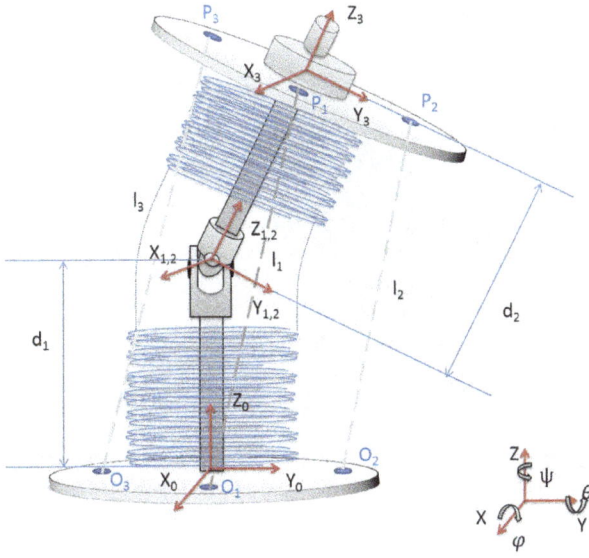

Figure 2. Neck mechanism with variables, local and global axes, cable lengths, upper and lower plate cable assembly points.

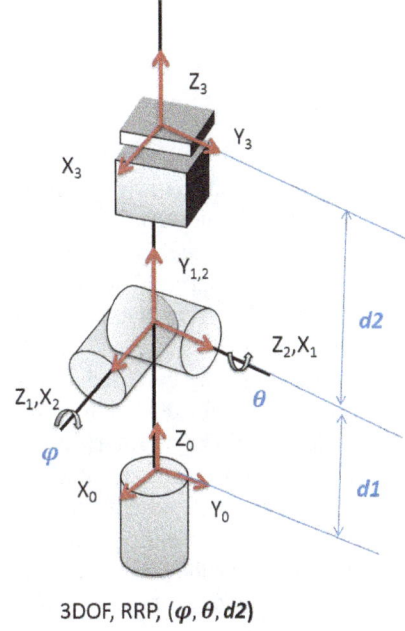

3DOF, RRP, (φ, θ, **d2**)

Figure 3. Reduced serial kinematic model of neck mechanism with RRP structure (pitch, roll and translation).

where the cables pass through at the lower plate can define the lower plane (O_1, O_2, O_3), the transformation matrix is used to define P_i according to O_i. The translation ($\mathbf{Tr}_1$) of constant length of the lower part of the shaft (d_1), the rotations ($\mathbf{Rot}_1$) of pitch and roll (θ, ϕ) and the translation ($\mathbf{Tr}_2$) of the upper plate (d_2) are taken into account in order to obtain this transformation. These transformations are given in Eq. (1), all together forming the transformation matrix $\mathbf{T}_{03}$ in Eq. (2).

$$\mathbf{Tr}_1 = \begin{bmatrix} 1 & 0 & 0 & 0 \\ 0 & 1 & 0 & 0 \\ 0 & 0 & 1 & d_1 \\ 0 & 0 & 0 & 1 \end{bmatrix},$$

$$\mathbf{Rot}_1 = \begin{bmatrix} c\theta & s\theta s\phi & s\theta c\phi & 0 \\ 0 & c\phi & -s\phi & 0 \\ -s\phi & c\phi s\phi & c\phi c\phi & d_1 \\ 0 & 0 & 0 & 1 \end{bmatrix},$$

$$\mathbf{Tr}_2 = \begin{bmatrix} 1 & 0 & 0 & 0 \\ 0 & 1 & 0 & 0 \\ 0 & 0 & 1 & d_2 \\ 0 & 0 & 0 & 1 \end{bmatrix} \tag{1}$$

$$\mathbf{T}_{03} = \begin{bmatrix} c\theta & s\theta s\phi & s\theta c\phi & s\theta c\phi d_2 \\ 0 & c\phi & -s\phi & -s\phi d_2 \\ -s\phi & c\phi s\phi & c\phi c\phi & c\theta c\phi d_2 + d_1 \\ 0 & 0 & 0 & 1 \end{bmatrix} \tag{2}$$

Then, the transformation matrix is used to calculate $P_i = [P_{i,x}, P_{i,y}, P_{i,z}]^T$ using $O_i = [O_{i,x}, O_{i,y}, O_{i,z}]^T$ via Eq. (3).

$$\begin{bmatrix} P_i \\ 1 \end{bmatrix} = \mathbf{T}_{03} \begin{bmatrix} O_i \\ 1 \end{bmatrix} \tag{3}$$

In order to relate the cable lengths (l_1, l_2, l_3) with upper plate position (θ, ϕ, d_2), the definition of the Euclidean distance between O_i and P_i is used. According to this definition, any cable length can be defined by the expression given in Eq. (4).

$$l_i = \sqrt{(P_{i,x} - O_{i,x})^2 + (P_{i,y} - O_{i,y})^2 + (P_{i,z} - O_{i,z})^2},$$
where $i = 1, 2, 3$ $\tag{4}$

If we could obtain the correct P_i coordinates using O_i from the kinematic solution, the nonlinear error function f_i defined by Eq. (5) must be zero.

$$f_i = l_i - \sqrt{(P_{i,x} - O_{i,x})^2 + (P_{i,y} - O_{i,y})^2 + (P_{i,z} - O_{i,z})^2}$$
$$= 0 \tag{5}$$

Using Eqs. (3) and (5), any variable can be obtained numerically using recursive Newton-Raphson algorithm given by Eq. (6).

$$q_k = q_{k-1} - \mathbf{J}^{-1} F \tag{6}$$

where $\mathbf{J}$ is the Jacobian matrix defined by Eq. (7) and q_k is the kth iteration of the solution for the vector of the generalised coordinates, while vector F is formed by equations f_i.

$$\mathbf{J} = \begin{bmatrix} \frac{\partial f_1}{\theta} & \frac{\partial f_1}{\phi} & \frac{\partial f_1}{d_2} \\ \frac{\partial f_2}{\theta} & \frac{\partial f_2}{\phi} & \frac{\partial f_2}{d_2} \\ \frac{\partial f_3}{\theta} & \frac{\partial f_3}{\phi} & \frac{\partial f_3}{d_2} \end{bmatrix} \qquad (7)$$

The solution of inverse kinematics is straightforward by using Eq. (5) when desired state variables are known. Additionally, a simple kinematic model is included, which minimizes the compression of the spring. The model assumes one cable is fixed at the initial position and takes only an orientation input. It is indeed possible to realize all the pitch and roll angles by manipulating only two cables. For each cable, a separate solution loop is defined according to the selected non-moving (i.e. idle) cable. The solution proceeds in the appropriate direction (i.e. towards minimizing approximation error) for finding the lengths of the remaining cables to realize the roll and pitch angles given in the inverse kinematic problem. According to this method, only two selected cables are manipulated at any motion command. This approach is also reflected at the simplified Jacobian matrix used in the inverse kinematic problem as given in Eq. (8).

$$\mathbf{J} = \begin{bmatrix} \frac{\partial f_1}{l_2} & \frac{\partial f_1}{l_3} & \frac{\partial f_1}{d_2} \\ \frac{\partial f_2}{l_2} & \frac{\partial f_2}{l_3} & \frac{\partial f_2}{d_2} \\ \frac{\partial f_3}{l_2} & \frac{\partial f_3}{l_3} & \frac{\partial f_3}{d_2} \end{bmatrix} = \begin{bmatrix} 0 & 0 & \frac{\partial f_1}{d_2} \\ 1 & 0 & \frac{\partial f_2}{d_2} \\ 0 & 1 & \frac{\partial f_3}{d_2} \end{bmatrix} \qquad (8)$$

The simplified kinematic model can be better understood by looking at the Fig. 4 where top view of the lower plate is given. The virtual lines between the points of O_i and the center divides the plate into three areas. Desired pitch and roll motions can be shown as a vector to determine the moving and fixed cables. Two cables neighbouring the area which includes the motion vector, are pulled to perform the desired motion. Besides, opposite cable length is set to default initial value. For example, to complete a 30° rotation in both pitch and roll axes, the distance between upper plate and the lower plate has to be decreased in the direction of vector $\boldsymbol{E}$. Thus, the neighbouring cables of this area which includes the vector are cable 1 and 3. The amount of pull or motor shaft rotation for these cables are calculated by using Eqs. (7) and (8). The length of cable 2 is set to default value.

The mechanism is implemented as a humanoid neck platform and it is used throughout the study. The same parameters with the implemented mechanism, which are given in Table 1, are used for workspace simulation. The positions of the midpoint and the orientations of the upper plate are calculated, using simplified forward kinematics, i.e., keeping one cable length constant at default value (95 mm) while changing the other two within 50–95 mm range, as seen in Fig. 5.

The workspace for the simplified kinematic algorithm is given in Fig. 5a. The workspace is obtained as a sum of three distinct cases. In each case, one of the cable is held at 95 mm. The midpoint of the upper plate is shown with different colours and symbols for each case. Figure 5b and c

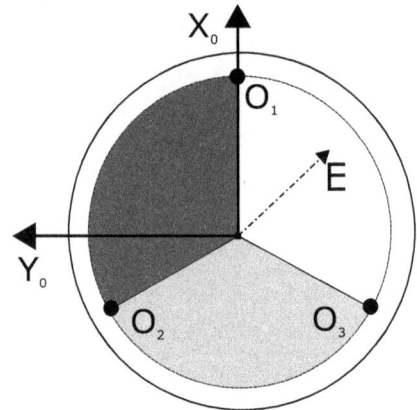

Figure 4. Top view of the lower plate and an example motion command vector.

Table 1. Simulation parameters in workspace analysis.

Variable	Value [unit]
Lower shaft length d_1	47 [mm]
Length of constant cable	95 [mm]
Distance between lower plate center and O_i points	40 [mm]
Distance between upper plate center and P_i points	40 [mm]

show $Y - Z$ and $X - Z$ views of the same figure, respectively. The cable numbers 2 and 3 are placed symmetrically with respect to x axis which results a symmetrical distribution in $Y - Z$ plane. Since the cable number 1 is on the x axis and there is no cable on the opposite side, the midpoints are distributed asymmetrically in $X - Z$ plane. Thus, single roll motions require at least two cables (2 and 3) have to be pulled. Even though, single positive pitch motion requires action of two cables similar to roll motion, only cable 1 is responsible for single negative roll motions. Figure 5d shows the range of motion in pitch and roll angles with respect to given variations of cable lengths. Obviously, each orientation value given in the range of motion (between −30 and 30°) is obtained by holding one cable length constant at default value.

Although, the range of motion (ROM) of the human neck differs from person to person regarding to their ages, genders and physical attributes, average values for pitch (flexion-extension), roll (lateral bending) and yaw (axial rotation) axes are reported around 60, 40 and 80°, respectively in the study of Ferrario et al. (2002). On the other hand, Bennett et al. (2002) shows that only a limited amount of the range (between 30 and 50 %) are utilised to complete daily tasks. The workspace of the humanoid neck platform is calculated between −30 and 30° yielding 60° in total for pitch and roll motions because of practical reasons such as strength of cables and stall torque of motors. Additionally, the lower plate blocks the upper plate when roll and pitch angles are increased to a certain value which is related with the plate dimensions. Thus, the main drawback of the mechanism can be

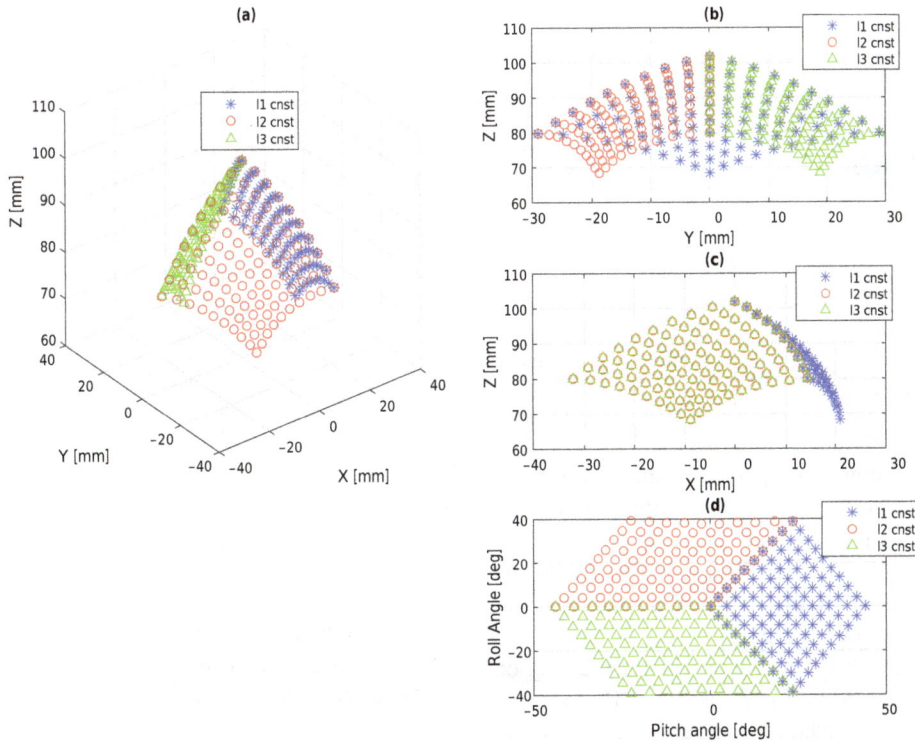

Figure 5. Workspace of the proposed mechanism in **(a)** 3-D, **(b)** $Y - Z$, **(c)** $X - Z$ view, **(d)** Pitch and Roll Angles.

considered as the motion boundaries when compared to the full ROM of humans. However, it is enough for performing daily activities when the proposed mechanism is used as a humanoid neck joint.

4   Dynamic Modelling

In dynamic modelling, the result of kinematic model (position, velocity and acceleration) obtained in Sect. 3 is used to define the dynamic motion of the mechanism under cable based forces. The dynamic modelling is composed of three sequential steps: (i) reduction of 3-D model into a 2-D model without information loss, (ii) force analysis on a bending helical spring using Castigliano's Theorem, and (iii) complete dynamic model.

4.1   Dimension Reduction of the Model

To decrease the complexity of computations and its cost during the analysis of helical spring under bending and compression effects, it is possible to reduce the dimension without information loss. Deformations of the helical spring can be considered and calculated in a 2-D model because of its cylindrical shape. In order to simplify the computation of bending of helical spring in 3-D coordinates, a different Euler angles convention is used as follows: Two rotations of mechanism around X_0 and Y_0 axes are defined in Euler $(X - Y - Z)$ convention and called roll and pitch angles, re-

spectively. However, these two angles can be defined in Euler $(Z - X - Y)$ convention. Thus, the $X - Y$ plane is rotated around Z_0 axis so that bending of the helical spring appears only in this plane. Then the bending or rotation around new x axis is called as deflection angle and denoted with ρ.

4.2   Helical Spring Analysis

The helical spring is subjected to bending effects rather than buckling. The cables are assumed to have constant lengths, hence no plastic/elastic deformation are allowed for the cables. The universal joint in the middle determines the bending point of the helical spring. Figure 6 shows geometrical relations and frames which are used to calculate the dynamic model of the system.

The frame $x_k y_k z_k$ is attached to the bottom center all of the coils. The variable k is the index number of the coils so that $x_0 y_0 z_0$ and $X_0 Y_0 Z_0$ are coincident on the lower plate. Successive coil frames are rotated around their x axes in equally so that z_{n-1} and Z_3 are tangent. The vector N_k stands for the origins of all coils.

The same idea is used for helical spring analysis as in Leech (1994) and each coil of the spring is separately analysed. In order to use Castigliano's Theorem, infinitesimal elements are defined in the helical spring. Figure 6 shows the helical spring and the frame definitions. Angular position of the infinitesimal elements on kth coil is defined as α on $x_k y_k$ plane.

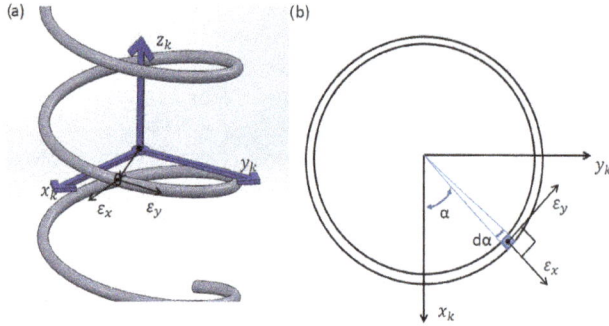

Figure 6. (a) Local frames on spring and infinitesimal element. **(b)** Top view of a spring coil.

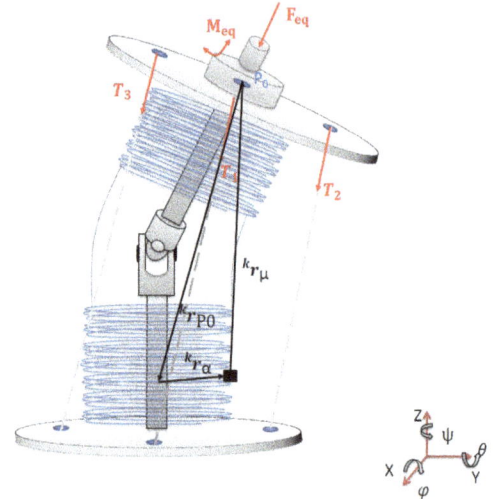

Figure 7. Spring and cable forces acting on the mechanism.

In order to define strength properties of the infinitesimal element on the spring coils, a 2-D frame is attached to all elements. The vector $\boldsymbol{\epsilon}_x$ is placed on tangential direction and the vector $\boldsymbol{\epsilon}_y$ is placed on normal direction. This frame and the other variables which are used to calculate the deformation of the spring are shown in Fig. 6.

The cable forces are shown in Fig. 7 which are acting on the upper plate are defined as $\boldsymbol{T}_1$, $\boldsymbol{T}_2$ and $\boldsymbol{T}_3$. These forces are unidirectional and are reduced to generalized forces definition which includes a force on $-\boldsymbol{z}_{n-1}$ direction which is $\boldsymbol{F}_{eq}$ and a moment around $\boldsymbol{x}_{n-1}$ axis which is $\boldsymbol{M}_{eq}$. To calculate the relation between the deformation of the spring and the generalized forces, the distances between upper plate center point and central points of the infinitesimal elements have to be defined. The position vector of the central point of the upper plate P_0 with respect to frame k is represented as $^k\boldsymbol{r}_{P_0}$. The position of an infinitesimal element "μ' on kth coil of the spring, $^k\boldsymbol{r}_\mu$, is defined as in Eq. (9),

$$^k\boldsymbol{r}_\mu = {}^k\boldsymbol{r}_\alpha + {}^k\boldsymbol{r}_{P_0} \tag{9}$$

where $^k\boldsymbol{r}_\alpha$ is described as the position of the infinitesimal element in the particular coil and is given in Eq. (10).

$$^k\boldsymbol{r}_\alpha = \begin{bmatrix} R\cos(\alpha) \\ R\sin(\alpha) \\ \left(N_{k,z} + \frac{N_{k+1,z} - (N_{k,z})\alpha}{2\Pi}\right) \end{bmatrix} \tag{10}$$

The moment vector, $^k\boldsymbol{M}_\mu$, which is resulting from the equivalent force vector and is acting on the element μ is calculated by using cross product given in Eq. (11).

$$^k\boldsymbol{M}_\mu = {}^k\boldsymbol{r}_\mu \times {}^k\boldsymbol{F}_{eq} \tag{11}$$

Next, the total moment acting on the element $(^k\boldsymbol{M}_{\mu t})$ is calculated as given in Eq. (12).

$$^k\boldsymbol{M}_{\mu t} = {}^k\boldsymbol{M}_\mu + {}^k\boldsymbol{M}_{eq} \tag{12}$$

In order to use the moment in Castigliano's Formula, it has to be defined in the element specific frame $\boldsymbol{\epsilon}_x \boldsymbol{\epsilon}_y$. Since the

given moments are defined in coil frame, a rotation is needed as α about $\boldsymbol{z}_k$ axis, using the rotation matrix $(\text{Rot}_z(\alpha))$. The rotated moment vector $(^k\boldsymbol{M}_c)$ can be obtained as

$$^k\boldsymbol{M}_c = \text{Rot}_z(\alpha)^k\boldsymbol{M}_{\mu t}. \tag{13}$$

Using the Castigliano's Theorem, bending $((U_{b1})_k, (U_{b2})_k)$ and torsional $((U_t)_k)$ strain energies are given in Eqs. (14), (15) and (16), respectively. The material properties which are modulus of elasticity and shear modulus are represented as E and G. The geometric quantities I, J and R are defined as area moment inertia, polar moment inertia and radius of the helical spring.

$$(U_{b1})_k = \int_0^{2\Pi} \frac{(^k M_{c,y})^2}{2EI} R \, d\alpha \tag{14}$$

$$(U_{b2})_k = \int_0^{2\Pi} \frac{(^k M_{c,z})^2}{2EI} R \, d\alpha \tag{15}$$

$$(U_t)_k = \int_0^{2\Pi} \frac{(^k M_{c,x})^2}{2GJ} R \, d\alpha \tag{16}$$

Summation of three strain energies give full strain energy of a single coil $((U)_k)$ and it is added for every coil to calculate the total strain energy of the spring (U), which are given in Eqs. (17) and (18), respectively.

$$(U)_k = (U_{b1}) + (U_{b2}) + (U_t)_k \tag{17}$$

$$(U) = \sum_{k=0}^{n-1} (U)_k \tag{18}$$

The Castigliano's Theorem states that derivatives of the strain energies equal to the deformations which are the compression distance Δd_2 and the 2-D bending angle ρ which is projection of pitch and roll angles on 2-D plane.

$$\Delta d_2 = \frac{\partial(U)}{\partial F_{eq}} \tag{19}$$

$$\rho = \frac{\partial(U)}{\partial M_{eq}} \tag{20}$$

As a result, the nonlinear force-deformation relation or stiffness equation of the spring can be written in given form in Eq. (20).

$$\begin{bmatrix} F_{eq} \\ M_{eq} \end{bmatrix} = [\mathbf{K}] \begin{bmatrix} \Delta d_2 \\ \rho \end{bmatrix} \tag{21}$$

The analysis is performed by using the Matlab Symbolic Toolbox for each Δd_2 (between 0° and 32 mm with 1 mm increments) and ρ (between 0 and 40° with 1° increments). The result of 2×2 $\mathbf{K}$ matrices are collected within a look-up table. Wire and spring radii are 1.5 and 16.5 mm, respectively. The number of active coils are 8 and the material of the spring is ASTM A227. In the dynamic model, the algorithm uses the appropriate $\mathbf{K}$ matrix values according to the generalised coordinates. With the definition of $\mathbf{K}$ matrix as in Eq. (22), the K_{12} and K_{21} elements are same due to the symmetrical structure of the stiffness matrix. The simulation results for K_{11}, K_{12} and K_{22} elements are given in Fig. 8.

$$[\mathbf{K}] = \begin{bmatrix} K_{11} & K_{12} \\ K_{21} & K_{22} \end{bmatrix} \tag{22}$$

K_{11} can be interpreted as the relation between the translational motion of the upper plate Δd_2 and the reaction force of the spring in the same direction. Since the stiffness matrix is symmetric, K_{12} and K_{21} are equal. K_{12} is the relation between combined rotation ρ and reaction force of the spring. Similarly, K_{21} connects Δd_2 and the reaction moment. The combined rotation and the reaction moment are connected by K_{22}. According to the Fig. 8, when both roll and pitch angles are zero, the translational motion has no effect on the change of the elements of stiffness matrix. Thus, this configuration of the mechanism can be called as a singular position for stiffness. On the other hand, all of the elements of the matrix decreases with the increase in the value of Δd_2, which means the spring gets softer with the compression. Compared to other parameters, K_{11} responds to the change of ρ differently. The decrease of K_{11} is obvious with the increasing values of ρ while the other parameters increase. So, the bending motion softens the spring for compression effect and it gets stiffer for rotation. Although the stiffness of the mechanism is adjusted with the translational motion of the upper plate, the bending angle has more effect on the stiffness of the mechanism. As a result of this, the stiffness value depends on the configuration of the mechanism. But, it also changes with the

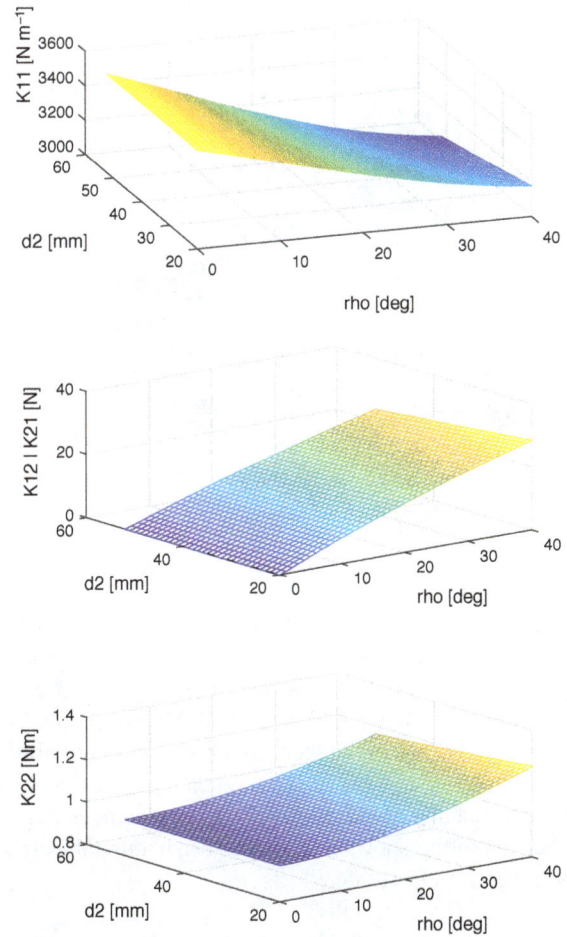

Figure 8. Stiffness matrix change with respect to bending angle and compression length.

variation of Δd_2 which gives variable stiffness ability to the mechanism.

Since finite element analysis (FEA) is commonly used tool for structural mechanics, this technique is used to validate the results of the stiffness analysis proposed in this study. In Fig. 9a and b, initial shape and the deformed shape of the mechanism are shown, respectively. The upper and the lower plates and the shaft are assumed to be rigid parts to isolate spring in analysis particularly and to simplify the model. Additionally, the spring is fixed to lower and upper plates. A revolute joint relation is defined between two shaft parts. Similarly, a translational joint constraint is described between the upper plate and the lower part of the shaft. A prescribed motion profile is applied to these relations as 10 mm translation and 40° rotation. As a result, reaction forces and moments of the spring are obtained during the motion.

FEA and proposed stiffness analysis technique are compared in Fig. 10. Both translation and rotation motions are applied as a linearly increasing function, i.e. ramp function. Thus, x axes of Fig. 10a and b are given as compression in millimetres and bending in degrees. The reaction forces

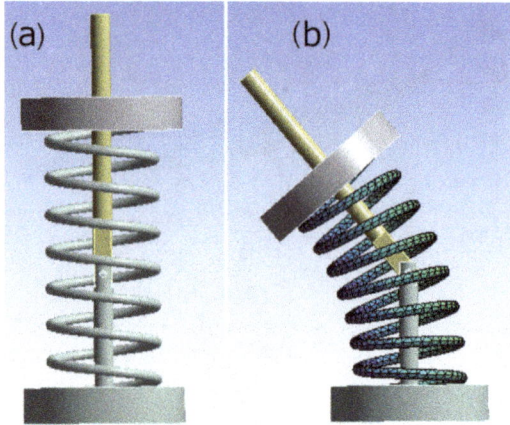

Figure 9. The models of the mechanism in finite element analysis. **(a)** Initial, **(b)** deformed.

and moments are calculated using FEA and stiffness values which are given in Fig. 8. The mean errors between FEA and the proposed method are 2.2 % for reaction force and 1.5 % for reaction moment. A convergence problem can be seen in the initial steps of FEA method. Moreover, quadratic function behaviour can be seen from both force and moment graphs which is a result of linearly changing stiffness values. Subsequently, the results of FEA demonstrates that linear helical spring shows a nonlinear behaviour subjected to bending and compression effects.

4.3 Complete Dynamic Model

The equation of motion of a serial manipulator is defined as in Eq. (23).

$$\mathbf{M}(q)\ddot{q} + \mathbf{C}(q,\dot{q})\dot{q} + \mathbf{G}(q) = \mathbf{B}(q)T + F_s(q) \qquad (23)$$

The vector q is the generalized coordinates vector $[\theta, \phi, d_2]^T$, $\dot{q}$ and $\ddot{q}$ are the velocity and acceleration vectors, respectively. $\mathbf{M}$ is the mass matrix, $\mathbf{C}$ includes Coriolis and centrifugal forces and $\mathbf{G}$ is the gravity matrix. There are no motors connected joints directly. Cable tensions are taken into account with a mapping matrix $\mathbf{B}$ which is given in Eq. (24). F_s is the generalized force vector of helical spring and s_i is the unit vector on the ith cable direction. The variables $P_{i,x}$, $P_{i,y}$, $P_{i,z}$ are the positions of connection points of upper plate and each cable.

$$\mathbf{B}(q) =$$
$$\begin{bmatrix} s_{1,z} & s_{2,z} & s_{3,z} \\ -P_{1,z}s_{1,y} + P_{1_y}s_{1,z} & -P_{2,z}s_{2,y} + P_{2_y}s_{1,z} & -P_{2,z}s_{2,y} + P_{2_y}s_{2,z} \\ P_{1,z}s_{1,x} - P_{1_x}s_{1,z} & P_{2,z}s_{2,x} - P_{2_x}s_{2,z} & P_{2,z}s_{2,x} - P_{2_x}s_{2,z} \end{bmatrix} \quad (24)$$

5 Experiments and Results

In this section, the implementation of the proposed mechanism is explained. First, the mechanical construction and

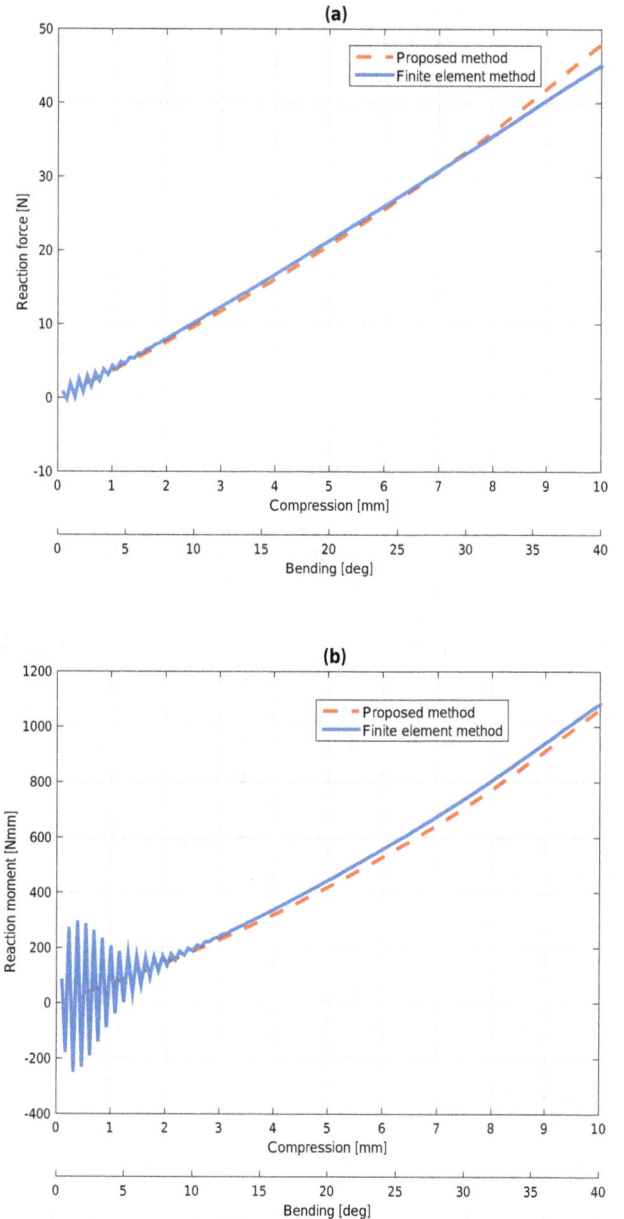

Figure 10. Behaviour of the spring under compression and bending effects. **(a)** Reaction force. **(b)** Reaction moment.

components are introduced. Then control structure, sensors and the software are given. In the experimental results section, the mechanism performs a controlled motion where the parameters are measured to validate the kinematic and dynamic models. The implemented mechanism is used as humanoid neck in UMAY project (Boyraz et al., 2013).

5.1 Experimental Setup

The mechanism is constructed in order to perform validation of both kinematic and dynamic models. Aluminum (7075) is used in upper and lower plates. The shafts are made of

Figure 11. UMAY neck mechanism.

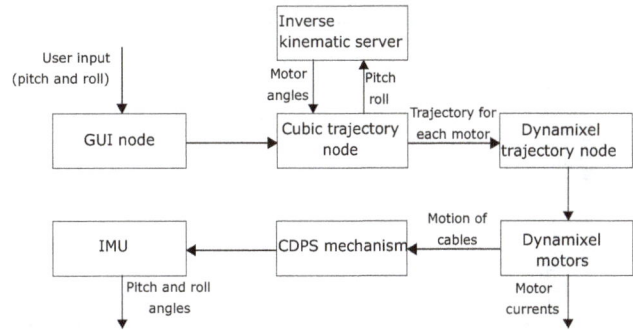

Figure 12. Control structure used in experiments.

chromed steel (AISI E 52100). Four Robotis Dynamixel MX-28 motors, which has 2.5 Nm stall torque at 12 V, are used. One of the motors are directly connected to shaft in order to perform yaw motion. Other three motors are connected to the cables with 14 mm radius capstans. Fishing lines (with 0.5 mm radius) are used as cable material which have a 100 N payload capacity. The friction force between the lower plate and the cables are reduced using PTFE tubes (with 2 mm inner radius) in O_1, O_2 and O_3 points. A picture of the implemented mechanism is given in Fig. 11 and a video is presented in the Supplement. The video is composed of randomly chosen pitch and roll motion combinations. Although the translational motion of the upper plate in zero positions of pitch and roll angles neither affects the stiffness nor the orientation, it is illustrated in the video to emphasize the motion. Three cables are pulled equal amount in length to obtain pure translational motion in the video.

5.2 Results

The experimental setup is equipped with Razor 9 DOF inertial measurement unit (IMU) which is attached on the upper plate along the x axis. The IMU is used to collect the ori-

entation data of output shaft in pitch and roll axes. IMU is used to create a truth table to validate the orientation outputs. The only feedback devices are the encoders of Dynamixel motors.

The whole position control structure used in experiments is programmed within Robot Operating System (ROS) which can be seen in Fig. 12. It is worth noting that the control scheme considers only kinematics of the mechanism. A graphical user interface (GUI) node is designed to collect data from the human user. The commands are sent to cubic trajectory node which enquires necessary motor angles from inverse kinematic server. Since the aim is to validate the models, the server calculates the necessary motor angle changes with the simplified solution of inverse kinematics as explained in Sect. 3. In this solution, one of the cables are held at a constant position. Therefore, the inverse kinematic server returns with the other cable lengths and d_2 parameter while ignoring the stiffness value. The cubic trajectory node takes these arguments as input and calculates a cubic polynomial for motor angles. The total trajectory time is given as 2 s for each trajectory which is a feasible amount of time for given workspace and motors. The cubic trajectory of motor shaft angles are sent to Dynamixel trajectory node.

In this section, all given data are obtained from a single one of several experiments. Thus, all figures are connected to each other. The experimental results are given for pitch/roll angles and related motors. The yaw axis motion and dynamics are excluded because they do not affect the motion of the upper plate and stiffness. The experiment is performed for 80 s which is long enough for an arbitrary duration to validate random pitch and roll input commands. A user gives random pitch and roll angle commands using GUI which is given in detail in this section. The experimental results are presented in two categories. First, in order to validate kinematic model, orientation of the output shaft and forward kinematics are compared. Second, motor load data and the inverse dynamics solution are examined.

Kinematic results of the experiment are given in Fig. 13 for pitch and roll axes. The user command is shown as step function in figures which starts a 2 s cubic trajectory. Forward

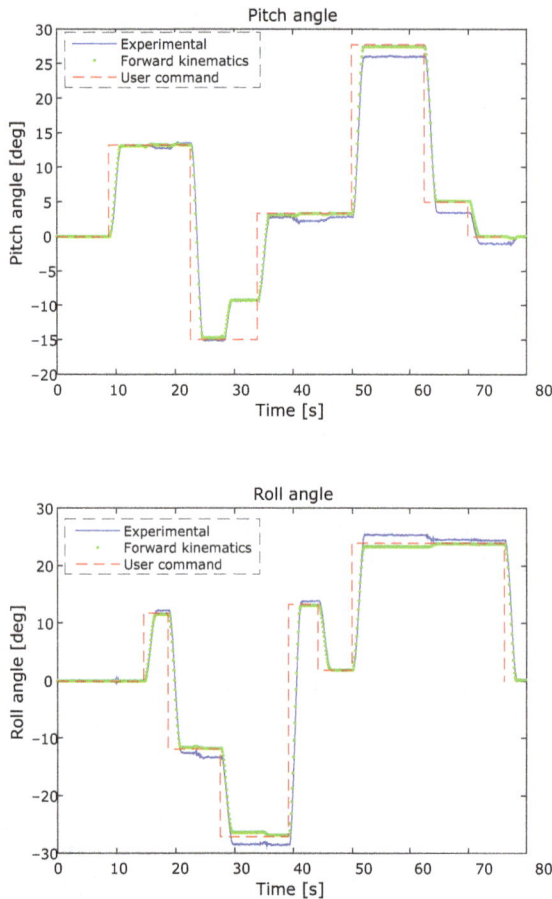

Figure 13. Angular positions of the output shaft in pitch and roll axes.

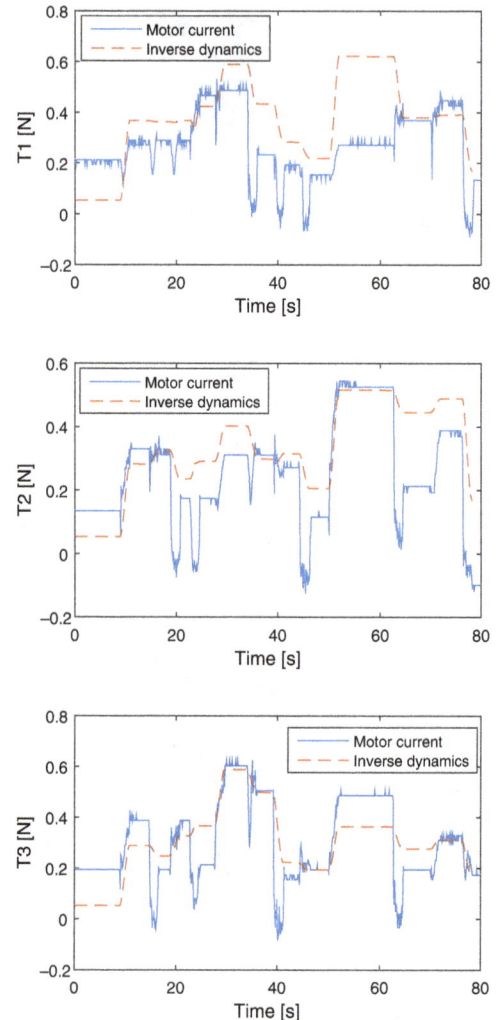

Figure 14. Cable tension estimations from inverse dynamics and motor currents.

kinematic solutions are calculated using motor shaft angles which are obtained from experiment.

Inverse dynamic algorithm is performed using user commands collected from GUI in order to estimate cable tensions for each motor. Additionally, the load data is used to estimate cable tensions which is provided by Dynamixel motors and obtained from motor current values. Two cable tension estimations are compared for front cable, left cable and right cable as can be seen in Fig. 14. The general behavior of these data for each motor are similar, but there are errors in some parts of these results because of three reasons. First, the motor current measurements are very noisy due to the use of pulse width modulation of the motor low level controller. Second, neglected effects such as friction in gearbox of motor, friction between cable and PTFE parts appears difference between torque estimation from motor current and inverse dynamics model result. Finally, DC motors have no load current which is not taken into account in this estimation.

6 Conclusions

A cable-driven parallel-series hybrid variable stiffness joint mechanism is presented in this paper. First, the mechanism is explained reporting advantages and disadvantages. Then, kinematic analysis of the mechanism is given considering it as a serial manipulator, simplifying the solution. By using the forward kinematics the workspace is computed. Then, dynamic model is presented. The stiffness analysis of helical spring between upper and lower plates is explained and for the spring which is used in the experimental setup, the look-up table of stiffness matrix according to compression and bending angle is calculated. To validate the results of the proposed helical spring analysis method, FEA is applied for a motion scenario and the results of both analysis are compared.

The proposed mechanism is implemented and operated to validate the models given in the paper. Experiments are studied for kinematics and dynamics. In kinematics part, the orientation of the output shaft is compared to forward kinematics computations. It is shown that forward kinematics can predict motion profile of the output shaft. In dynamics section, two estimations of cable tensions, one from motor current and the other one from inverse dynamics calculations, are examined. It can be seen that, the results from the inverse dynamic calculations are similar to the motor current values.

In future work, the closed-loop stiffness control and hybrid position-force control techniques are going to be applied to the mechanism. With its large workspace, smooth motion profiles, lightweight mechanical structure and variable stiffness properties, the possible application area to this mechanism could be humanoid robots. It is indeed used as a humanoid neck mechanism in UMAY project. Humanoid robots interact with their environments and humans. Force interaction with humans requires safe operation conditions. After applying force control techniques, it can be used in humanoid joints where the force interaction occurs such as arms and hands in a safe manner as a result of compliant nature of the mechanism.

Competing interests. The authors declare that they have no conflict of interest.

Acknowledgements. The project is supported by Scientific Research Projects Unit of Istanbul Technical University Young Researcher Award.

Edited by: M. Wojtyra

References

Alici, G. and Shirinzadeh, B.: Topology optimisation and singularity analysis of a 3-SPS parallel manipulator with a passive constraining spherical joint, Mech. Mach. Theory, 39, 215–235, 2004.

Bennett, S. E., Schenk, R. J., and Simmons, E. D.: Active range of motion utilized in the cervical spine to perform daily functional tasks, Clinical Spine Surgery, 15, 307–311, 2002.

Boyraz, P., Yigit, C. B., and Bicer, H. O.: UMAY 1: A modular humanoid platform for education and rehabilitation of children with autism spectrum disorders, IEEE, 9th Asian Control Conference (ASCC), 1–6, 2013.

Cao, D. and Tucker, R. W.: Nonlinear dynamics of elastic rods using the Cosserat theory: modelling and simulation, Int. J. Solids Struct., 45, 460–477, 2008.

Conrad, B. L., Jung, J., Penning, R. S., and Zinn, M. R.: Interleaved continuum-rigid manipulation: An augmented approach for robotic minimally-invasive flexible catheter-based procedures, IEEE International Conference on Robotics and Automation (ICRA), 718–724, 2013.

Ferrario, V. F., Sforza, C., Serrao, G., Grassi, G., and Mossi, E.: Active range of motion of the head and cervical spine: a three-dimensional investigation in healthy young adults, J. Orthop. Res., 20, 122–129, 2002.

Gao, B., Xu, J., Zhao, J., and Xi, N.: Combined inverse kinematic and static analysis and optimal design of a cable-driven mechanism with a spring spine, Adv. Robotics, 26, 923–946, 2012a.

Gao, B., Xu, J., Zhao, J., Xi, N., Shen, Y., and Yang, R.: A humanoid neck system featuring low motion-noise, J. Intell. Robot. Syst., 67, 101–116, 2012b.

Gravagne, I. A., Rahn, C. D., and Walker, I. D.: Large deflection dynamics and control for planar continuum robots, IEEE/ASME Transactions on Mechatronics, 8, 299–307, 2003.

Grioli, G., Wolf, S., Garabini, M., Catalano, M., Burdet, E., Caldwell, D., Carloni, R., Friedl, W., Grebenstein, M., Laffranchi, M., Lefeber, D., Stramigioli, S., Tsagarakis, N., van Damme, M., Vanderborght, B., Albu-Schaeffer, A., and Bicchi, A.: Variable stiffness actuators: The user's point of view, Int. J. Robot. Res., 34, 727–743, 2015.

Ham, R. V., Sugar, T. G., Vanderborght, B., Hollander, K. W., and Lefeber, D.: Compliant actuator designs, IEEE Robot. Autom. Mag., 16, 81–94, 2009.

Hollander, K. W., Sugar, T. G., and Herring, D. E.: Adjustable robotic tendon using a "jack spring", in: Rehabilitation Robotics, IEEE, 9th International Conference on ICORR 2005, 113–118, 2005.

Jones, B. A. and Walker, I. D.: Kinematics for multisection continuum robots, IEEE T. Robot., 22, 43–55, 2006.

Kim, J. S., Jeong, J. H., and Park, J. H.: Inverse kinematics and geometric singularity analysis of a 3-SPS/S redundant motion mechanism using conformal geometric algebra, Mech. Mach. Theory, 90, 23–36, 2015.

Leech, A. R.: A study of the deformation of helical springs under eccentric loading, PhD thesis, Naval Postgraduate School, Monterey, California, 1994.

Lim, W., Yeo, S., Yang, G., and Mustafa, S.: Kinematic analysis and design optimization of a cable-driven universal joint module, in: Advanced Intelligent Mechatronics, IEEE/ASME International Conference on AIM 2009, 1933–1938, 2009.

Lim, W. B., Yang, G., Yeo, S. H., and Mustafa, S. K.: Modular Cable-Driven Robotic Arms for Intrinsically Safe Manipulation, Service Robots and Robotics: Design and Application: Design and Application, p. 274, IGI Global, Hershey PA, 2012.

Liu, X.-J., Wang, J., and Pritschow, G.: A new family of spatial 3-DoF fully-parallel manipulators with high rotational capability, Mech. Mach. Theory, 40, 475–494, 2005.

Mizuuchi, I., Tajima, R., Yoshikai, T., Sato, D., Nagashima, K., Inaba, M., Kuniyoshi, Y., and Inoue, H.: The design and control of the flexible spine of a fully tendon-driven humanoid "Kenta", IEEE/RSJ International Conference on Intelligent Robots and Systems, 3, 2527–2532, 2002.

Nori, F., Jamone, L., Sandini, G., and Metta, G.: Accurate control of a human-like tendon-driven neck, IEEE, 7th IEEE-RAS International Conference on Humanoid Robots, 371–378, 2007.

Potkonjak, V., Svetozarevic, B., Jovanovic, K., and Holland, O.: The puller-follower control of compliant and noncompliant antagonistic tendon drives in robotic systems, Int. J. Adv. Robot. Syst., 8, 143–155, 2011.

Tunay, I.: Spatial continuum models of rods undergoing large deformation and inflation, IEEE T. Robot., 29, 297–307, 2013.

Vanderborght, B., Albu-Schaeffer, A., Bicchi, A., Burdet, E., Caldwell, D. G., Carloni, R., Catalano, M., Eiberger, O., Friedl, W., Ganesh, G., Garabini, M., Grebenstein, M., Grioli, G., Haddadin, S., Hoppner, H., Jafari, A., Laffranchi, M., Lefeber, D., Petit, F., Stramigioli, S., Tsagarakis, N., Van Damme, M., Van Ham, R., Visser, L. C., and Wolf, S.: Variable impedance actuators: A review, Robot. Auton. Syst., 61, 1601–1614, 2013.

Walker, I. D.: Continuous backbone "continuum" robot manipulators, ISRN Robotics, 2013, 726506, doi:10.5402/2013/726506, 2013.

Wendlandt, J. M. and Sastry, S. S.: Design and control of a simplified Stewart platform for endoscopy, Proceedings of the 33rd IEEE Conference on Decision and Control, 1, 357–362, 1994.

Woehrmann, M., Doerbaum, M., Ponick, B., and Mertens, A.: Design of a fully actuated electromagnetic bending actuator for endoscopic applications, in: Innovative Small Drives and Micro-Motor Systems, GMM/ETG Symposium, 9, 1–6, 2013.

Yang, G., Lin, W., Pham, C., and Yeo, S. H.: Kinematic design of a 7-DOF cable-driven humanoid arm: a solution-in-nature approach, IEEE/ASME International Conference on Advanced Intelligent Mechatronics, Proceedings, 444–449, 2005.

5

Comparison of parallel or convergent proximal schanz screw placement of pertrochanteric fixator in intertrochanteric fracture model

Arif Gok[1], Sermet Inal[2], Ferruh Taspinar[3], Eyyup Gulbandilar[4], and Kadir Gok[5]

[1]Amasya University, Technology Faculty, Mechanical Engineering, 05000 Amasya, Turkey
[2]Dumlupinar University, School of Medicine, Department of Orthopaedics and Traumatology,
Campus of Evliya Celebi, 43100 Kutahya, Turkey
[3]Dumlupinar University, School of Health Science, Department of Physiotherapy and Rehabilitation,
43100 Kutahya, Turkey
[4]Eskisehir Osmangazi University, Faculty of Engineering & Architecture, Department of Computer
Engineering, Meselik Campus, 26480 Eskisehir, Turkey
[5]Manisa Celal Bayar University, Hasan Ferdi Turgutlu Teknoloji Fakültesi, Mechanical and Manufacturing
Engineering, 45400 Manisa, Turkey
Correspondence to: Arif Gok (arif.gok@amasya.edu.tr)

Abstract. Intertrochanteric femoral fractures are serious traumas among elderly patients. In these patients, external fixator is a preferable method for the fixation of fractures. Therefore, this study was planned to compare the parallel and convergent proximal schanz screw placement of pertrochanteric fixator in the intertrochanteric femoral fractures with respect to biomechanical forces that stabilize the fracture line and to present their clinical importance. A commercial finite element based program, AnsysWorkbench was used to investigate the biomechanical parameters of the femoral intertrochanteric fractures and different placement of implants. The von Mises stress, von Mises strain and shear stress on the proximal and distal surface of the fracture line were lower in the convergent pertrochanteric fixator. Proximal schanz screws in convergent configuration pertrochanteric fixator had greater stress and strain values than proximal schanz screws in parallel configuration pertrochanteric fixator. The distance between the proximal schanz screws on the fracture line was measured as 12 mm in convergent configuration pertrochanteric fixator, and as 3.5 mm in parallel configuration pertrochanteric fixator. The angle between the proximal schanz screws in the convergent configuration was measured as 12.88°. The effect of convergent and parallel configuration pertrochanteric fixators on axial loading demonstrated that convergent configuration pertrochanteric fixator was safer in this respect.

1 Introduction

Intertrochanteric femoral fractures (ITFs) are generally associated with low energy traumas among elderly patients and with high energy traumas among young patients. Especially in elderly patients, and depending on age-related concomitant conditions such as osteoporosis, these fractures can lead to high morbidity and mortality rates. For this reason, ITFs are considered among the difficult-to-treat group of fractures (Healey and Msorman, 1993; Laskin et al., 1979). These

fractures occur especially among patients with a history of diabetes, heart disease or chronic obstructive pulmonary disease. For elderly patients with these concurrent conditions, the administration as well as the duration of anesthesia involves certain risks (Christodoulou and Sdrenias, 2000). For this reason, it is generally described that treatment procedures involving external fixation methods should be preferred for such patients. External fixation methods ensure treatment with minimal surgical trauma and blood loss, and a short duration of anesthesia and surgery (Dhal and Singh,

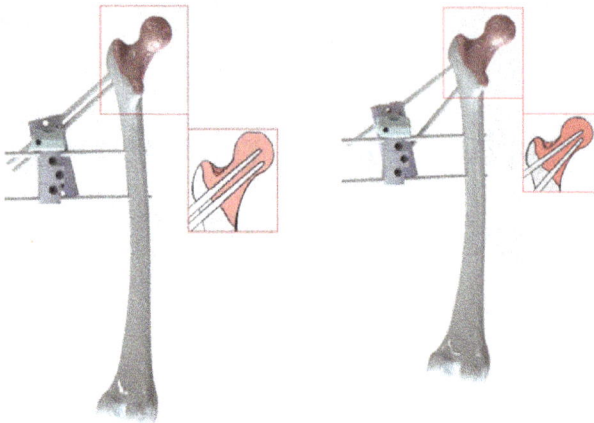

Figure 1. Images of PTF with different configuration.

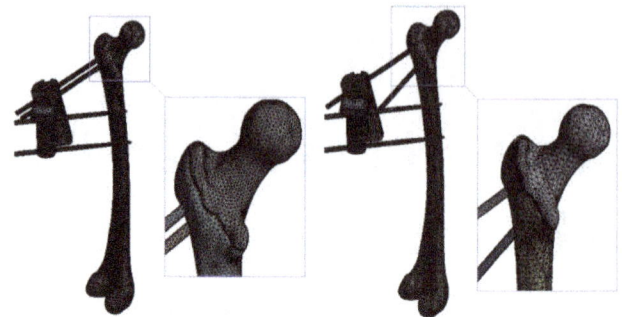

Figure 2. Mesh structure of the model.

Table 1. Biomechanical properties of the femur.

Parameters	Value
Density ($kg\,m^{-3}$)	2100
Young's modulus (GPa)	17
Yielding strength (MPa)	135
Tensile strength (MPa)	148
Poisson's ratio	0.35

1996; Kamble et al., 1996; Moroni et al., 2005; Vossinakis and Badras, 2001, 2002a). Earlier mobilization of patients allows the prevention of postoperative complications such as urinary infections, pneumonia, decubitus ulcers and bone infections (Badras et al., 1997; Eksioglu et al., 2000).

External fixators have been used since 1950. However, despite the limited number of patients treated with external fixators in the literature, complications such as pin site infections, varus deformities and shortness have been described (Barros et al., 1995; Christodoulou and Sdrenias, 2000; Kamble et al., 1996). To ensure early rehabilitation with these fractures, the most important goal is to preserve the reduction position of the fracture (Irfan, 1997; Pervez et al., 2004; Vossinakis and Badras, 2002b). For this reason, the effectiveness of external fixators on fracture stability should be considered as a key point. Aside from comparing the stabilizing effect of different external fixators applied together with other internal fixation materials, it is also necessary to investigate the effectiveness of different implantation configurations. Thus, by identifying which configurations would ensure stronger stabilization by fixators implanted for hip fractures, it becomes possible to predict the quality of mobilization. As a result, possible complications can be prevented. In our study, our aim was biomechanically to investigate the effectiveness of Pertrochanteric Fixators (PTFs) applied with two different configurations of schanz screws inserted to the femoral head in ITFs.

2 Materials and methods

Modeling using Finite Element Method (FEM) of the femur

A commercial finite element based program, MSC Patran/Mentat/Marc is used to investigate the biomechanical parameters of the femoral fracture and implants (Mahaisavariya et al., 2006). The human femoral model is scanned using 3 dimension (3-D) scanner and point cloud is obtained. After that, 3-D model of femur is created using

point cloud data by Geomagic Studio 10 programme. The femoral ITF is created using SolidWorks programme as seen in Fig. 1. The modeling of the implants are also modeled in 3-D using the SolidWorks 2013 programme. These models are imported into the AnsysWorkbench to prepare the FEM and the mesh generation of the FEM is created. The mesh generation of the models is created using the tetrahedrons element type as seen in Fig. 2. The generated finite element model has 147 783 nodes and 96 290 elements. While the element size of the model is selected as 4 mm, the contact regions are selected as 2 mm.

The femur is fixed from the distal condylar articular face. The Pertrochanteric Fixator (PTF) implantations are performed similar to the surgical implantation technique used in routine orthopaedic surgeries. Two proximal schanz screws of fixators are applied in two different configurations, the first one is performed in parallel and the second one is applied in convergent way (Fig. 1). The distal schanz screws of both PTFs are applied in the same configuration; perpendicular to the femoral shaft and parallel to each other. Contact types among the parts of the implants and bone are defined as a frictional contact. These contact interactions are assumed between the different parts of the models. Friction coef?cients are taken as 0.46 for bone–bone interactions and 0.42 for bone–implant interactions. For boundary conditions, the loading vector (350 N by z-axis) is applied through the orthogonal plane to the femoral head while the distal end of the femur is fixed (Goffin et al., 2013). The FEM is applied to both PTF configurations. All materials used for PTF were stainless steel from the ANSYS Material Library. The biomechanical properties are summarized in Table 1 (Tu et al.,

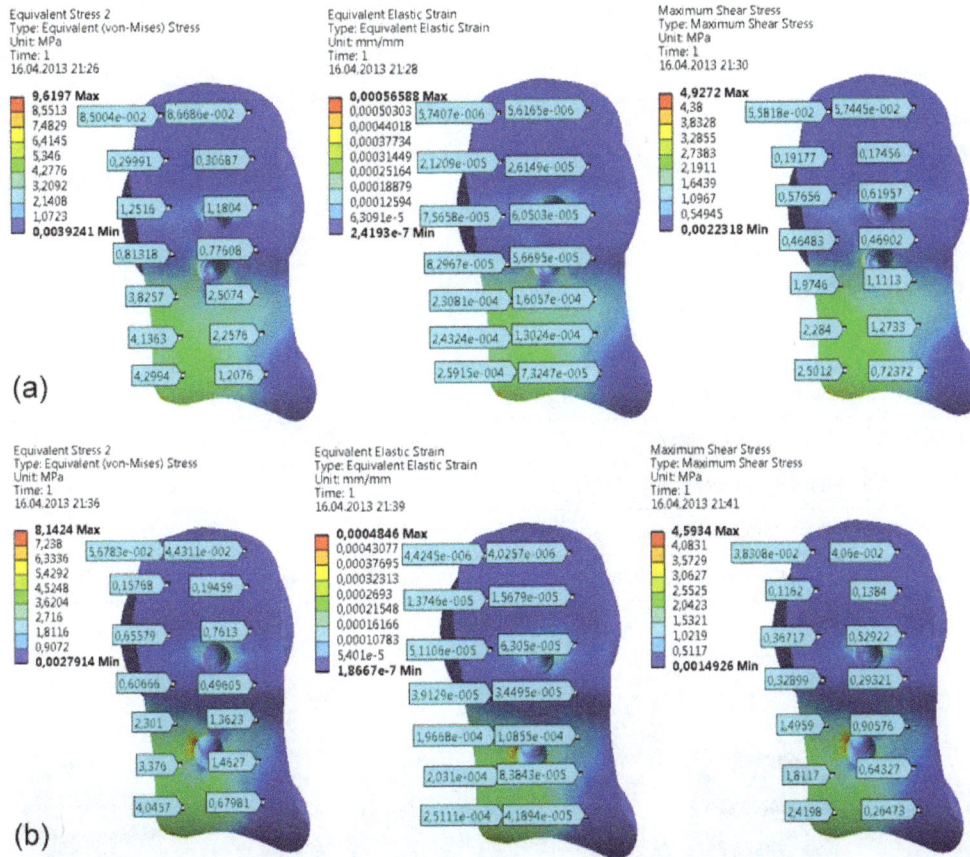

Figure 3. The biomechanical parameters on the proximal surface of the fracture line for both configurations.

2009). After the boundary conditions, biomechanical properties of the materials for each component were identified and simulations were solved. The Von-Mises stress, strain, shear stress values are evaluated at the upper and lower surfaces of the fracture line and implants for each configuration. Also the distances between the proximal schanz screws at the fracture line for each configuration are measured. The angle between proximal schanz screws in convergent technique is also evaluated. The parameters are compared and the research question to this study was whether different proximal schanz screw placement of PTFs might have biomechanical advantages during axial loading for patients to be mobilized.

3 Results

The Von Mises stress, strain and shear stress values on the upper and lower surfaces of the fracture line were found to be lower in convergent configuration PTF in comparison to parallel configuration PTF (Figs. 3 and 4). It was also determined that proximal schanz screws in convergent configuration PTF had greater stress and strain values than proximal schanz screws in parallel configuration PTF (Figs. 5 and 6). The distance between the proximal schanz screws on the fracture line was measured as 12 mm in convergent con-

figuration PTF, and as 3.5 mm in parallel configuration PTF. The angle between the proximal schanz screws in the convergent configuration was measured as 12.88° (Table 2).

4 Discussion

For the past 40 years, through various cadaver studies, model and clinical trials, researchers have investigated new materials and techniques that would ensure early and safe rigid fixation and load transfer (Audigé et al., 2003; Bong et al., 2004; Bridle et al., 1991). Studies in recent years have demonstrated that, in comparison to past studies, better results are now being obtained with external fixators used for the treatment of trochanteric fractures (Christodoulou and Sdrenias, 2000; Vossinakis and Badras, 2001, 2002a). This study was performed in order to compare the biomechanical effects of convergent and parallel configuration PTFs during axial loading by using FEM. It was demonstrated in this context that PTF with convergent configuration ensured better stabilization of the fracture line.

An evaluation of the literature revealed no studies on the biomechanical properties of external fixators used in different configurations for hip fractures, and the effect of these configurations on clinical outcome. The majority of the stud-

Table 2. The biomechanical parameters at the fracture line.

Variables	Parallel	Convergent
Von Mises stress		
The proximal surface	1.64 ± 1.5	1.14 ± 1.24
The distal surface	0.80 ± 0.7	0.68 ± 0.61
Von Mises strain		
The proximal surface (10^{-5})	10.1 ± 8.6	7.9 ± 8
The distal surface (10^{-5})	4.6 ± 3.9	4.5 ± 3.18
Shear		
The proximal surface (10^{-5})	0.88 ± 0.82	0.66 ± 0.73
The distal surface (10^{-5})	0.49 ± 0.43	0.40 ± 0.33
Maximum Von Mises stress (schanz)	65.915	68.35
Maximum Von Mises strain (schanz)	0.00034	0.00035
The angle between proximal schanz screws (degree)	0	12.88
The distance between the proximal schanz screws on the fracture line (mm)*	3.5	12

* mm: milimetres.

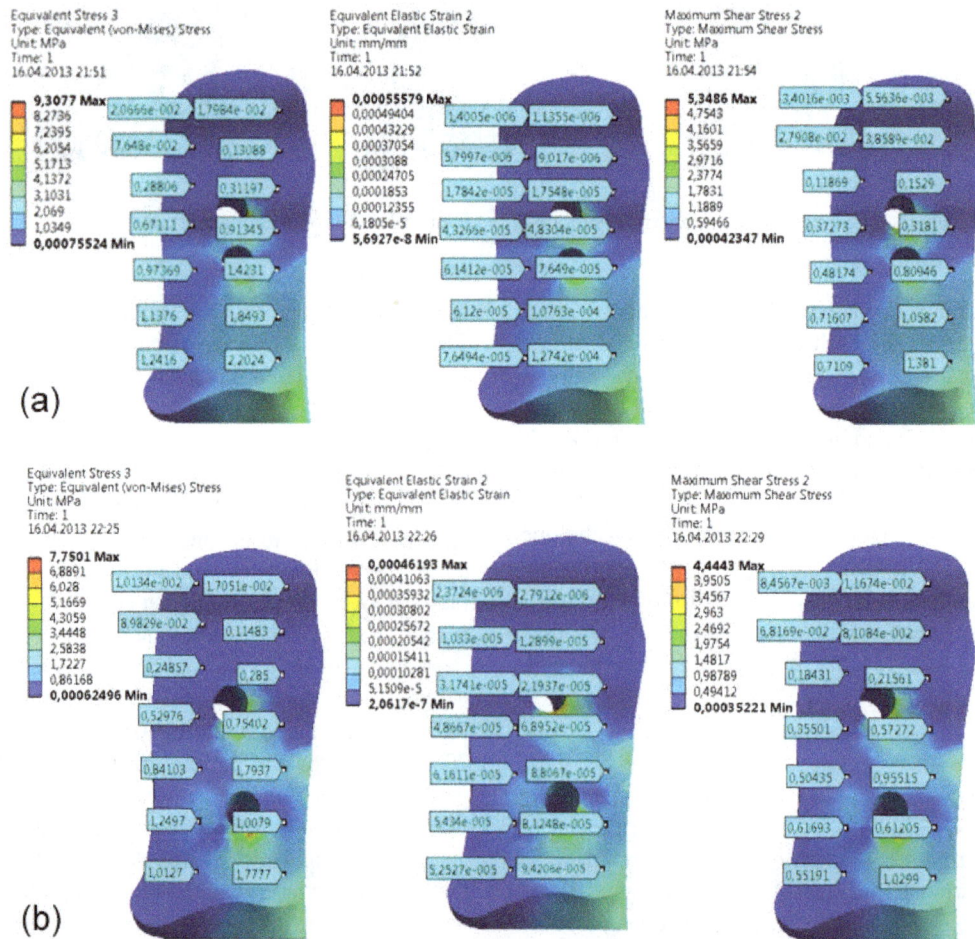

Figure 4. The biomechanical parameters on the distal surface of the fracture line for both configurations.

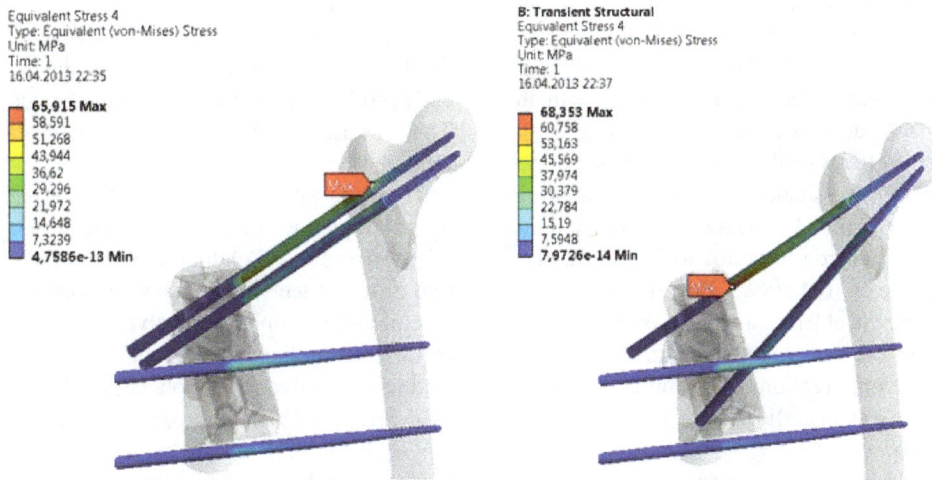

Figure 5. Demonstration of the maximum stress of proximal schanz screws.

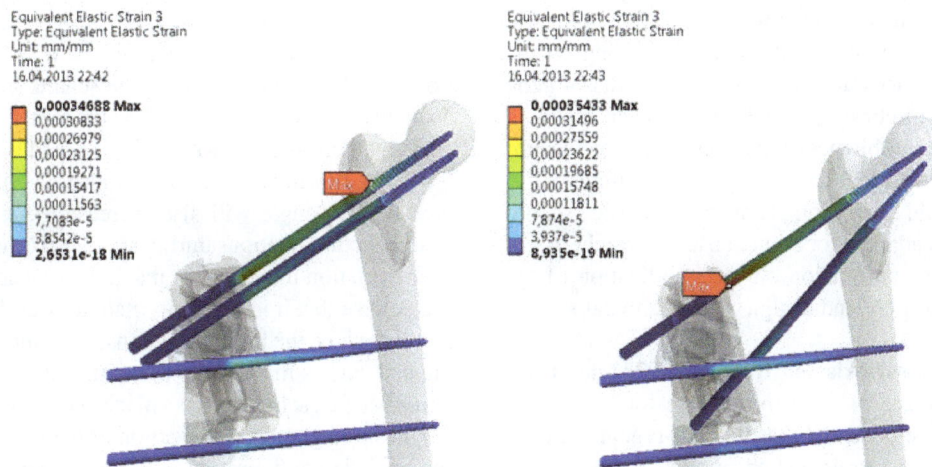

Figure 6. Demonstration of the maximum strain of proximal schanz screws.

ies on external fixators used for hip fractures demonstrate minimal blood loss, reduced surgery risk and earlier mobilization for this treatment method. They also show a very low incidence of postoperative complications, and a low rate of morbidity and mortality associated with these complications (Kazakos et al., 2007; Ozkaya et al., 2008).

Vekris et al. (2011) clinically compared parallel and convergent configured PTFs used for the treatment of ITFs. Based on the study results, they suggested that parallel and convergent configured PFTs provided the same clinical results, and that the parallel configuration was preferable due to its easier applicability. No biomechanical analyses were conducted in this study; only a clinical interpretation of the results was performed.

In the study of Eksioglu et al. (2000), it was described that internal fixators were insufficient for early rehabilitation, and that they did not allow walking with full weight

bearing until union occurs. In addition to this, they emphasized that delayed and difficult rehabilitation was associated with an increase in early mortality rates. For this reason, Eksioglu reported that external fixators applied for ITFs should be considered as semi-conservative method. In recent years, the increase in the number of studies performed on external fixators has led the authors of this article to biomechanically investigate different configurations that can be applied with external fixators. To our knowledge, there is currently no other biomechanical research in the literature regarding this subject.

As lateral fixation devices lead to compression in larger bones, they also tend to reduce the shear forces between surfaces (Scarante et al., 1993). Closed reduction protects the fracture hematoma and ensures rapid healing (Vossinakis and Badras, 2002b). In patients with trochanteric fractures they treated with external fixators, Aly et al. (2004) described that external fixation provided good protection for the reduction, and also corrected the varus deformity while causing compression on the fracture line. In addition to this, although certain mechanical problems have been reported with the external fixators, it was described that the elasticity of the schanz screws and the effects of the tension band supported the mechanical stability (Vossinakis and Badras, 2002a). In the study of Özdemir et al. (2003), it was described that union in the treatment of trochanteric fractures with external fixators required an average of 10.9 weeks. However, this recovery period was identified in elderly and high risk patients; it is hence possible to consider and plan studies involving young patients and different configurations. The ability to transfer with full weight bearing at an earlier stage might potentially result in a shorter recovery period.

Moroni et al. (2005) investigated various screw properties in their study, and determined that external fixators with hydroxyapatite-coated screws could potentially be used for osteoporotic elderly patients. Certain researchers have also described that the schanz screws in external fixators could be used in different numbers and angles depending on the width of the femoral neck and the preferences of the surgeon (Girgin et al., 1993; Subaşı et al., 1998). However, it should be considered that the biomechanical responses of the implants will also be different depending on the differences in application. As such, these biomechanical properties should be investigated and adapted for clinical use. Researchers will thus be able to have an opinion on the stabilization of the fracture line, and to plan and design further clinical studies accordingly.

Baumgaertner (Vossinakis and Badras, 2002b) described that fracture fixation is of great importance for bone healing. In our study, we observed that the convergent configuration had lower stress, strain and shear stress values on the upper and lower surface of the fracture line in comparison to the parallel configuration. Furthermore, it was observed that the stress and strain in the proximal schanz screws were greater in the convergent configuration. This demonstrated that convergent configuration PTF assumed a greater portion of the loads on the fracture line, and that they hence provided greater stability. On the other hand, these same biomechanical parameters were observed as being lower in schanz screws in parallel configuration PTF. This confirmed that the parallel configuration bore less load and hence assumed a smaller portion of the loads on the fracture line. Although Vekris et al. (2011) described the parallel configuration as a preferable method in terms of its ease of applicability, they were not able to biomechanically demonstrate this configuration's resistance against axial loading. The conclusion we reached with regards to biomechanics was different from the one reached by Vekris et al. (2011) in their clinical study.

Especially in severe osteoporotic patients, we believe that it is necessary to investigate the different configurations of external fixators in order to prevent and overcome mechanical complications such as the shortness of extremity and

varus deformity. This is because these patients are generally elderly patients with poor overall health and bone quality. For this reason, positive differences that can be biomechanically produced and proposed are of considerable importance. This will assist in the planning of the ideal configuration and treatment method for the patient.

There are various stability studies describing the use of diagonal or parallel Kirschner wires in the surgical treatment of elbow fractures. Tachdjian (John Anthony, 2002) describes two different techniques for supracondylar elbow fractures. One of these techniques involves the application of the wires on the fracture line in a diagonal configuration, while second technique involves the application of the wires in a parallel configuration. Furthermore, as an aspect that is more important than the diagonal or parallel application of the wires, it was described that increasing the distance between the Kirschner wires passing over the fracture line led to greater stability. In our study, the distance between the schanz screws over the fracture line was 3.5 times greater in the convergent configuration in comparison to the parallel configuration. Our results are thus in parallel with Tachdjian's observations. We believe that placing the schanz screws sufficiently apart such that they remain inside the femur neck will also increase the distance between the pins, and thereby strengthen the stability. However, it should also be considered that increasing the angle will also increase the likelihood of mechanical complications during surgery. The elbow joint is not an articulation that bears as much direct load as the hip joint. We believe that it is very important to bear this consideration in mind, since the hip joint is constantly subject to axial loading and carries more weight than the elbow joint. According to our results, schanz screws placed convergently on a simple ITF ensured greater stabilization of the fracture line. For this reason, this configuration could be preferred to the parallel configuration in osteoporotic patients.

Although we identified studies in the literature in which the schanz screws of external fixators were applied at different angles to the femoral head of cases with ITFs, we nevertheless did not encounter any study describing the effects of applications at different angles on the biomechanical stability across the fracture line. For this reason, there is no clear information on which configuration is more advantageous. We also did not encounter any study similar to our own that biomechanically compared PTFs of different configurations. As it involved a biomechanical analysis performed by using FEM method (a method known to be reliable), our study was considered to be reliable and valid. However, problems such as the inability to perform three dimensional scans on the human femur, the assumption that the femur has an equally distributed density, the inability to show the proximal trabecular structure and the calcar, cortical, spongiform bony structures, and the inability to reflect the extent of osteoporosis (if applicable) represented the limitations of this study.

5 Conclusion

An investigation of the effect of convergent and parallel configuration PTFs on axial loading demonstrated that convergent configuration PTF was safer in this respect. We believe that this observation will contribute to future clinical applications.

Competing interests. The authors declare that they have no conflict of interest.

Edited by: D. Pisla

References

Aly, T. A., Hafez, K., El-nor, T. A., and, Osama, A.: Treatment of Trochanteric Fractures By External Fixator in Patients With High Unacceptable Operative Risk, Pan Arab. J. Orth. Trauma, 8, 157–162, 2004.

Audigé, L., Hanson, B., and Swiontkowski, M. F.: Implant-related complications in the treatment of unstable intertrochanteric fractures: meta-analysis of dynamic screw-plate versus dynamic screw-intramedullary nail devices, Int. Orthop., 27, 197–203, 2003.

Badras, L., Skretas, E., and Ed, V.: Traitement des fractures pertrochantériennes par fixateurexterne, Rev. Chir. Orthop., 84, 461–465, 1997.

Barros, J. W., Ferreira, C. D., Freitas, A. A., and Farah, S.: External fixation of intertrochanteric fractures of the femur, Int. Orthop., 19, 217–219, 1995.

Bong, M. R., Patel, V., Iesaka, K., Egol, K. A., Kummer, F. J., and Koval, K. J.: Comparison of a Sliding Hip Screw with a Trochanteric Lateral Support Plate to an Intramedullary Hip Screw for Fixation of Unstable Intertrochanteric Hip Fractures: A Cadaver Study, J. Trauma Acute Care Surg., 56, 791–794, 2004.

Bridle, S., Patel, A., Bircher, M., and Calvert, P.: Fixation of intertrochanteric fractures of the femur. A randomised prospective comparison of the gamma nail and the dynamic hip screw, J. Bone Joint Surg., 73-B, 330–334, 1991.

Christodoulou, N. A. and Sdrenias, C. V.: External Fixation of Select Intertrochanteric Fractures With Single Hip Screw, Clin. Orthop. Rel. Res., 381, 204–211, 2000.

Dhal, A. and Singh, S. S.: Biological fixation of subtrochanteric fractures by external fixation, Injury, 27, 723–731, 1996.

Eksioglu, F., Gudemez, E., Cavusoglu, T., and Sepici, B.: Treatment of intertrochanteric fractures by external fixation, Bull. Hosp. Joint. Dis., 59, 131–135, 2000.

Girgin, O., Öztan, L., Özlü, K., and Ög, B.: İntertrokanterik femur kırı klarının eksternal fiksatör ile tedavisi, 15–19 May 1993, Nevşehir, Türkiye, 607–610, 1993.

Goffin, J. M., Pankaj, P., and Simpson, A. H.: The importance of lag screw position for the stabilization of trochanteric fractures with a sliding hip screw: A subject-specific finite element study, J. Orthop. Res., 31, 596–600, 2013.

Healey, H. S. and Msorman, L.: Osteoporosis, in: Manual of Rheumatology and Outpatient Orthopaedic Disorders, Little Brown, Boston, 1993.

Irfan, Ö.: Kalça kırı klarında prognozu etkileyen risk faktörleri, Acta Orthop. Traumatol. Turc., 31, 374–377, 1997.

John Anthony, H.: Upper Extremity Injuries, in: Tachdjian's Pediatric Orthopaedics, W. B. Saunders Company, USA, 2002.

Kamble, K. T., Murthy, B. S., Pal, V., and Ráo, K. S.: External fixation in unstable intertrochanteric fractures of femur, Injury, 27, 139–142, 1996.

Kazakos, K., Lyras, D. N., Verettas, D., Galanıs, V. P. I., and Kostas, X.: External fixation of intertrochanteric fractures in elderly high-risk patients, Acta Orthop. Belg., 73, 44–48, 2007.

Laskin, R. S., Gruber, M. A., and Zimmerman, A. J.: Intertrochanteric Fractures of the Hip in the Elderly: A Retrospective Analysis of 236 cases, Clin. Orthop. Rel. Res., 141, 188–195, 1979.

Mahaisavariya, B., Sitthiseripratip, K., and Suwanprateeb, J.: Finite element study of the proximal femur with retained trochanteric gamma nail and after removal of nail, Injury, 37, 778–785, 2006.

Moroni, A., Faldini, C., Pegreffi, F., Hoang-Kim, A., Vannini, F., and Giannini, S.: Dynamic Hip Screw Compared with External Fixation for Treatment of Osteoporotic Pertrochanteric Fractures, A Prospective, Random. Study, 87, 753–759, 2005.

Özdemir, H., Dabak, T. K., Ürgüden, M., and Gür, S.: A different treatment modality for trochanteric fractures of the femur in surgical high-risk patients: a clinical study of 44 patients with 21-month follow-up, Arch. Orthop. Trauma Surg., 123, 538–543, 2003.

Ozkaya, U., Parmaksızoglu, A. S., Gul, M., Kabukcuoglu, Y., Ozkazanlı, G., and Basilgan, S.: Management of osteoporotic pertrochanteric fractures with external fixation in elderly patients, Acta Orthop. Traumatol. Turc., 42, 246–251, 2008.

Pervez, H., Parker, M. J., and Vowler, S.: Prediction of fixation failure after sliding hip screw fixation, Injury, 35, 994–998, 2004.

Scarante, B., Ranellucci, M., and Feas, L.: The dynamic axial fixator in the treatment of pertrochanteric fractures of the femur, Int. J. Orthop. Trauma, 3, 58–60, 1993.

Subaşı, M., Atlıhan, D., Katırcı, T., Dindar, N., Aşık, Y., and Hasan, Y.: İntertrokanterik femur kırı klarının eksterna lfiksatör ile tedavisi, Acta Orthop. Traumatol. Turc., 32, 40–43, 1998.

Tu, Y. K., Liu, Y. C., Yang, W. J., Chen, L. W., Hong, Y. Y., Chen, Y. C., and Lin, L. C.: Temperature Rise Simulation During a Kirschner Pin Drilling in Bone, 11–13 June 2009, 3rd International Conference on Bioinformatics and Biomedical Engineering, ICBBE 2009, Beijing, China, 1–4, 2009.

Vekris, M. D., Lykissas, M. G., Manoudis, G., Mavrodontidis, A. N., Papageorgiou, C. D., Korompilias, A. V., Kostas-Agnantis, I. P., and Beris, A. E.: Proximal screws placement in intertrochanteric fractures treated with external fixation: comparison of two different techniques, J. Orthop. Surg. Res., 6, 48–60, 2011.

Vossinakis, I. C. and Badras, L.: Management of pertrochanteric fractures in high-risk patients with an external fixation, Int. Orthop., 25, 219–222, 2001.

Vossinakis, I. C. and Badras, L. S.: The external fixator compared with the sliding hip screw for pertrochanteric fractures of the femur, J. Bone Joint Surg., 84-B, 23–29, 2002a.

Vossinakis, I. C. and Badras, L. S.: The Pertrochanteric External Fixator Reduced Pain, Hospital Stay, and Mechanical Complications in Comparison with the Sliding Hip Screw, J. Bone Joint Surg., 84, 1488–1501, 2002b.

6

Configuration synthesis of generalized deployable units via group theory

Tuanjie Li[1], Jie Jiang[2], Hangjia Dong[1], and Lei Zhang[1]

[1]School of Electromechanical Engineering, Xidian University, Xi'an, 710071, China
[2]Engineering College, Honghe University, Mengzi, 661100, China

Correspondence to: Tuanjie Li (tjli888@126.com)

Abstract. The generalized deployable mechanism is composed of generalized links and generalized kinematic pairs. The generalized links include the flexible members, springs, and cables etc. The generalized kinematic pairs include the preloaded kinematic pairs and flexible hinges etc. The generalized deployable mechanism consists of generalized deployable units. Based on group theory, this paper presents a brief and effective configuration synthesis method for generalized deployable units which is composed of generalized links and generalized kinematic pairs expressed by group. The permutation group is used to obtain all the permutation types of generalized kinematic pairs and generalized links. The configuration matrix of generalized deployable unit is established through combining permutation group, and the topological configurations of generalized deployable units are created. The combining rules of groups are developed. The method of isomorphism detection is used to ensure the uniqueness of configurations. The configurations of generalized deployable units including four and six generalized links are generated respectively.

1 Introduction

The deployable mechanism is proposed by Pinero in the 1960s (Pinero, 1962). Deployable mechanisms, also named as deployable structures (Pellegrino, 2001), can vary their shape automatically from a compact, packaged configuration to an expanded, operational configuration. The first properly engineered deployable structures were used as stabilization booms on early spacecraft. Later on, more complex structures were devised for solar arrays, communication reflectors and telescopes (Li, 2014; Tibert, 2002; Mruthyunjaya, 2003). In other fields there have been a variety of developments, including retractable roofs for stadia, foldable components for cars, portable structures for temporary shelters and exhibition displays. Obviously, the requirements that have to be met by a deployable mechanism in its operational configuration (e.g. providing shelter from rain, in the case of umbrella, or forming an accurate reflective surface, in the case of a deployable reflector antenna for telecommunications) are different from the requirements in the package configuration (usually, this should be as small as possible). But an essen-

tial requirement is that the deployment and retraction transformation process should be possible without any damage, and should be autonomous and reliable (Pellegrino, 2001). For this purpose, the deployable mechanisms are composed of same deployable units due to modular design. The deployable and foldable transformation of deployable units depends on the elastic deformation of springs, flexible components or cable driven. Thus, the deployable unit not only includes the traditional rigid links and kinematic pairs, but also includes the flexible components, springs, cables and the preloaded kinematic pairs and flexible hinges etc, and thus defined as the generalized deployable unit which is composed of generalized links and generalized kinematic pairs. The generalized links include flexible members, springs, and cables etc. The generalized kinematic pairs include preloaded kinematic pairs and flexible hinges etc.

With the development of deployable mechanisms, it is possible to evolve new configurations that meet the requirements of engineering applications. The configuration synthesis is the effective method to address the key problem. The configuration synthesis of traditional mechanisms has been one of hotspot issues. The early configuration synthesis mainly depended on the experience and inspiration. Until the 1960s, the topological graph theory was introduced to describe the mechanisms and lots of configuration synthesis methods were advanced (Davies, 1966). For the deployable mechanisms, Gantes (1991) investigated the geometric modeling and design methodology of deployable structures featuring stable and stress-free states in both the deployed and the collapsed configurations. Warnaar and Chew (1995a, b) studied the generation of deployable truss modules with the aid of graph theory. Chen and You (2009) conducted lots of researches in the field of monocyclic constraint mechanisms on the basis of Bennett and Myard mechanisms, and generated some new space deployable mechanisms. Ding et al. (2013) proposed a novel deployable mechanism based on polyhedral linkages, and discussed the mobility and the singularities. Lu et al. (2014) proposed a family of novel deployable prism mechanisms based on the Hoekens straight-line linkages. Lu et al. (2015) researched a network of type III Bricard linkages, and performed two methods of connecting linkages. As an expansion of traditional deployable mechanisms consisted of rigid links and kinematic pairs, the generalized deployable mechanism, which is composed of generalized links and generalized kinematic pairs, is a system that can vary their shape automatically from a compact, packaged configuration to an expanded, operational configuration. The generalized links, as the expansion of the traditional rigid links, contain all members which are capable to transform motions and forces, including the flexible members, springs, and cables etc. The generalized kinematic pairs, as the expansion of the traditional pairs, contain all kinematic pairs which connect different links and offer motion constraints, including not only traditional prismatic pairs and revolute pairs but also preloaded prismatic pairs, preloaded revolute pairs, flexible hinges etc. The combination of different generalized links and generalized kinematic pairs may generate all sorts of configurations with different performances. The structural complexity of generalized deployable mechanisms results in that the topological representation and configuration synthesis cannot be addressed by the existing methods. Proceeding from the generalized links and generalized kinematic pairs, we develop the configuration synthesis of generalized deployable units based on the finite group theory. The configurations of generalized deployable units including four and six generalized links are created.

2 Graphical representation and configuration matrix of generalized deployable units

The generalized deployable units include cables, flexible components, and preloaded kinematic pairs etc. The adjacent matrix of graph theory is often used to describe topological relationship between links and kinematic pairs. Thus, the configuration matrix of generalized deployable units is

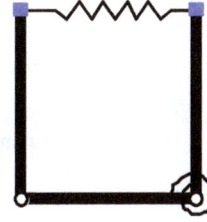

Figure 1. The four-link generalized deployable unit.

defined to describe the topological relationship between generalized links and generalized kinematic pairs. The diagonal elements in configuration matrix represent the type of generalized links, and the nondiagonal elements represent the type of generalized kinematic pairs. The weighted values in the configuration matrix are defined to express the types of generalized links and generalized kinematic pairs, as shown in Tables 1 and 2.

In order to express the relation between the generalized links and generalized kinematic pairs, the configuration matrix is introduced to describe the generalized deployable units. The configuration matrix and the generalized deployable unit are one to one correspondence. The diagonal elements of configuration matrix represent generalized links, and other elements represent the relation between two generalized links. We have

$$\mathbf{A} = \left(a_{ij}\right)_{n \times n}, \tag{1}$$

where, $\mathbf{A}$ is the configuration matrix of a generalized deployable unit, n is the number of generalized links, and

$$a_{ij} = \begin{cases} w_1 & \text{If } i \neq j, \text{ the weighted value of generalized} \\ & \text{kinematic pair connecting two links.} \\ w_2 & \text{If } i = j, \text{the weighted value of generalized link.} \\ 0 & \text{Otherwise.} \end{cases}$$

For example, a four-link generalized deployable unit is shown in Fig. 1, its configuration matrix is

$$\mathbf{A} = \begin{bmatrix} 3 & 9 & 0 & 9 \\ 9 & 1 & 7 & 0 \\ 0 & 7 & 1 & 5 \\ 9 & 0 & 5 & 1. \end{bmatrix}$$

3 Configuration synthesis method

The basic idea of configuration synthesis method is the link replacement in the traditional kinematic chains or frame structures. The configurations of traditional kinematic chains or frame structures are called as the initial configurations, and the corresponding matrix is called as the initial matrix. The main diagonal elements of initial matrix constitute a vector called as the links generator a. The non-zero elements

Table 1. Types of generalized links.

Generalized links	Weighted values (w_2)	Graphical representation
Rigid link	1	
Flexible link	2	
Spring	3	
Cable	4	

Table 2. Types of generalized kinematic pairs.

Generalized kinematic pairs	Weighted values (w_1)	Graphical representation
Revolute pair	5	
Preloaded hinge	7	
Flexible hinge	8	
Fixed joint	9	

of non-main-diagonal elements of initial matrix constitute a vector called as the kinematic pairs generator b. The vector groups of links and kinematic pairs are obtained from their generators by the defined generation relation. The matrix groups of links and kinematic pairs are established by the defined mapping relation. The configuration matrix groups are generated by combining the matrix groups of links and kinematic pairs. The generalized deployable units are created after removing the unreasonable configurations such as the locally rigid configurations, the unreasonable stress configurations and isomorphism configurations.

3.1   Vector groups of links and kinematic pairs

The vector groups of links and kinematic pairs are established by the defined generation relation for the generators. All the elements of link generator are 1, which means all the links are rigid. All the elements of kinematic pair generator are 5, which means all the kinematic pairs are revolute pairs. The generation relation of links is defined as follows.

$$\varphi(i, h) = [1 \to i, h], \qquad i = 1, 2, 3, 4, \tag{2}$$

where, i denotes as the generation relation number and the weighted values of links. The h stands for the hth element in the generator. Equation (2) transforms the weighted value of the hth element in the generator from 1 (rigid link) to i (other type).

Similarly, the generation relation of kinematic pairs is defined as follows.

$$\psi(i_Y, h) = [5 \to i_Y, h], \qquad i_Y = 5, 7, 8, 9, \tag{3}$$

where, i_Y represents the generation relation number and the weighted value of kinematic pairs.

For example, the generator $a = [1, \ 1, \ 1]$ means that all the links are rigid. The operation of $\varphi(2, 4) \to a$ means the generation relation $\varphi(2, 4) = [1 \to 2, 4]$ is applied to a, and a is then transformed into $[1, 1, \ 2]$. That is, the four rigid links are transformed into three rigid links and one flexible link.

The vector subgroups of links and kinematic pairs can be obtained while the generation relations are operated for the generators which is realized with a recursive function. The vector subgroups of links are generated as follows:

$$G_1 = \varphi(i, 1) \to a \tag{4}$$
$$G_k = \varphi(i, k) \to G_{k-1}(h), \ k > 1, \tag{5}$$

where k is the subgroup number.

The vector subgroups of kinematic pairs are generated as follows.

$$Y_1 = \psi(i_Y, 1) \to b \tag{6}$$
$$Y_k = \psi(i_Y, k) \to Y_{k-1}(h), \ k \gg 1. \tag{7}$$

A series of subgroups of links and kinematic pairs are obtained in accordance with the above recursive functions. Every subgroup is a part of configurations, so all the configurations are the union of subgroups. The same configuration may exist in the different subgroups, which can be found by the intersection of subgroups. Thus, the vector groups of links and kinematic pairs are derived respectively as follows.

$$G_s = \bigcup_{k=1}^{m} G_k - \bigcup_{\ell=1, \delta=1, \ell \neq \delta}^{m(m-1)} \left\{ G_\delta \bigcap G_\ell \right\} \tag{8}$$

$$Y_s = \bigcup_{k=1}^{n} Y_k - \bigcup_{\ell=1, \delta=1, \ell \neq \delta}^{n(n-1)} \left\{ Y_\delta \bigcap Y_\ell \right\}, \tag{9}$$

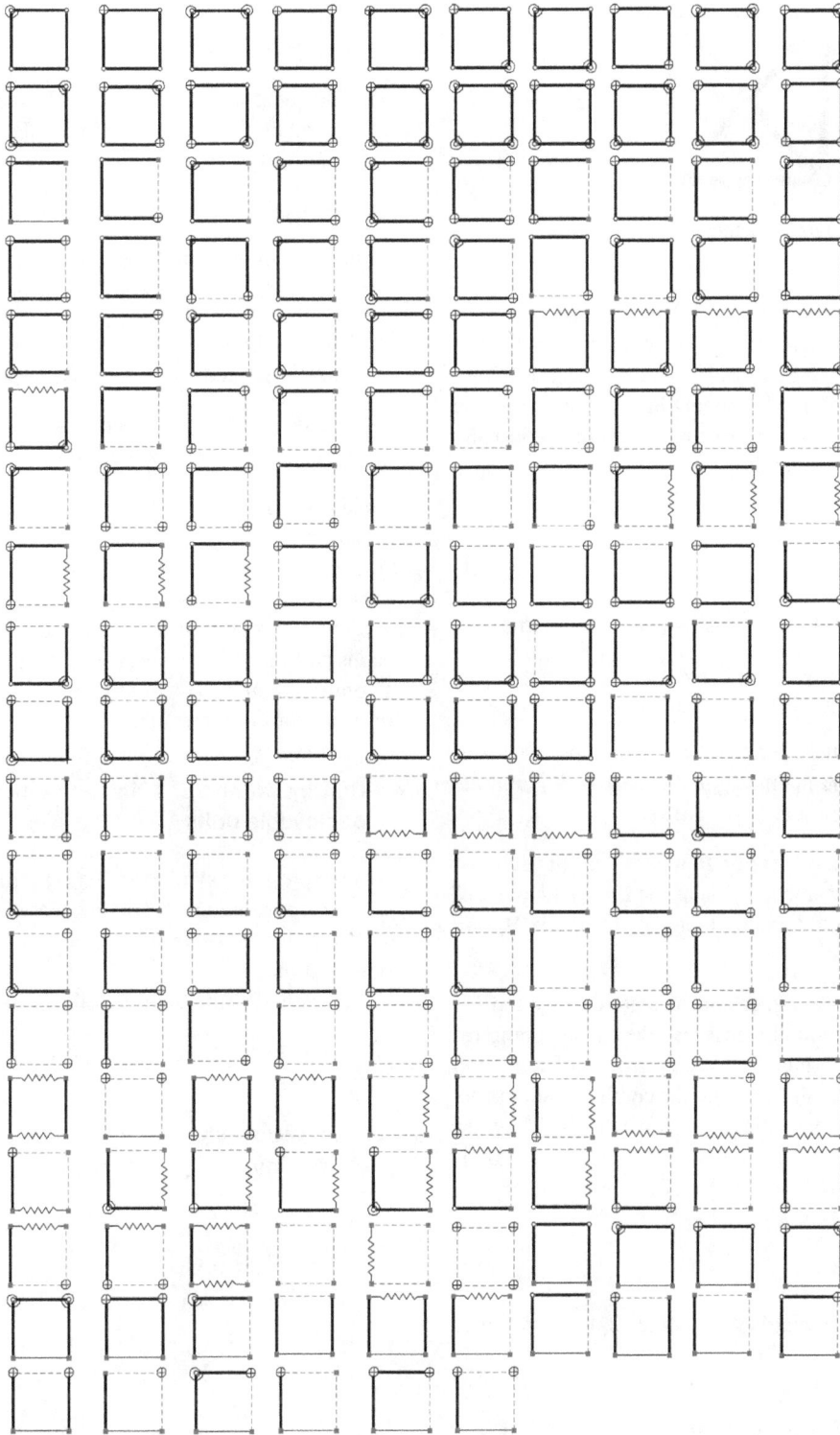

Figure 2. Configurations of 4-link generalized deployable unit.

where, m denotes the number of link subgroups, and n means the number of kinematic pair subgroups.

The first items in right side of Eqs. (8) and (9) are the union of all subgroups, and the second items are the intersections of any two subgroups. The subtraction of two items elimilates isomorphic configurations.

The permutation group is used to obtain the all permutation types of generalized kinematic pairs and generalized

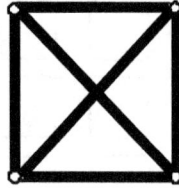

Figure 3. Planar 6-bar frame structure.

links. A permutation group of a set A denoted as $S(A)$, is a set of permutations of A that forms a group under composition of functions (Kurzweil and Stellmacher, 2004; Richard, 2009). It is the full permutation of elements in a group. Thus, the full-permutation vector groups of links and kinematic pairs are derived respectively as follows:

$$G = S(G_s) \tag{10}$$
$$Y = S(Y_s). \tag{11}$$

3.2 Configuration matrix groups of links and kinematic pairs

First, the two mapping relationships are defined as follows:

1. $\mathbf{A}_p = J[G(p)]$, the operator J means the pth vector of G is placed at the main diagonal of matrix $\mathbf{A}_p$, and the others elements of matrix $\mathbf{A}_p$ are zeros.

2. $\mathbf{B}_q = H[Y(q)]$, the operator H means the qth vector of Y are placed the same location as the kinematic pairs of initial matrix, and the other elements of matrix $\mathbf{B}_q$ are zeros.

It can be noted that the dimension of matrixes $\mathbf{A}_p$ and $\mathbf{B}_q$ is the same as that of initial matrix. By the two mapping relationships, the full-permutation vector groups of links and kinematic pairs are transformed into the configuration matrix groups of links and kinematic pairs, respectively. They are $\mathbf{AL} = \{\mathbf{A}_1, \cdots, \mathbf{A}_p, \cdots, \mathbf{A}_v\}$ and $\mathbf{BK} = \{\mathbf{B}_1, \cdots, \mathbf{B}_q, \cdots, \mathbf{B}_w\}$. The addition of $\mathbf{A}_p$ and $\mathbf{B}_q$ can create a configuration matrix, that is

$$\mathbf{C}_{pq} = \mathbf{A}_p + \mathbf{B}_q (p = 1, \cdots, v; \ q = 1, \cdots, w). \tag{12}$$

Then, the configuration matrix group is generated as follows.

$$\mathbf{C} = \{\mathbf{C}_{11}, \cdots, \mathbf{C}_{1q}, \cdots, \mathbf{C}_{pq}, \cdots, \mathbf{C}_{vw}\}. \tag{13}$$

It can be seen that an element of group $\mathbf{C}$ corresponds to a configuration matrix of a generalized deployable unit.

3.3 Removing the unreasonable configurations and isomorphism detection

The unreasonable configurations may exist in the configuration matrix group. In order to remove the unreasonable configurations, we propose some synthesis rules as follows.

1. The bars are replaced sequentially by flexible links, springs, or cables, respectively.

2. The rotational and cylindrical kinematic pairs are replaced by the preloaded hinges, flexible hinges, or fixed joints, respectively.

3. The fixed joint cannot connect two bars.

4. The preloaded hinge is composed of a torsional spring and a pair, which can only connect two bars.

5. The flexible hinge can connect two bars, two flexible links, a bar and a flexible link, respectively.

6. The connections between two springs, two cables, or a spring and a cable cannot exist.

The above synthesis rules can be performed based on the information of configuration matrix of generalized deployable units. In addition, Li et al. (2011) have proposed the method of the powers of the adjacency matrix for identifying the isomorphism of topological graphs and weighted multicolored graphs conveniently, correctly and uniquely. Thus the powers of configuration matrix are used to identify the isomorphism configurations of generalized deployable units.

4 Configurations of 4-link and 6-link generalized deployable units

First, the proposed synthesis method is illustrated by the configuration generation of 4-link generalized deployable unit by taking the planar four-bar kinematic chain as the initial configuration.

The synthesis process is as follows:

1. The links generator is $a = [1, 1, 1, 1]$, and the kinematic pairs generator is $b = [5, 5, 5, 5]$.

2. The vector subgroups of generalized links can be obtained from Eqs. (4) and (5) as

$$G_1 = \{[1,1,1,1],[2,1,1,1],[3,1,1,1],[4,1,1,1]\} \tag{14}$$

$$\begin{aligned} G_2 = \{&[1,1,1,1],[1,2,1,1],[1,3,1,1],[1,4,1,1],\\ &[2,1,1,1],[2,2,1,1],[2,3,1,1],[2,4,1,1],\\ &[3,1,1,1],[3,2,1,1],[3,3,1,1],[3,4,1,1],\\ &[4,1,1,1],[4,2,1,1],[4,3,1,1],[4,4,1,1]\} \end{aligned}$$
$$\cdots \tag{15}$$

Meanwhile, the vector subgroups of kinematic pairs can be obtained from Eqs. (6) and (7) as

$$Y_1 = \{[5,5,5,5],[7,5,5,5],[8,5,5,5],[9,5,5,5]\} \tag{16}$$

$$\begin{aligned} Y_2 = \{&[5,5,5,5],[5,7,5,5],[5,8,5,5],[5,9,5,5],\\ &[7,5,5,5],[7,7,5,5],[7,8,5,5],[7,9,5,5],\\ &[8,5,5,5],[8,7,5,5],[8,8,5,5],[8,9,5,5],[9,5,5,5], \end{aligned}$$

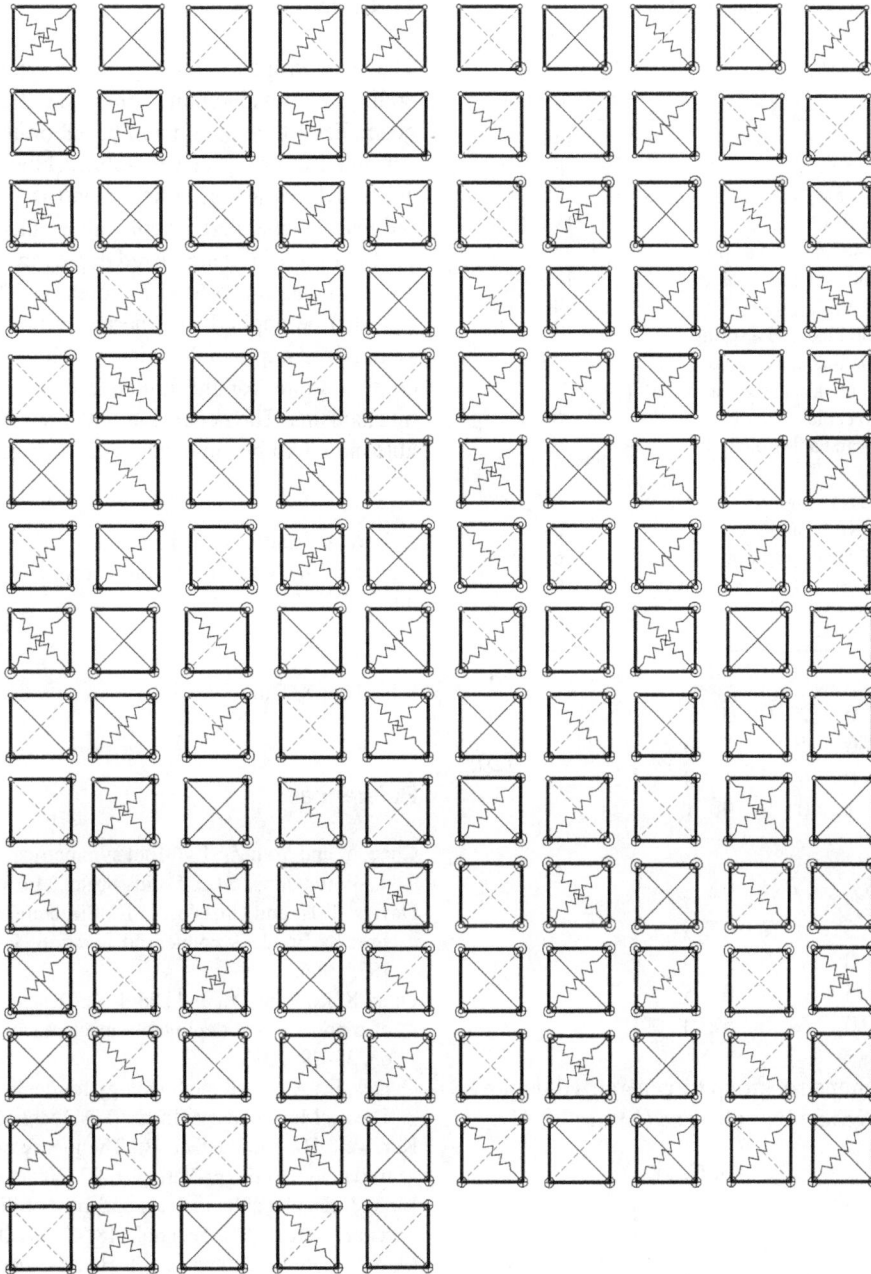

Figure 4. Configurations of 6-link generalized deployable unit with the rigid peripheral four links.

$$[9, 7, 5, 5], [9, 8, 5, 5], [9, 9, 5, 5]\}$$

$$\cdots \tag{17}$$

3. The isomorphic configurations of generalized links and kinematic pairs can be eliminated according to Eqs. (8) and (9), and the vector groups of generalized links and kinematic pairs are obtained as follows.

$$G_{\mathrm{s}} = \{[1, 1, 1, 1], [2, 1, 1, 1], [3, 1, 1, 1], [4, 1, 1, 1],$$
$$[1, 2, 1, 1], [1, 3, 1, 1], [1, 4, 1, 1], [2, 1, 1, 1],$$
$$[2, 2, 1, 1], [2, 3, 1, 1], [2, 4, 1, 1], [3, 1, 1, 1],$$

$$[3, 2, 1, 1], [3, 3, 1, 1], [3, 4, 1, 1], \cdots \} \tag{18}$$

$$Y_{\mathrm{s}} = \{[5, 5, 5, 5], [7, 5, 5, 5], [8, 5, 5, 5], [9, 5, 5, 5],$$
$$[5, 7, 5, 5], [5, 8, 5, 5], [5, 9, 5, 5],$$
$$[7, 7, 5, 5], [7, 8, 5, 5], [7, 9, 5, 5],$$
$$[8, 7, 5, 5], [8, 8, 5, 5], [8, 9, 5, 5], \cdots \}. \tag{19}$$

4. The configuration matrix groups of links and kinematic pairs are established respectively, as

$$\mathbf{AL} = \{A_1, \cdots A_l, \cdots A_{m1}\}$$

Table 3. Configurations of 6-link generalized deployable unit.

Types of peripheral four links	Number of configurations
Four rigid links	145
Three rigid links and a spring	40
Three rigid links and a cable	40
Two rigid links and two flexible links	180
Two rigid links, a flexible links and a spring	99
Two rigid links, a flexible link and a cable	102
A rigid link and three flexible links	68
A rigid link, two flexible links and a spring	77
A rigid link, two flexible link and a cable	77
A rigid link, a flexible link, a cable and a spring	15
Two rigid links and two cables	6
A rigid link, two cables and a flexible link	6
Four flexible links	38
Three flexible links and a spring	19
Two flexible links and two springs	6
Two flexible links, a cable and a spring	6
Two flexible links and two cables	6
Total	930

$$= \left\{ \begin{matrix} 1 & 0 & 0 & 0 & 2 & 0 & 0 & 0 \\ 0 & 1 & 0 & 0 & 0 & 1 & 0 & 0 \\ 0 & 0 & 1 & 0 & 0 & 0 & 1 & 0 \\ 0 & 0 & 0 & 1, & 0 & 0 & 0 & 1 \end{matrix} \cdots \right\}. \tag{20}$$

$$\mathbf{BK} = \{B_1, \cdots, B_q, \cdots, B_w\}$$

$$= \left\{ \begin{matrix} 0 & 5 & 0 & 5 & 0 & 7 & 0 & 5 \\ 5 & 0 & 5 & 0 & 7 & 0 & 5 & 0 \\ 0 & 5 & 0 & 5 & 0 & 5 & 0 & 5 \\ 5 & 0 & 5 & 0, & 5 & 0 & 5 & 0 \end{matrix} \cdots \right\}. \tag{21}$$

5. The configuration matrix group of generalized deployable units is derived by Eqs. (12) and (13), as

$$\mathbf{C} = \{C_{11}, \cdots, C_{1q}, \cdots, C_{pq}, \cdots, C_{vw}\}$$

$$= \left\{ \begin{matrix} 1 & 5 & 0 & 5 & 1 & 7 & 0 & 5 \\ 5 & 1 & 5 & 0 & 7 & 1 & 5 & 0 \\ 0 & 5 & 1 & 5 & 0 & 5 & 1 & 5 \\ 5 & 0 & 5 & 1, & 5 & 0 & 5 & 1 \end{matrix} \cdots \right\}. \tag{22}$$

All the 186 configurations of 4-link generalized deployable unit are created, as shown in Fig. 2.

Then, the planar 6-bar frame structure, as shown in Fig. 3, is regarded as the initial configuration. All the 930 configurations of 6-link generalized deployable unit are created, as shown in Table 3 according to the types of peripheral four links. If the peripheral four links are rigid, the number of configurations is 145, as shown in Fig. 4. Some potential applications of synthesized configurations need to develop in the future. Obviously, the second configuration in Fig. 4 can be assembled into the deployable mechanism of Astromesh antenna (Thomson, 1999).

5 Conclusions

The paper proposes a configuration synthesis method of generalized deployable units composed of generalized links and generalized kinematic pairs based on the group theory. The permutation group of generalized kinematic pairs and generalized links are established and the topological configurations of generalized deployable units are generated. The configurations of generalized deployable units including four and six generalized links are created. The proposed method can also be applied to generate the other configurations of generalized deployable units, which will provide the enough configurations for the innovate design of large deployable mechanisms. The networking method of generalized deployable units will become keystone of research in the future.

Acknowledgements. The authors gratefully acknowledge the support of the National Natural Science Foundation of China (grant no. 51375360).

Edited by: X. Ding

References

Chen, Y. and You, Z.: Two-fold symmetrical 6R foldable frame and its bifurcations, Int. J. Solids Struct., 46, 4504–4514, 2009.

Davies, T. H. and Crossley, F. E.: Structural analysis of plane linkages by Franke's condensed notation, J. Mechanisms, 1, 171–183, 1966.

Ding, X. L., Yang, Y., and Dai, J. S.: Design and kinematic analysis of a novel prism deployable mechanism, Mech. Mach. Theory, 63, 35–49, 2013.

Gantes, C.: A design methodology for deployable structures, Ph.D. Thesis, Massachusetts Institute of Technology, 1991.

Kurzweil, H., Stellmacher, B.: The theory of finite groups: An introduction, Springer-Verlag, 1–92, 2004.

Li, T. J.: Deployable analysis and control of deployable space antenna, Aerosp. Sci. Technol., 18, 42–47, 2014.

Li, T. J., Cao, W. Q., and Yan, T. H.: Applications of graph theory in mechanism analysis, in: Emerging topics on differential geometry and graph theory, edited by: Lucas, B. and Francois, R., New York, Nova Science Publishers, 1–34, 2011.

Lu, S. N., Zlatanov, D., Ding, X. L., and Molfino, R.: A new family of deployable mechanisms based on the Hoekens linkage, Mech. Mach. Theory, 73, 130–153, 2014.

Lu, S. N., Zlatanov, D., Ding, X. L., Zoppi, M., and Guest, S. D.: A Network of Type III Bricard Linkages, ASME 2015 International Design Engineering Technical Conferences and Computers and Information in Engineering Conference, Boston, Massachusetts, USA, Paper No. DETC2015-47139, 2015.

Mruthyunjaya, T. S.: Kinematic structure of mechanisms revisited, Mech. Mach. Theory, 38, 279–320, 2003.

Pellegrino, S.: Deployable structures, Springer, 1–36, 2001.

Pinero, E. P.: Expandable space framing, Progressive Architecture, 43, 154–155, 1962.

Richard, A. B.: Introductory combinatorics, 5th ed., Pearson, New York, 1–52, 2009.

Thomson, M. W.: The Astromesh deployable reflector, IEEE Antennas Prop., 3, 1516–1519, 1999.

Tibert, G.: Deployable tensegrity structures for space applications, Ph.D. Thesis, Sweden, Royal Institute of Technology, 2002.

Warnaar, D. B. and Chew, M.: Kinematic synthesis of deployable-foldable truss structure using graph theory, part 1: graph generation, J. Mech. Design, 117, 112–116, 1995a.

Warnaar, D. B. and Chew, M.: Kinematic synthesis of deployable-foldable truss structure using graph theory, part 2: generation of deployable truss module design concepts, J. Mech. Design, 117, 117–122, 1995b.

7

Development of a parallel robotic system for transperineal biopsy of the prostate

Doina Pisla[1], **Paul Tucan**[1], **Bogdan Gherman**[1], **Nicolae Crisan**[2], **Iulia Andras**[2], **Calin Vaida**[1], and **Nicolae Plitea**[1]

[1]CESTER – Research Center for Industrial Robots Simulation and Testing, Technical University of Cluj-Napoca, Cluj-Napoca, Romania

[2]University of Medicine and Pharmacy, Cluj-Napoca, Romania

Correspondence to: Calin Vaida (calin.vaida@mep.utcluj.ro)

Abstract. Prostate cancer is the second deadliest form of cancer, even though it is less invasive and easily curable in early stages, due to the lack of an efficient and accurate diagnosis strategy. To date, the standard diagnosis procedure involves a blind biopsy with a high rate of false negative results. In order to overcome these limitations, the paper proposes the development of a novel parallel robotic structure for transperineal prostate biopsy that enables an accurate diagnosis through ultrasound-guided targeted tissue sampling. The robotic system consists of two parallel modules, each with 5 degrees of freedom (DOFs): one module guiding the transrectal ultrasound probe (TRUS) and the other guiding the biopsy gun. The two modules are designed to work together in order to help the physician with the tissue sampling of the prostate. The singular configurations of both robotic modules are analyzed and solutions for avoiding them are provided. The experimental model of the robotic structure is described along with the initial test results, which evaluate the robot accuracy for several medically relevant sets of coordinates.

1 Introduction

In 2015, the American Cancer Society (ACS, 2016) provided statistical data regarding cancer occurrence. According to these data, 1.6 million cases of cancer were estimated in 2015, 848 200 cases in male patients and 810 170 cases in female patients. The most common form of cancer in female patients is breast cancer with an occurrence rate of 29 % (234 950 cases), while for male patients the most common cancer is prostate cancer with an occurrence rate of 26 % (220 532 cases). As indicated by the ACS, the next in rank after prostate cancer is lung and bronchial cancer with an occurrence rate of 14 % (118 748 cases); almost half are prostate cancer cases. Incidence rates of prostate cancer have changed substantially over the past 20 years, rapidly increasing from 1988 to 1992, declining sharply from 1992 to 1995, remaining stable from 1995 to 2000, and decreasing (on average) from 2000 to 2011. This erratic trend primarily reflects changing patterns in the utilization of prostate-specific anti-gen (PSA) blood testing for the detection of prostate cancer (ACS, 2016).

In the matter of cancer mortality, lung cancer is by far the leading cause of death among men (28 %), followed by prostate (9 %) and colon and rectal cancer (8 %). Biopsy, as a medical procedure, focuses on removing a body tissue sample in order for it to be examined with a microscope. Currently this is the most efficient way to determine the malignancy of tumors. The main method used in prostate cancer diagnosis is the core needle biopsy, which is usually performed by a urologist (Jemal et al., 2009).

The equipment used in prostate biopsy includes an image-acquiring tool, also known as a transrectal ultrasound probe (TRUS), and a biopsy gun that samples the prostate (Pondman et al., 2008) in a sequence illustrated in Fig. 1. In the first step, the tip of the needle is positioned proximal to the prostate, then a button is pushed on the biopsy gun that automatically performs the next three steps.

Depending on the insertion of the biopsy needle, a biopsy of the prostate can be done in two ways. In the case of a

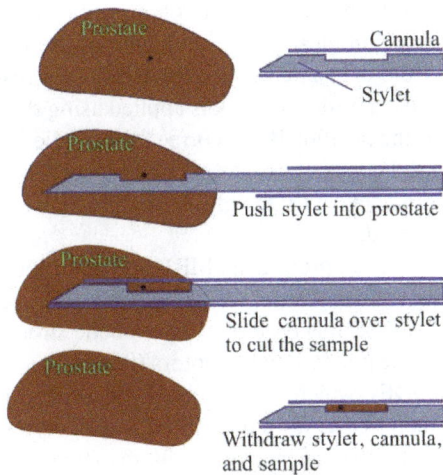

Figure 1. Tissue sampling procedure in four steps.

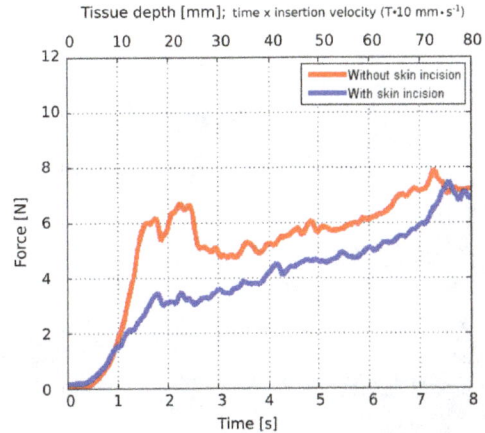

Figure 2. Experimental results of the needle incision procedure with and without skin incision (Vaida et al., 2017).

transrectal biopsy, the patient is seated in a lateral decubitus position; the TRUS probe is lubricated with a special anesthetic gel and inserted into the rectum. The ultrasound probe provides real-time imaging of the entire prostate gland, enabling the physician to select the prostate area to be sampled. Attached to the TRUS probe is a special guiding tool for the biopsy gun that will preserve the orientation of the needle in the echography plane, allowing for the continuous monitoring of the needle location. When the desired location is reached, the firing mechanism of the biopsy gun is actuated and the needle collects the tissue sample. After the sampling, the needle is extracted and the tissue is placed into a special container for further analysis. Typically a 12-core biopsy is performed in an attempt to cover the entire volume of the prostate, but the procedure is carried out in a blind manner due to the limited image quality and manual positioning accuracy of the system. Another drawback specific to the transrectal approach is the fact that the needle, in order to collect the tissue sample, has to pass through the rectum wall, thus easily generating serious infection causing the patient to be hospitalized for a certain period of time after the biopsy (Taneja et al., 2013).

Transperineal biopsy of the prostate implies a different protocol for the procedure. The patient is positioned in the gynaecological position (Free Education Network, 2016). The TRUS probe is lubricated and inserted into the rectum, as in transrectal biopsy, and used to visualize and guide the needles to the sampling positions in real time. The needles are inserted through the perineum in two possible scenarios: on multiple individual trajectories passing through the skin each time or on trajectories with a common point at which an incision is made through the skin (Taneja et al., 2013). The two approaches were compared experimentally and a detailed analysis is presented in Vaida et al. (2017). The results, presented in Fig. 2, point out an important difference between the two approaches. When the needle passes

through the skin, there is a force spike that does not appear in the second case. This indicates that the second approach offers better accuracy in avoiding possible needle bending and deflection from the ideal trajectory. From a medical point of view, the experimental data we collected was certified by the European Institute of Oncology in Milan, Italy, where doctors use the skin incision approach to perform transperineal biopsies of the prostate (de Cobelli et al., 2015).

In a comparison between the two approaches, the advantage of the transperineal approach (Fig. 3) is better access over the prostate gland, especially for the sampling of the apex of the prostate (an area that cannot be reached transrectally). In the case of the transrectal approach, the needles are inserted through the base of the prostate (far from the apex) and the length of the sampled tissue has to be at least 10 mm to obtain proper material for analysis. At the same time, the transperineal approach eliminates the septic risk raised by the transrectal approach because the needle does not penetrate the intestines in order to collect the tissue (Pepe and Aragona, 2014).

The most commonly used procedure in prostate cancer diagnosis is the TRUS-guided prostate biopsy. However, this procedure has significant disadvantages. In general, the biopsy cores are clustered together, which raises the issue of optimal sampling of the entire prostate. Moreover, the precise localization of possible lesions and the resampling of a region of interest are not possible with the standard freehand TRUS biopsy technique. Since the cancer detection rate is correlated with the quality of biopsy core sampling, an improved core sampling technique should maximize the prostate cancer detection rate, which in turn leads to better disease management (Kaye et al., 2015).

One possible solution to the standard TRUS biopsy limitations is the use of a robot (which enhances precision), together with three-dimensional reconstruction software. A robotic ultrasound application may provide image guid-

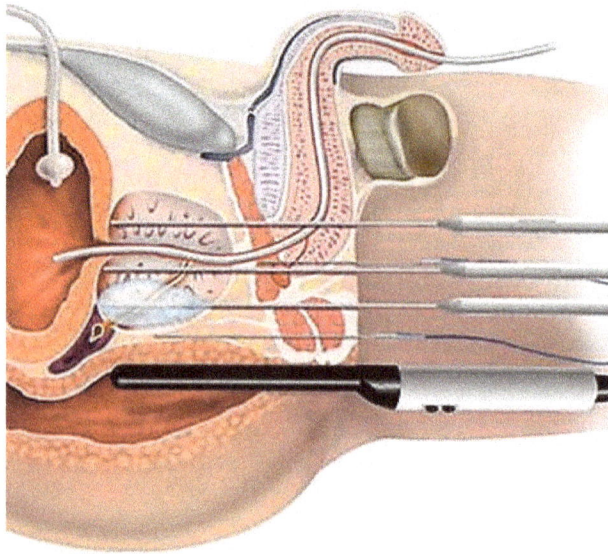

Figure 3. Transperineal approach to prostate biopsy (Avantgarde, 2016).

ance for common prostate cancer treatment methods (e.g., brachytherapy and radical prostatectomy; Kaye et al., 2015). Research centers around the world are trying to provide solutions for medical requirements in an attempt to improve lives through the construction of robotic systems (Gherman et al., 2016; Berceanu and Tarnita, 2010; Tarnita, 2016) or by studying and improving existing systems (Tarnita and Marghitu, 2013; Ottaviano et al., 2014; Berceanu et al., 2010).

In 2007, Cheng proposed a robotic system that allows for the transperineal approach to prostate biopsy (Cheng, 2007). In this system, the needle is placed at an angle with respect to the endorectal probe and set to collect different samples by modifying this angle. The robotic system is made up of two mechanical modules: the first module is for needle guidance, while the second module is for imagistic sampling. Both modules are mounted on a mobile platform with 5 DOFs. The needle is inserted manually and guided by the urologist, while the sampling depth is controlled by a switch mounted on the biopsy gun. The orientation mechanism of the needle and the switch are actuated. The imagistic sampling module is made up of an endorectal probe and its support mechanism that executes an active translation. The 5 DOF system offers sufficient dexterity to position the imagistic sampling module, but the tissue sampling is done manually.

Another solution for the transperineal approach is proposed by Long et al. (2012). The system includes a guiding module for the biopsy gun mounted on one side of the imagistic module. The system positions the needle on a specified trajectory and fixes the sampling depth. The robot has 7 DOFs. A TRUS probe is attached to the ultrasound apparatus. The guiding system of the needle is calibrated before the endorectal probe procedure. The robot positions the module of the biopsy gun close to the perineum of the patient and the first needle is inserted. Once the needle has reached the sampling area, a position check is applied using the endorectal probe. If the position is not correct, the needle is retracted and reinserted correctly. This system does not have an automated guiding device for the ultrasound probe, which is thus manually guided.

Stoianovici et al. propose an MRI-safe robot for biopsy of the prostate (Stoianovici et al., 2014). For this system, the transrectal approach is preferred. The robotic structure is actuated with pneumatic stepper motors (PneuStep). The structure has 3 DOFs, which is considered sufficient to guide the TRUS probe; 1 DOF is implemented separately and represents a new method of biopsy needle insertion. The robotic structure assists the urologist by automatically orienting the biopsy gun to the targeted biopsy area and fixing the sampling depth using MRI. The robot is attached to the MRI table using a fixture plate after the patient has been seated in the necessary position for the medical procedure (lateral decubitus). The endorectal extension includes an MRI coil and a set of markers to memorize the patient position. The guided biopsy needle passes through the endorectal extension at an angle calculated by the robotic system. The system is provided with a rotation joint around the endorectal extension, a rotation joint for needle guidance, and a translation joint for needle insertion. The insertion of the needle is done manually, and only the depth of sampling and the orienting angles are fixed by the endorectal probe. The disadvantage of the system is the transrectal approach, which is a procedure that creates complications, as mentioned previously.

Another robotic system for transperineal prostate biopsy is proposed by Vaida et al. (2015). The robotic solution consists of a modular parallel structure with two independent kinematic chains using a specific RCM (remote center of motion). The TRUS guidance module has 4 DOFs; with respect to entry points, there are three rotations and a translation along the longitudinal axis of the probe. The module for manipulating the biopsy gun has only 3 DOFs since the rotation of the needle around its longitudinal axis is unnecessary. The first module of the TRUS probe is positioned on the biopsy table and the biopsy gun module is mounted on top of the first in an upside-down position with the same kinematic appearance. The fourth active joint is replaced with a simple actuation device that triggers the biopsy gun. This system requires high-quality machining for the components of the RCM mechanism, which is sometimes difficult to achieve.

The robotic structure presented in this paper combines a series of technologies to help the urologist perform a biopsy in a short period of time with a reduced number of samples. The robotic system proposed in this paper is designed to work as a fusion system between previously obtained MRI results (from an MRI scan of the prostate area) and real-time images provided by the endorectal probe in order to validate the sampling points of the biopsy. After the MRI scan of the suspected cancer patient has been performed, the sampling points are marked by the urologist. The marked sampling points are then referenced in the robot reference system and the endorectal probe is inserted. After the volumetric image of the prostate is obtained, the sampling points are identified and the biopsy is performed. The advantage of this robotic system is that after the patient has been seated and calibrated with the robot reference system, the surgeon's only task is to supervise the system, check the validity of the sampling points, and comply with the safety features of the system (every procedure performed by the robot has to be approved by the urologist).

The paper is organized as follows. Section 2 presents the novel parallel robotic system designed for prostate biopsy with its two modules and the inverse kinematic equations. Section 3 presents the singularities in the robotic system. In Sect. 4, the workspace of the robotic structure is presented. The simulation results of the parallel robot for transperineal biopsy are presented in Sect. 5. The experimental model of the robotic structure is presented in Sect. 6 followed by a set of experimental results in Sect. 7. The conclusions are presented in Sect. 8, followed by acknowledgements and references.

2 BIO-PROS-1 parallel robot

In order to define the specific tasks of the robotic system for transperineal prostate biopsy, a medical protocol (for robotic-assisted prostate biopsy) was developed in collaboration with a team of oncologists from the University of Medicine and Pharmacy in Cluj-Napoca as a stepwise procedure.

1. The exact tumor location and size are determined with an MR imaging device, where a special smaller probe will be inserted through the urethra into the prostate as a fixed marker.

2. The MRI data are processed in order to establish the coordinates for the target points for the needle, calculated relative to the special probe.

3. The patient is positioned in the gynecological position.

4. Anesthesia is administered; the transperineal approach involves a local anesthetic of the perineum, the area through which the needle is inserted.

5. The robot is positioned relative to the patient and the coordinate systems of the patient and robot are correlated with all the necessary transformations (from this moment on, the robot will remain fixed with respect to the patient).

6. The TRUS probe is inserted for real-time image acquisition along with the same special smaller probe placed inside the prostate. The probe will be used for the MRI–TRUS fusion to ensure that the coordinates determined

as target points will be properly defined during the biopsy.

7. The exact locations of the points are recalculated based on the MRI–TRUS fusion image (the prostate is modified in shape due to the different positions of the patient inside the MRI and during the procedure).

8. The sampling is performed using the target point coordinates determined in step 7, with real-time monitoring of the enhanced images using one of the two main options for the definition of the insertion points locations.

9. A needle-guiding template is used, which allows the needles to be driven on parallel trajectories in different areas of the prostate.

10. A single entry port is also used, which is created by a very small incision in the skin situated on the median line of the body to drive the needles on concurrent trajectories (Vaida et al., 2017).

11. The biopsy gun is driven to the insertion point by the robotic system with the final orientation (calculated based on the insertion target coordinates pairs).

12. The needle actuation module is engaged to insert the needle (on a linear path) until the target point is reached.

13. The needle actuation module activates the sampling function of the biopsy gun, thereby sampling the tissue.

14. The needle actuation module retracts the needle on the same trajectory until the needle is out of the patient, at which time the robot arm moves away from the patient to enable the unloading of the gun and the sample retrieving.

15. The samples are stored in special containers (small bottles with identification tags for the sampling area).

   After the sampling, anatomopathological analysis of the tissue is performed to determine the possible presence and spread of cancerous cells in the prostate.

The robotic system should guide both the TRUS and the biopsy gun, and hence there are two main modules, one for each instrument (Fig. 4). The kinematics of the BIO-PROS-1 parallel robotic structure have been presented in Pisla et al. (2015).

The main requirements given by urologists are high accuracy, a medically relevant workspace, safe positioning and orientation of the medical instrument, and performing the biopsy in a targeted way under real-time ultrasound guidance. These led to the development of a robotic system consisting of two modules (robotic structures) that work together (Plitea et al., 2015a). The first module, designed for TRUS probe guidance (illustrated in Fig. 5), has a modular structure consisting of a parallel module with $M = 3$ DOFs of family

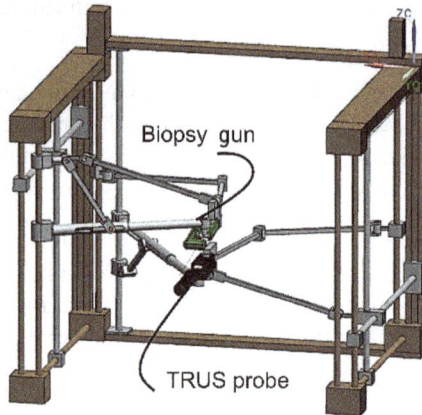

Figure 4. Simplified CAD model (concept) of BIO-PROS-1.

$F = 1$ (the number of imposed constraints for 1 DOF; common to all links of the mechanism) with three active joints (q_1, q_2, q_3). The second parallel module is $M = 3$ DOFs of family $F = 1$ with two active joints (q_4, q_5); it works in cylindrical coordinates. A fixed coordinate system ($OXYZ$) is defined and attached to the frame of the robot (Fig. 5), where the OZ axis represents the translational axis of the (q_3) translational joint, and the OY axis is parallel to the (q_1) translational joint. With respect to this coordinate system, the first module is positioned in the OYZ plane, while the second module is at a distance (a_4) with the active joints (q_4, q_5) moving on an axis parallel to OZ. The second robotic module, used for guiding the biopsy gun, has the same kinematic structure as the first; however, as can be seen from Fig. 6, it has both modules in the same plane. With respect to the fixed coordinate system, $OXYZ$ is parallel to the OYZ plane at a distance (X_C). Both robotic modules have active translational joints actuated along axes parallel to the OY (q_1 and q'_1) and OZ ($q_2, q_3, q_4, q_5, q'_2, q'_3, q'_4, q'_5$) of the fixed frame of the robot. Each pair of modules (composing the TRUS guiding module and the biopsy-gun guiding module) is connected through a pair of Cardan joints that hold the mobile platform, which further integrates the active instruments. The geometric parameters of the TRUS-probe guiding module are as follows:

$R_1, R_2, R_3, c_4, d_4, h$ — fixed link lengths;

e, e_1, e_2, e_3, e_4 — distances between the rotational axes

     of passive joints;

c — distance between the two Cardan joints (A_1 and A_2).

The geometric parameters of the biopsy-gun guiding module are as follows:

$R'_1, R'_2, R'_3, c'_4, d'_4, h'$ — fixed link lengths;

$e', e'_1, e'_2, e'_3, e'_4$ — distances between the rotational axes

     of passive joints;

c' — distance between the two Cardan joints (A'_1 and A'_2).

The inverse geometric model of the TRUS guiding module is determined based on simple analytical equations: knowing the TRUS position, the tip point $E(X_E, Y_E, Z_E)$ and its orientation, and the angles ψ and θ. The generalized coordinates are presented in Eqs. (1) to (5):

$$q_1 = Y_E - (h + c) \cdot \sin(\theta) \cdot \sin(\psi) - e, \qquad (1)$$

$$q_2 = Z_E + (h + c) \cdot \cos(\theta) + e \qquad (2)$$

$$+ \sqrt{R_1^2 - (X_E - (h + c) \cdot \sin(\theta) \cdot \cos(\psi) - e_1 - e_2)^2},$$

$$q_3 = Z_E + (h + c) \cdot \cos(\theta), \qquad (3)$$

$$q_4 = Z_E + h \cdot \cos(\theta) - e_3, \qquad (4)$$

$$q_5 = Z_E - h \cdot \cos(\theta) - e_3 + \sqrt{d_4^2 - (R_4 - c_4 - e_4)^2}. \qquad (5)$$

The generalized equations for the direct kinematic model of the TRUS guiding module are

$$\begin{cases} X_E = X_{A_1} + d \cdot \sin(\theta) \cdot \cos(\psi) \\ Y_E = Y_{A_1} + d \cdot \sin(\theta) \cdot \sin(\psi) \\ Z_E = Z_{A_1} - d \cdot \cos(\theta) \end{cases}, \qquad (6)$$

$$\theta = \mathrm{acos}\left(\frac{Z_{A_1} - Z_{A_2}}{c} \right) \qquad (7)$$

$$\psi = \mathrm{atan2}\left(Y_{A_2} - Y_{A_1}, X_{A_2} - X_{A_1} \right),$$

where

$$\begin{cases} X_{A_1} = e_1 + e_2 + \sqrt{R_1^2 - (q_2 - q_3 - e)^2} \\ Y_{A_1} = q_1 - e \\ Z_{A_1} = q_3 \end{cases}. \qquad (8)$$

By solving the system of equations (Eq. 9), the expressions for $X_{A_2}, Y_{A_2}, Z_{A_2}$ can be obtained:

$$\begin{cases} c^2 - \left(Z_{A_1} - Z_{A_2}\right)^2 \\ \quad = \left(Y_{A_2} - Y_{A_1}\right)^2 + \left(X_{A_2} - X_{A_1}\right)^2 \\ \left(X_{A_2} - e_5\right)^2 + \left(a_4 - Y_{A_2}\right)^2 \\ \quad = \left(e_4 + c_4 + \sqrt{d_4^2 - (q_5 - q_4)^2}\right)^2 \\ Z_{A_2} = q_4 + e_3 \end{cases}. \qquad (9)$$

Equation (9) is a quadratic equation leading to a double solution for X_{A_2} and Y_{A_2}, which define the two possible working modes of the robotic system.

The equations that characterize the inverse kinematic model of the biopsy gun module for BIO-PROS-1 are described in Eqs. (10) to (14):

$$q'_1 = Y'_E - (h' + c') \cdot \sin(\theta') \cdot \sin(\psi') - e', \qquad (10)$$

$$q'_2 = Z'_E + (h' + c') \cdot \cos(\theta') + e' \qquad (11)$$

$$+ \sqrt{R_1'^2 - (X'_E - (h' + c') \cdot \sin(\theta') \cdot \cos(\psi') - e'_1 - e'_2)^2},$$

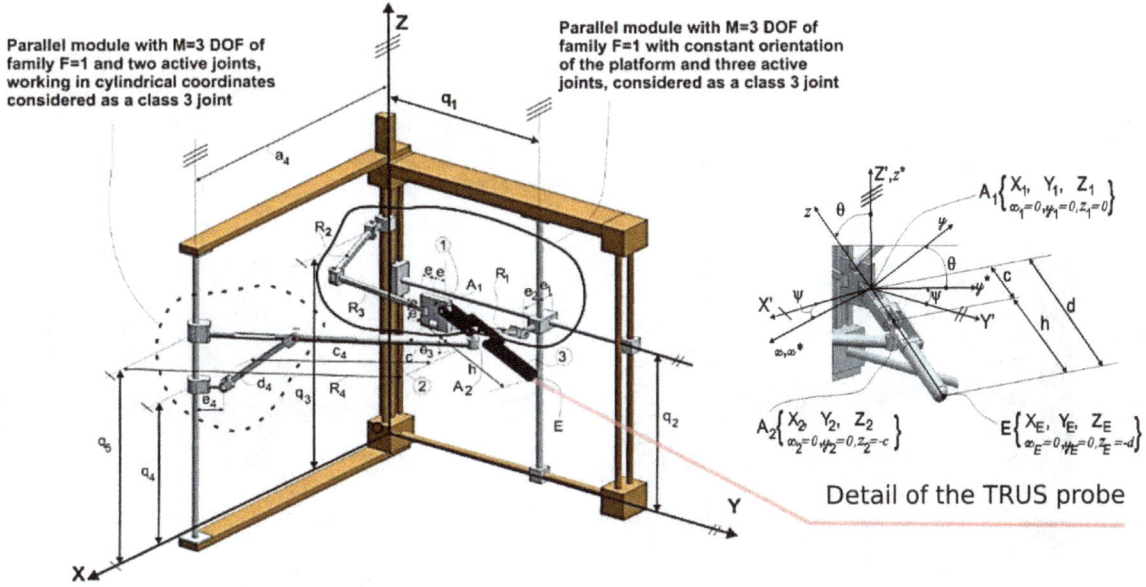

Figure 5. Kinematic model of the TRUS-probe guiding module.

$$q'_3 = Z'_E + (h' + c') \cdot \cos(\theta'), \tag{12}$$

$$q'_4 = Z'_E + h' \cdot \cos(\theta') - e'_3, \tag{13}$$

$$q'_5 = Z'_E - h' \cdot \cos(\theta') - e'_3 \tag{14}$$
$$+ \sqrt{d'^2_4 - (R'_4 - c'_4 - e'_4)^2}.$$

Using the direct kinematic model, the equations that characterize the needle tip coordinates and the orientation of the biopsy gun are presented in Eqs. (15) to (18):

$$\begin{cases} X_{E'} = X_{A'_1} + d' \cdot \sin(\theta') \cdot \cos(\psi') \\ Y_{E'} = Y_{A'_1} + d' \cdot \sin(\theta') \cdot \sin(\psi') \\ Z_{E'} = Z_{A'_1} - d' \cdot \cos(\theta') \end{cases}, \tag{15}$$

$$\theta' = \mathrm{acos}\left(\frac{Z_{A'_1} - Z_{A'2}}{c'}\right) \tag{16}$$

$$\psi' = \mathrm{atan2}\left(Y_{A'2} - Y_{A'_1}, X_{A'2} - X_{A'_1}\right),$$

where

$$\begin{cases} X'_{A_1} = X_C - e'_1 - e'_2 - \sqrt{R'^2_1 - (q'_2 - q'_3 - e')^2} \\ Y'_{A_1} = q'_1 - e' \\ Z'_{A_1} = q'_3 \end{cases}. \tag{17}$$

By finding the solutions for the system of equations (Eq. 18), the expressions for X'_{A_2}, Y'_{A_2}, Z'_{A_2} are obtained:

$$\begin{cases} c'^2 - (Z'_{A_1} - Z'_{A_2})^2 \\ \quad = (Y'_{A_2} - Y'_{A_1})^2 + (X'_{A_2} - X'_{A_1})^2 \\ (X_C - e'_5 - X'_{A_2})^2 + Y'^2_{A_2} \\ \quad = \left(c'_4 + e'_4 + \sqrt{d'^2_4 - (q'_5 - q'_4)^2}\right)^2 \\ Z'_{A_2} = q'_4 + e'_3 \end{cases}. \tag{18}$$

An analysis of Eq. (18) reveals that in the case of the second module, the coordinates for X'_{A_2} and Y'_{A_2} also have a double solution, leading to the two working modes of the robot module.

3 Singularity analysis of the parallel robot BIO-PROS-1

The singularities in the parallel manipulators can be studied using different mathematical techniques, such as analyzing the Jacobian matrix (of the loop closure equations) rank and condition number (Merlet, 2006), using screw theory (Zlatanov et al., 2002), using a study parameterization of the Euclidian displacement group (Walter and Husty, 2010), or using the augmented Jacobian matrix (Joshi and Tsai, 2002). A parallel robot in singular configurations can instantaneously lose the ability to transmit motion and become uncontrollable (Plitea, 2015b) in two possible cases: the mechanism loses its stiffness by gaining DOFs, or the mechanism locks by losing DOFs (Podder et al., 2010). Identification and avoidance of the singular loci is crucial in ensuring kinematic accuracy and robotic system stability.

Patient safety is one critical aspect regarding the use of a robotic system for medical procedures such as biopsies.

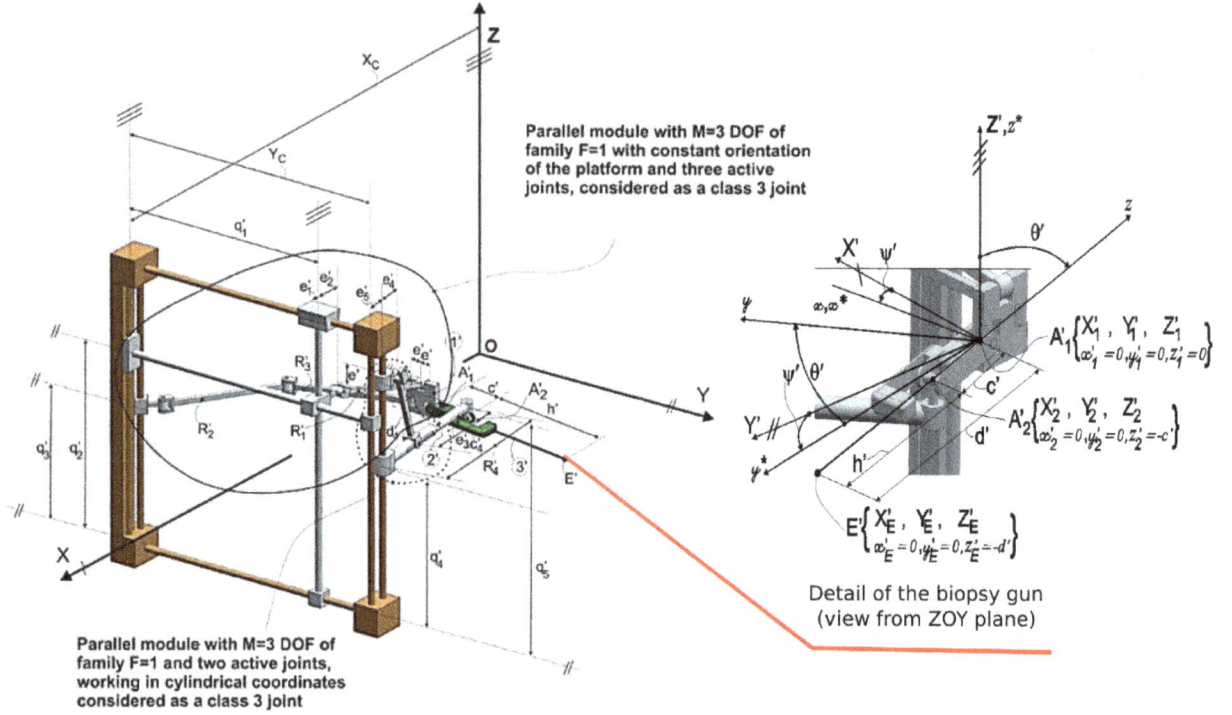

Figure 6. Kinematic model of the biopsy-gun guiding module.

Singularity analysis therefore has great importance in order to identify and avoid singular loci in the robot workspace (Gherman et al., 2010). The method used in this paper for the singularity analysis is based on evaluating the determinants of the Jacobian matrices **A** and **B**, which are obtained from the inverse and direct geometric models of the robot (Gosselin and Angeles, 1990; Gosselin and Wang, 1997). This groups them into three categories: serial singularities (the determinant of the Jacobian matrix for the inverse kinematic problem is zero and the robot losses degrees of mobility); parallel singularities (the determinant of the Jacobian matrix for the direct kinematic problem is zero and the robot gains degrees of freedom, becoming uncontrollable); and architectural singularities (the determinant of the Jacobian matrix for both kinematic problems are zero and the end effector can be moved while all the active joints are locked).

Starting from the geometric models that describe the relations between the coordinates of the active joints $q'_1, q'_2, q'_3, q'_4, q'_5$ and the coordinates of the tip of the needle, $E = (X_E, Y_E, Z_E, \psi_E, \theta_E)$, as shown in Pisla et al. (2015). The implicit functions can be defined by Eq. (19) for the TRUS probe module and Eq. (20) for the biopsy gun module:

$$\begin{cases} f_1: Y_E - d \cdot \sin(\theta) \cdot \sin(\psi) - e - q_1 = 0; \\ f_2: Z_E + d \cdot \cos(\theta) - 2 \cdot e \\ \quad - \sqrt{R_1^2 - (X_E - d \cdot \sin(\theta) \cdot \cos(\psi) - e_1 - e_2)^2} \\ \quad - q_2 = 0; \\ f_3: Z_E + d \cdot \cos(\theta) - e - q_3 = 0; \\ f_4: Z_E + h \cdot \cos(\theta) - e_3 - q_4 = 0; \\ f_5: \left(c_4 + e_4 + \sqrt{d_4^2 - (q_5 - q_4)^2} \right)^2 \\ \quad - (a_4 - X_E + h \cdot \cos(\psi) \cdot \sin(\theta))^2 \\ \quad + (Y_E - h \cdot \sin(\psi) \cdot \sin(\theta) - c_4 - e_4)^2 - q_5 = 0, \end{cases}$$ (19)

$$\begin{cases} f_1: Y'_E - d' \cdot \sin(\theta') \cdot \sin(\psi') - e' - q'_1 = 0; \\ f_2: Z'_E + d' \cdot \cos(\theta') + 2 \cdot e' \\ \quad + \sqrt{R_1'^2 - (X_C - X'_E + d' \cdot \sin(\theta') \cdot \cos(\psi') - e'_1 - e'_2)^2} \\ \quad - q'_2 = 0; \\ f_3: Z'_E + d' \cdot \cos(\theta') - q'_3 = 0; \\ f_4: Z'_E + h' \cdot \cos(\theta') - e'_3 - q'_4 = 0; \\ f_5: \left(c'_4 + e'_4 + \sqrt{d_4'^2 - (q'_5 - q'_4)^2} \right)^2 \\ \quad - (X_C - X_E + h \cdot \cos(\psi) \cdot \sin(\theta) - e_5)^2 \\ \quad - (Y_C - Y_E - h \cdot \sin(\psi) \cdot \sin(\theta))^2 = 0. \end{cases}$$ (20)

The equation describing the kinematic model for the velocities of the ultrasound-guiding robotic module is presented in Eq. (21),

$$\mathbf{A} \cdot \dot{X} + \mathbf{B} \cdot \dot{q} = 0,$$ (21)

where $\dot{X}$ represents the velocity vectors of the ultrasound tip and $\dot{q}$ the velocity vector of the active joints:

$$\dot{X} = \left[\dot{X}_E, \dot{Y}_E, \dot{Z}_E, \dot{\psi}, \dot{\theta} \right]^{\mathrm{T}}, \quad \dot{q} = \left[\dot{q}_1, \dot{q}_2, \dot{q}_3, \dot{q}_4, \dot{q}_5 \right]^{\mathrm{T}}.$$ (22)

A and **B** are Jacobian matrices obtained using Eqs. (23) and (24):

$$\mathbf{A} = \begin{pmatrix} 0 & 1 & 0 & \frac{\partial f_1}{\partial \psi} & \frac{\partial f_1}{\partial \theta} \\ \frac{\partial f_2}{\partial X_E} & 0 & 1 & \frac{\partial f_2}{\partial \psi} & \frac{\partial f_2}{\partial \theta} \\ 0 & 0 & 1 & 0 & \frac{\partial f_3}{\partial \theta} \\ 0 & 0 & 1 & 0 & \frac{\partial f_4}{\partial \theta} \\ \frac{\partial f_5}{\partial X_E} & \frac{\partial f_5}{\partial Y_E} & 0 & \frac{\partial f_5}{\partial \psi} & \frac{\partial f_5}{\partial \theta} \end{pmatrix}, \quad (23)$$

$$\mathbf{B} = \begin{pmatrix} -1 & 0 & 0 & 0 & 0 \\ 0 & -1 & 0 & 0 & 0 \\ 0 & 0 & -1 & 0 & 0 \\ 0 & 0 & 0 & -1 & 0 \\ 0 & 0 & 0 & \frac{\partial f_5}{\partial q_4} & \frac{\partial f_5}{\partial q_5} \end{pmatrix}, \quad (24)$$

where

$$\frac{\partial f_1}{\partial \psi} = -d \cdot \cos(\psi) \cdot \sin(\theta), \quad (25)$$

$$\frac{\partial f_1}{\partial \theta} = -d \cdot \sin(\psi) \cdot \cos(\theta), \quad (26)$$

$$\frac{\partial f_2}{\partial X_E} = \frac{e_1 - X_E + e_2 + d \cdot \cos(\psi) \cdot \sin(\theta)}{\sqrt{R_1^2 - (e_1 - X_E + e_2 + d \cdot \cos(\psi) \cdot \sin(\theta))^2}}, \quad (27)$$

$$\frac{\partial f_2}{\partial \psi} = d \cdot \sin(\psi) \cdot \sin(\theta) \quad (28)$$
$$\cdot \frac{(e_1 - X_E + e_2 + d \cdot \cos(\psi) \cdot \sin(\theta))}{\sqrt{R_1^2 - (e_1 - X_E + e_2 + d \cdot \cos(\psi) \cdot \sin(\theta))^2}},$$

$$\frac{\partial f_2}{\partial \theta} = -d \cdot \sin(\theta) \quad (29)$$
$$- \frac{d \cdot \cos(\psi) \cdot \cos(\theta) \cdot (e_1 - X_E + e_2 + d \cdot \cos(\psi) \cdot \sin(\theta))}{\sqrt{R_1^2 - (e_1 - X_E + e_2 + d \cdot \cos(\psi) \cdot \sin(\theta))^2}},$$

$$\frac{\partial f_3}{\partial \theta} = -d \cdot \sin(\theta), \quad (30)$$

$$\frac{\partial f_4}{\partial \theta} = -h \cdot \sin(\theta), \quad (31)$$

$$\frac{\partial f_5}{\partial X_E} = 2 \cdot (a_4 - X_E + h \cdot \cos(\psi) \cdot \sin(\theta)), \quad (32)$$

$$\frac{\partial f_5}{\partial Y_E} = 2 \cdot (h \cdot \sin(\psi) \cdot \sin(\theta) - Y_E), \quad (33)$$

$$\frac{\partial f_5}{\partial \psi} = 2 \cdot h \cdot \sin(\psi) \cdot \sin(\theta) \cdot (a_4 - X_E + h \cdot \cos(\psi) \quad (34)$$
$$\cdot \sin(\theta)) + 2 \cdot h \cdot \cos(\psi) \cdot \sin(\theta) \cdot (Y_E - h \cdot \sin(\psi) \cdot \sin(\theta))$$

$$\frac{\partial f_5}{\partial \theta} = 2 \cdot h \cdot \sin(\psi) \cdot \cos(\theta) \cdot (Y_E - h \cdot \sin(\psi) \cdot \sin(\theta)), \quad (35)$$
$$+ 2 \cdot h \cdot \cos(\psi) \cdot \cos(\theta) \cdot (a_4 - X_E + h \cdot \cos(\psi) \cdot \sin(\theta)),$$

$$\frac{\partial f_5}{\partial q_4} = \frac{-2 \cdot (q_4 - q_5) \cdot \left(c_4 + e_4 + \sqrt{d_4^2 - (q_4 - q_5)^2}\right)}{\sqrt{d_4^2 - (q_4 - q_5)^2}}, \quad (36)$$

$$\frac{\partial f_5}{\partial q_5} = \frac{2 \cdot (q_4 - q_5) \cdot \left(c_4 + e_4 + \sqrt{d_4^2 - (q_4 - q_5)^2}\right)}{\sqrt{d_4^2 - (q_4 - q_5)^2}}. \quad (37)$$

The determinant of the **A** Jacobian matrix of the inverse kinematic model for the TRUS probe module is defined by the expression

$$\det(A) = \frac{2 \cdot c^2 \cdot t_1 \cdot t_2}{t_3}, \quad (38)$$

where

$$t_1 = \sin(\theta)^2 \cdot [Y_E \cdot \cos(\psi) - X_E \cdot \sin(\psi) + a_4 \cdot \sin(\psi)], \quad (39)$$
$$t_2 = e_1 - X_E + e_2 + (c + h) \cdot \cos(\psi) \cdot \sin(\theta), \quad (40)$$
$$t_3 = \sqrt{R_1^2 - (X_E - (h + c) \cdot \sin(\theta) \cdot \cos(\psi) - e_1 - e_2)^2}. \quad (41)$$

The singularity condition is satisfied when $\det(A) = 0$. This determinant vanishes in the following cases.

a. $c = 0$.

The term "c" is a geometric parameter that defines the distance between the two Cardan joints, always taking values greater than zero. If "c" becomes zero, the two Cardan joints would superpose, changing the geometry of the robot. This case is purely theoretical and will not occur.

b. $\sin(\theta) = 0$, leading to $\theta = 0$ or $\theta = \pi$.

This expression implies that the TRUS probe is placed vertically, parallel to the OZ axis and oriented with the tip of the probe downwards ($\theta = 0$) or with the tip of the probe upwards ($\theta = \pi$). Because of the relative position between the robot and the patient, the TRUS probe will always work close to the horizontal plane $\theta = \pi/2$ (Fig. 7). Furthermore, in the design parameters of the robot, the geometric dimensions of the elements do not allow for the vertical positioning of the TRUS probe. Thus, this singularity is eliminated in the design stage.

c. $(Y_E \cdot \cos(\psi) - X_E \cdot \sin(\psi) + a_4 \cdot \sin(\psi)) = 0$,

leading to

$$\tan(\psi) = \frac{Y_E}{X_E - a_4}. \quad (42)$$

Figure 8 represents this type of configuration, when $\psi = $ atan2$(Y_E, X_E - a_4)$, translated in the positive direction of the OX axis. Practically, this configuration implies that links

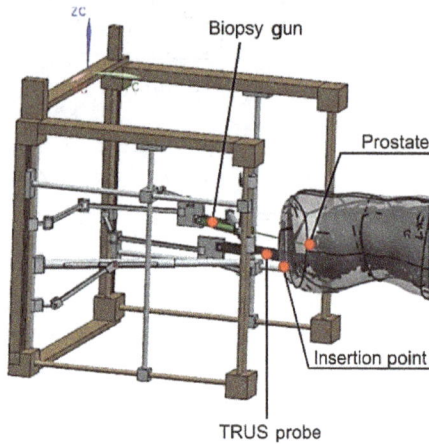

Figure 7. Working mode of BIO-PROS-1.

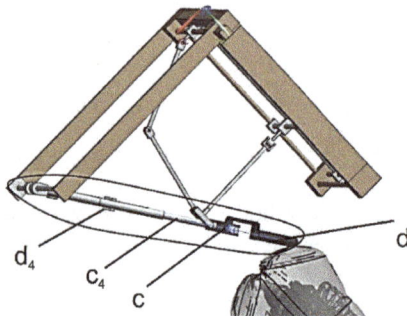

Figure 8. Singular position (c) of the TRUS probe module.

c_4, d_4, c, and h are in the same plane. This position represents the plane that separates the two working modes of the robotic module emerging from the double solution for Eq. (9). To avoid this singular configuration, the following condition must be implemented into the robot control: $\psi < \text{atan2}(Y_E, X_E - a_4)$. This will ensure that the robot will not change its working mode during functioning.

    d. $e_1 - X_E + e_2 + (c+h) \cdot \cos(\psi) \cdot \sin(\theta) = 0$,

leading to the expression

$$X_{A_1} - e_1 - e_2 = 0. \tag{43}$$

This configuration is presented in Fig. 9 and implies that the link R_1 is positioned into a parallel plane with the YOZ plane. To avoid this configuration when implementing the control module of the robot, the next condition can be implemented:

$$X_{A_1} > e_1 + e_2. \tag{44}$$

This configuration (Fig. 10) appears when the link R_1 is normal to the YOZ plane (or is positioned into a parallel plane to XOY). As in the previous case, given the fact that the singularity is on the boundary of the module workspace, the fol-

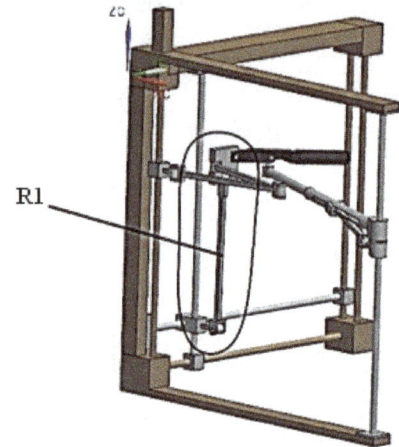

Figure 9. Singular position (d) of the TRUS probe module.

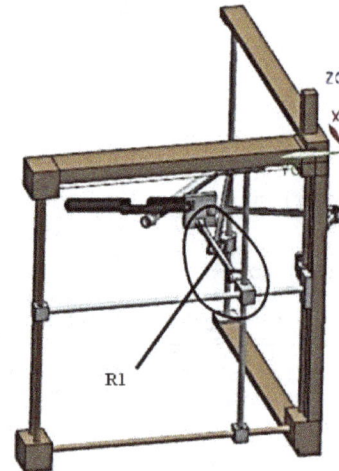

Figure 10. Singular position (e) of the TRUS probe module.

lowing mechanical constraint can be imposed on the control module of the robot:

$$R_1^2 > \left(X_{A_1} - e_1 - e_2\right)^2. \tag{45}$$

In conclusion, the singularity cases (a), (b), (d), and (e) are all at the workspace boundary. This is also demonstrated in the figures that represent each singularity, where the robot links are either fully extended or fully retracted relative to the $OXYZ$ frame. The only singular pose that has to be avoided (since it is located inside the robot workspace) during the robot operation is described in case (c).

    Next, the determinant of the direct kinematic model is analyzed. The general expression is

$$\det(B) = -\frac{2(q_4 - q_5)\left(c_4 + e_4 + \sqrt{d_4^2 - (q_4 - q_5)^2}\right)}{\sqrt{d_4^2 - (q_4 - q_5)^2}}. \tag{46}$$

The cases for which this determinant becomes zero are detailed below.

a. $q_4 - q_5 = 0$.

This expression implies that $q_4 = q_5$, meaning that the two active joints are overlapping. This singularity case is eliminated in the design phase since the robot mechanical structure will prohibit the values of the two active joints from becoming equal. By mounting a position sensor on one of the joints, the possibility of collision is also eliminated.

b. $d_4^2 - (q_4 - q_5)^2 = 0$.

The determinant of matrix **B** becomes zero when the d_4 link is in the vertical position (parallel to the OZ axis). This position can be eliminated by imposing the condition that the distance between joints q_4 and q_5 is lower than d_4 on the control module of the robot.

c. $c_4 + e_4 + \sqrt{d_4^2 - (q_4 - q_5)^2} = 0$.

Considering the condition imposed for the singularity (b), the left-hand side of this equation contains only positive terms. This means that this equality will never occur.

As a general overview, for the determinant of matrix **B**, the singular cases appear only at the boundary of the workspace; thus, through adequate design and control, they can be easily avoided without any negative influence on the behavior of the robot.

The equations for the biopsy gun module are very close to the ones for the first module, with very little variation in expression. They will be presented briefly. The determinant of the inverse kinematic problem for the biopsy gun module is given by the expression

$$\det(A') = \frac{2 \cdot c'^2 \cdot t_1 \cdot t_2}{t_3}, \tag{47}$$

where

$$t_1 = \sin\theta'^2 \cdot (Y_C - Y'_E) \cdot \cos\psi'^2 - (X_C - X'_E - e'_5) \cdot \sin\psi', \tag{48}$$

$$t_2 = (X_C - e'_1 - X'_E - e'_2 + (c' + h') \cdot \cos\psi' \cdot \sin\theta', \tag{49}$$

$$t_3 = \sqrt{R_1'^2 - (X_C - X'_E + (h' + c') \cdot \sin(\theta') \cdot \cos(\psi') - e'_1 - e'_2)^2}. \tag{50}$$

There are also five cases for which this expression becomes zero.

a. $c' = 0$.

The term c' is a geometric parameter (the distance between the two Cardan joints) and has a positive dimension imposed by the architecture of the robot.

b. $\sin(\theta') = 0$, leading to $\theta' = 0$ or $\theta' = \pi$.

This expression implies that the biopsy gun is placed vertically, parallel to the OZ axis oriented with the tip of the

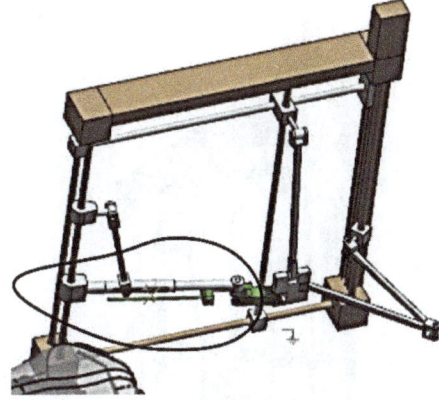

Figure 11. Singular position (c) of the biopsy gun module.

needle downwards $\theta' = 0$) or with the tip of the needle upwards ($\theta' = \pi$). Because of the relative position between the robot and the patient, this position cannot be reached, as seen in Fig. 10; the working position of the probe is close to the horizontal plane.

c. $((Y_C - Y'_E) \cdot \cos(\psi') - (X'_C - X'_E - e'_5) \cdot \sin(\psi'))$
$= 0$,

leading to

$$\tan(\psi') = \frac{Y'_C - Y'_E}{X'_C - X'_E - e'_5}. \tag{51}$$

Figure 11 represents this type of configuration translated in the positive direction of the OX axis. Practically, this configuration implies that links c'_4, d'_4, c', and h' are in the same plane. This configuration defines the boundary between the two working modes of this robotic module (defined by the double solution in Eq. 18).

d. $X_C - e'_1 - X'_E - e'_2 + (c' + h') \cdot \cos(\psi') \cdot \sin(\theta') = 0$,

leading to

$$X'_{A_1} - e'_1 - e'_2 = 0. \tag{52}$$

This configuration is illustrated in Fig. 12 and implies that the link R'_1 is positioned into a parallel plane with the YOZ plane. To avoid this configuration when developing the control module of the robot, the next condition can be implemented:

$$X'_{A_1} > e'_1 + e'_2. \tag{53}$$

This will reduce to the expression

$$R_1'^2 - (X'_{A_1} - e'_1 - e'_2.)^2 = 0. \tag{54}$$

This configuration appears when the link R'_1 is normal to the YOZ plane (or is positioned into a parallel plane

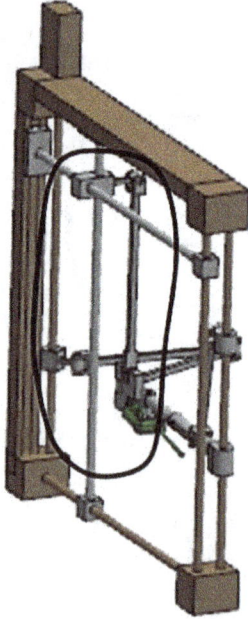

Figure 12. Singular position (d) of the biopsy gun module.

to XOY), represented in Fig. 13. As in the previous case, given the fact that the singularity is at the boundary of the workspace of the module, the following mechanical constraint can be imposed on the control module of the robot:

$$R_1'^2 > \left(X'_{A_1} - e'_1 - e'_2\right)^2. \tag{55}$$

The determinant of matrix **B** for the biopsy gun is the same as the determinant of matrix **B** for the endorectal probe module. The same singularities are obtained with all poses located on the workspace boundary:

$$\det(B') = -\frac{2\left(q'_4 - q'_5\right)\left(c'_4 + e'_4 + \sqrt{d_4'^2 - \left(q'_4 - q'_5\right)^2}\right)}{\sqrt{d_4'^2 - \left(q'_4 - q'_5\right)^2}}. \tag{56}$$

4   Workspace analysis of the parallel robot BIO-PROS-1

In the case of the ultrasound probe guidance, the robot workspace depends on the insertion point into the anus (I) and thus on the relative position of the patient to the robot. Once this point's coordinates are defined (either by visual guidance or by using an external marker and calibrated table for the robot), the ultrasound probe will have a spherical motion around this point that acts as a remote center of motion (RCM)), reaching a large number of target points and orientations. For the workspace generation, an inverse kinematic model was used. An initial volume, in the shape of a parallelepiped, is defined to obtain a range of values for each coordinate. Further on, each point is tested; if validated,

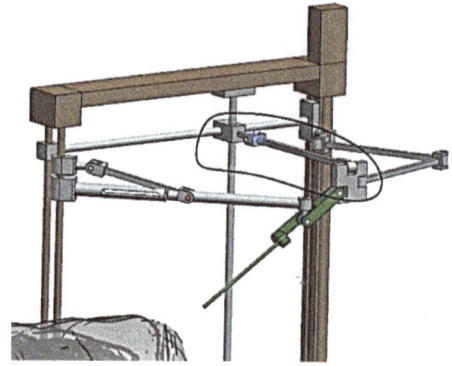

Figure 13. Singular position (e) of the biopsy gun module.

it is saved as a point in the robot workspace. The same approach is used for the biopsy gun module, by considering the approach with a single insertion point for each sampling. Supplementary conditions are introduced between the relative positions of the two modules based on the restrictions imposed by the medical procedure. The conditions imposed refer to the following:

1. the maximum and minimum stroke values for the robot active joints (q_1, q_2, q_3, q_4, q_5 for the TRUS probe module and $q'_1, q'_2, q'_3, q'_4, q'_5$ for the biopsy gun module);

2. the distance $\overline{IT}$ maximum value (the depth value of the probe introduced into the body) should be less than or equal to the probe length;

3. the distance $\overline{I'T'}$ that represents the length of the needle inserted into the body must not exceed the total length of the needle;

4. the value intervals of the orientation angles should be $60° < \psi < 120°$ and $45° < \theta < 135°$ (according to medical experts);

5. singular configurations and singularity points have to be avoided by imposing the mathematic conditions $\mathrm{abs}(\det(A) \neq 0)$ and $\mathrm{abs}(\det(\mathbf{B}) \neq 0)$ (Pisla, 2015);

6. the robotic structure should avoid positions in which one or more links of the robots would collide ($Z'_E > Z_E$ and $Z'_{A_1} > Z_{A_1}$);

7. and the workspaces of the TRUS-probe guiding module and the biopsy gun guiding module should not intersect.

Using this algorithm, the generated workspace is presented in Fig. 14, allowing a broad range of motions for the ultrasound probe for a selected insertion point into the colon. The workspace of the TRUS-probe guiding module is defined by pairs of points: I (the TRUS probe insertion point inside the rectum defined by the anus, which will remain fixed during

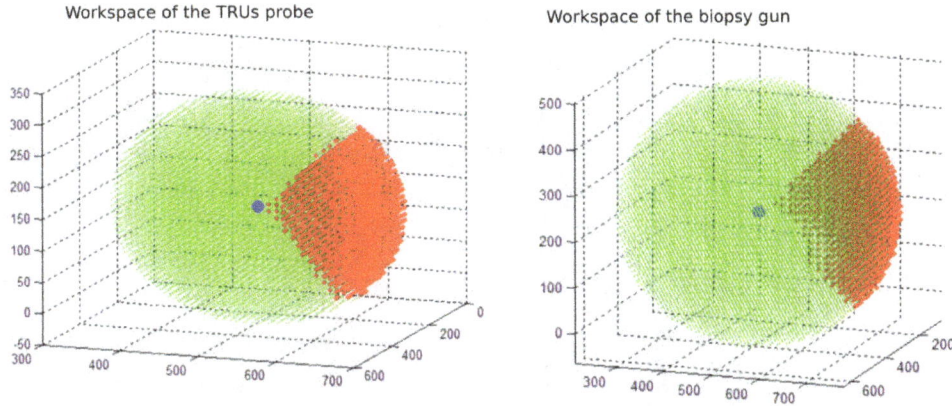

Figure 14. Ultrasound probe and biopsy gun workspace for a given insertion point.

the procedure) and T (the target point of the probe inside the colon in order to provide proper imaging of the target point for the sampling, which will have a range of coordinates inside the body, limited by the insertion point coordinates and the probe length). In the case of the biopsy-needle guidance robot, the workspace is largely dependent on the pair of insertion target points: I' (the needle insertion point into the perineum) and T' (the target point into the prostate tumor). Once these points have been chosen and their coordinates become known, the needle will be positioned at the insertion point with its final orientation (the angles ψ and θ). Afterwards it will be inserted on a linear trajectory up to the target point. Figure 14 presents the robot workspace for an insertion point I (for the ultrasound probe) and I' (for the biopsy gun), for which all the possible target locations for the two instruments, based on the set of restrictions, have been computed.

The geometrical parameters have the following constructively chosen values as a result of designing the robotic structure on a real scale.

a. For the ultrasound-probe guidance robot:

$R_1 = 305\,\text{mm}$; $e = 27.5\,\text{mm}$; $e_1 = 65\,\text{mm}$; $e_2 = 20\,\text{mm}$;

$e_3 = 30\,\text{mm}$;

$e_4 = 60\,\text{mm}$; $c = 80\,\text{mm}$; $h = 185\,\text{mm}$; $c_4 = 365\,\text{mm}$;

$d_4 = 190\,\text{mm}$;

$a_4 = 655\,\text{mm}$.

b. For the biopsy-needle guidance robot:

$R'_1 = 305\,\text{mm}$; $e' = 27.5\,\text{mm}$; $e'_1 = 65\,\text{mm}$;

$e'_2 = 20\,\text{mm}$; $e'_3 = 30\,\text{mm}$;

$e'_4 = 60\,\text{mm}$; $e'_5 = 40\,\text{mm}$; $c' = 80\,\text{mm}$; $h' = 185\,\text{mm}$;

$c'_4 = 365\,\text{mm}$;

$d'_4 = 190\,\text{mm}$; $a_4 = 655\,\text{mm}$; $X_C = 760\,\text{mm}$;

$Y_C = 600\,\text{mm}$.

The insertion points have the following coordinates: $I(X_I = 360\,\text{mm}; Y_I = 520\,\text{mm}; Z_I = 150\,\text{mm})$, $I'(X'_I = 365\,\text{mm}; Y'_I = 500\,\text{mm}; Z'_I = 220\,\text{mm})$. These are indicated by the blue points on the figure; the red represents the valid points with respect to the given insertion point (I and I'), and the green represents the points that cannot be reached from the insertion points due to the mechanical configuration of the structure. As presented above, the workspace was computed inside a parallelepiped shape defined by the range of all active joints; because all computations were made using the RCM, all the points are shaped into a sphere with a center in the insertion points (I and I'; the insertion point is fixed and the motion is around this point).

5 Simulation of parallel structure BIO-PROS-1

By using the inverse kinematic models presented in Pisla et al. (2015), a medically relevant scenario has been simulated in a program created using Matlab. The motions are similar for both robots (TRUS and the biopsy gun). The robotic system and the patient are positioned and the robot coordinate system is correlated with that of the patient based on several markers. After both robotic modules have been initialized (the coordinates of all the joints are known), the sampling procedure begins. First the TRUS probe moves towards the insertion point. This point is marked on the body; for the TRUS probe, this is the anus. After the insertion point for the probe has been reached the probe moves towards a preset target point inside the rectum, which represents the point that gives the best view of the sampling area of the prostate. To reach the target point, given the fact that the probe moves inside the patient, the motion parameters (speed and acceleration) are significantly lower than the motion parameters of the motion towards the insertion point of the probe.

After the probe has reached the target point, the biopsy gun module guides the needle towards the insertion point. This point is located on the perineum and is usually is marked by

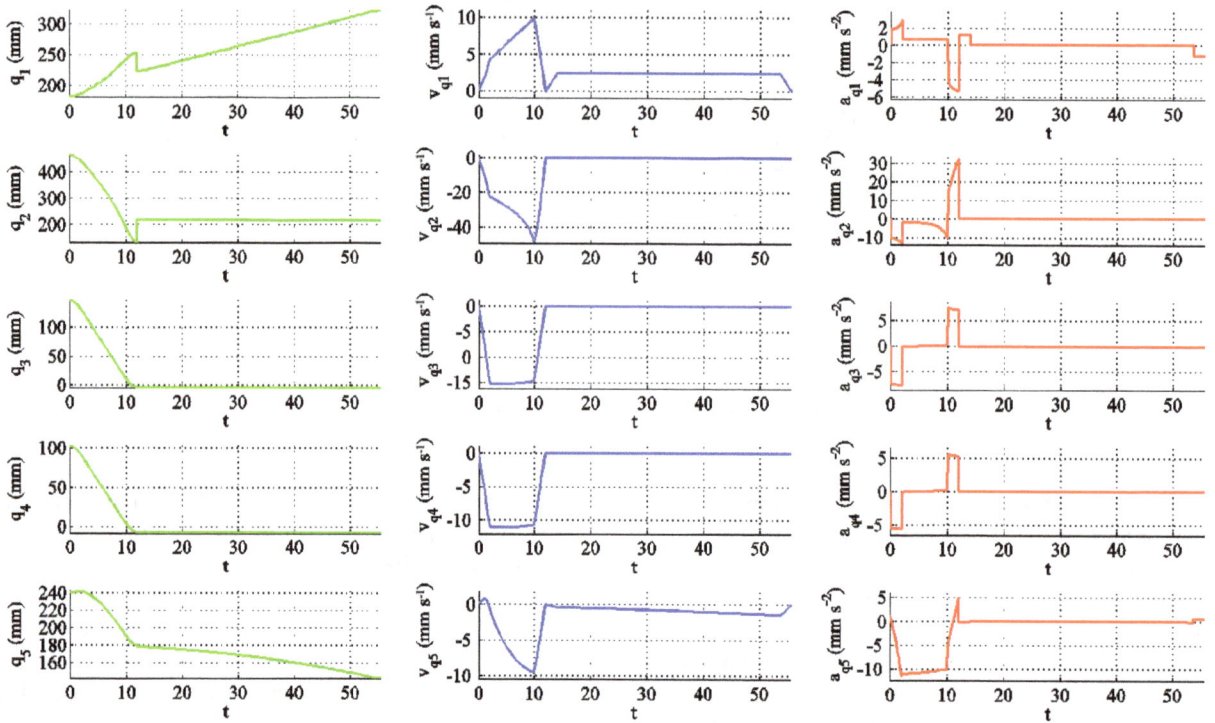

Figure 15. BIO-PROS-1 active coordinate time diagram for positions, speeds, and accelerations for the TRUS-probe guiding module.

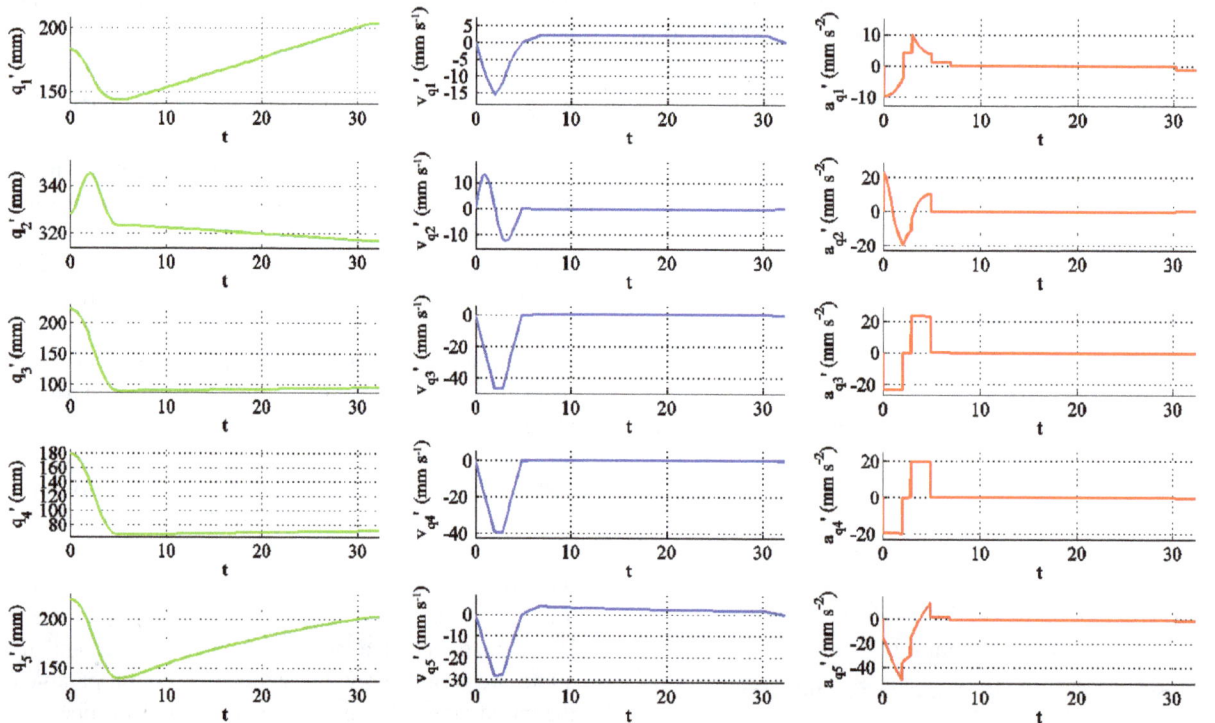

Figure 16. BIO-PROS-1 active coordinate time diagram for positions, speeds, and accelerations for the biopsy-gun guiding module.

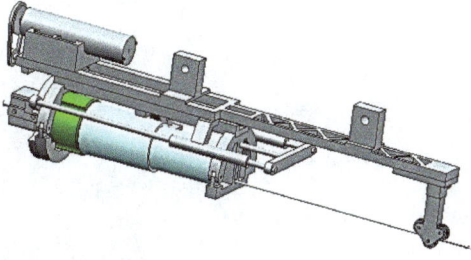

Figure 17. Biopsy gun holder.

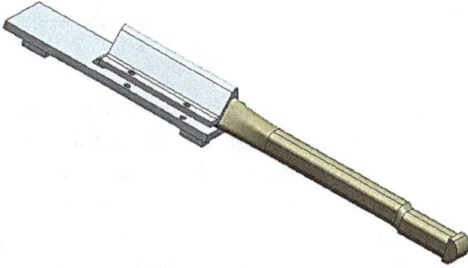

Figure 18. TRUS probe holder.

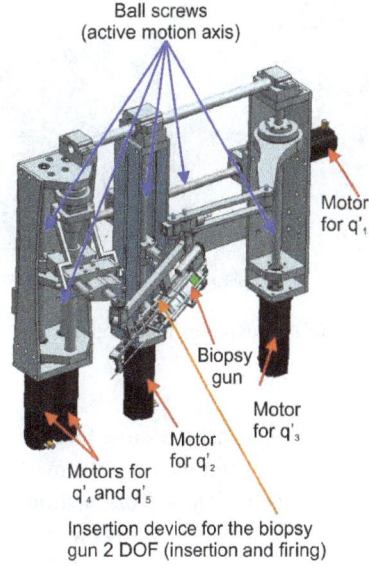

Figure 19. Biopsy gun module.

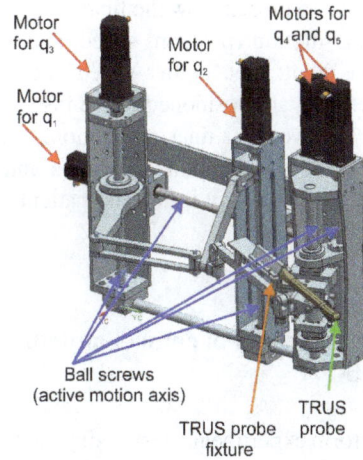

Figure 20. TRUS probe module.

a small cross incision to reduce the tissue resistance force (as described in the Introduction). Based on the pairs of coordinate points (insertion target), the robot calculates the needle orientation. When the insertion point is reached, the needle has the final orientation to enable its motion inside the body on a linear path. The target point is represented by the sampling area on the prostate. The motion towards the target point of the needle is a strictly linear motion along the needle axis; as in the case of the probe, the motion parameters are lowered. After the target point of the needle has been validated via an ultrasound image provided by the TRUS probe, the biopsy gun is triggered and the needle samples the prostate. With the sampled tissue inside the cannula of the biopsy gun (see Fig. 1), the needle is retracted on the same linear trajectory.

For the simulation, the starting points and orientations were chosen as follows.

– Starting pose for the ultrasound probe:

$$\begin{cases} X_C = 180\,\text{mm}; \\ Y_C = 470\,\text{mm}; \\ Z_C = 100\,\text{mm}; \end{cases} \begin{cases} \psi = 90°; \\ \theta = 80°. \end{cases} \tag{57}$$

– Starting pose for the biopsy needle:

$$\begin{cases} X'_C = 380\,\text{mm}; \\ Y'_C = 570\,\text{mm}; \\ Z'_C = 160\,\text{mm}; \end{cases} \begin{cases} \psi' = 90°; \\ \theta' = 80°. \end{cases} \tag{58}$$

The two pairs of insertion target points for the two robots are as follows.

– For the ultrasound probe:

$$\begin{cases} X_I = 370\,\text{mm}; \\ Y_I = 530\,\text{mm}; \\ Z_I = 85\,\text{mm}; \end{cases} \begin{cases} X_T = 370\,\text{mm}; \\ Y_T = 600\,\text{mm}; \\ Z_T = 110\,\text{mm}. \end{cases} \tag{59}$$

– For the biopsy gun:

$$\begin{cases} X'_I = 390\,\text{mm}; \\ Y'_I = 530\,\text{mm}; \\ Z'_I = 120\,\text{mm}; \end{cases} \begin{cases} X'_T = 380\,\text{mm}; \\ Y'_T = 590\,\text{mm}; \\ Z'_T = 125\,\text{mm}. \end{cases} \tag{60}$$

The orientation for the two guided elements can be derived from the insertion and target point coordinates:

$$\psi_{IT} = \text{atan2}\,(Y_T - Y_I,\, X_T - X_I),$$
$$\theta_{IT} = \text{atan2}\left(\sqrt{(Y_I - Y_T)^2 + (X_I - X_T)^2},\, Z_I - Z_T\right). \tag{61}$$

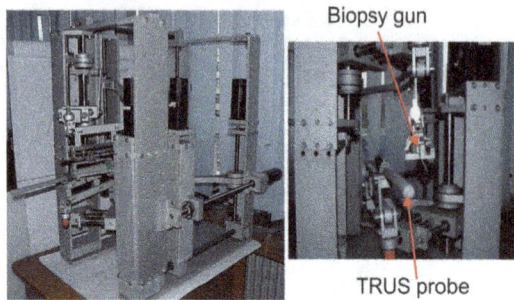

Figure 21. BIO-PROS-1 experimental model.

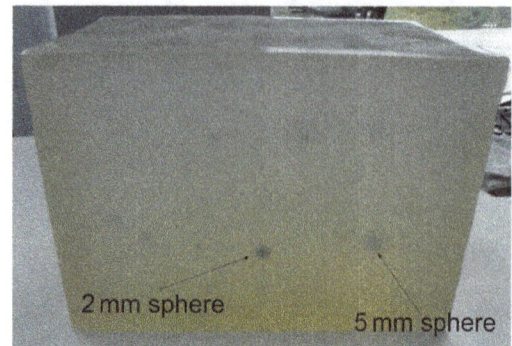

Figure 22. BIO-PROS-1 testing: ballistic gel cube with spheres.

The motion parameters are the same for both robotic modules: $v_{\max} = 20 \, \text{mm s}^{-1}$ and $a_{\max} = 10 \, \text{mm s}^{-2}$ for the first stage of the motion (position and orientation of the guided elements at the insertion points) and $v_{\max} = 2.4 \, \text{mm s}^{-1}$ and $a_{\max} = 1.2 \, \text{mm s}^{-2}$ for the second stage during which the probe and needle are inserted into the body (at the speeds required by urology specialists for safety reasons).

Figures 15 and 16 present how the final motion is achieved by assessing the motion (position, speed, and acceleration) for each active joint: at a higher speed and acceleration, the guided elements are positioned at the insertion points in the body and achieve their final orientation. Afterwards, at a slower speed and acceleration, the probe and needle are guided on a linear trajectory inside the patient (Pisla et al., 2015).

6 Experimental model of parallel structure BIO-PROS-1

In order to perform experimental tests using the BIO-PROS-1 robotic structure, an experimental model of the robotic system was designed. The main purpose of the robotic structure is to manipulate the two instruments used in the transperineal prostate biopsy: the biopsy gun (Fig. 17) (Vaida et al., 2017) and the TRUS probe (Fig. 18). The first step in the development of the experimental model was the design of the biopsy gun module and the geometrical parameters of the mobile platform to guide an ultrasound probe. In order to design the two instruments, real models of the instruments were analyzed. For the biopsy gun, the Bard Monopty 22 mm (Bard Biopsy, USA) was selected and the Endocavity Biplane E14CL4b (BK Ultrasound, USA) was selected for the TRUS probe.

Analyzing the properties (mass, size) of the two instruments resulted in the design elements of the robot (Figs. 19 and 20). Both modules were linked together through a frame and assembly organs. Each active joint was materialized through a ball screw axis (10 in total, 5 for each module) and each axis was actuated using a stepper motor (10 in total). For each passive rotational joint, radial-axial bearings

were selected to fulfil the motion requirements, and bearing housings were chosen for passive translational joints.

The result of the experimental design was a rigid and modular robotic structure able to fulfil the requirements for a robotic-assisted transperineal prostate biopsy. The overall dimensions of the structure resulted from combining the dimensions given by medical staff and the dimensions from computing the design parameters of the real model in order for the structure to manipulate the instruments in a safe environment.

The final result of the development of the parallel robotic structure BIO-PROS-1 can be seen in Fig. 21. The material used for the frame is an aluminum alloy, which is both rigid and low weight. For the motion axes, chrome steel was an applicable solution, while some parts were constructed using a rapid metal casting process.

7 Experimental data for the biopsy task

In order to evaluate the robot accuracy, a cube made of ballistic gel was created with spheres of different diameters placed inside in a well-defined pattern (Fig. 22). The robot task was to insert the biopsy needle inside a sphere. The silicone cube has an overall size of $150 \times 100 \times 100$ mm and the coordinates of the spheres are as follows, also listed in the Table 1.

A set of 10 consecutive runs was made for each sphere to demonstrate that the robot can reach each of them. This initial set of experimental runs validated the robot accuracy in the range of 2 mm, which is the size of the smallest sphere reached inside the cube. For the spheres with diameters of 2 mm, there were some target misses in the first runs, but following calibration the spheres were reached each time.

For the second part of the measurements, the robot was evaluated using an external measurement system, the FaroArm Edge, which is portable (Faro Technologies, UK). The measurement was achieved in four main steps for both the TRUS probe and the biopsy gun.

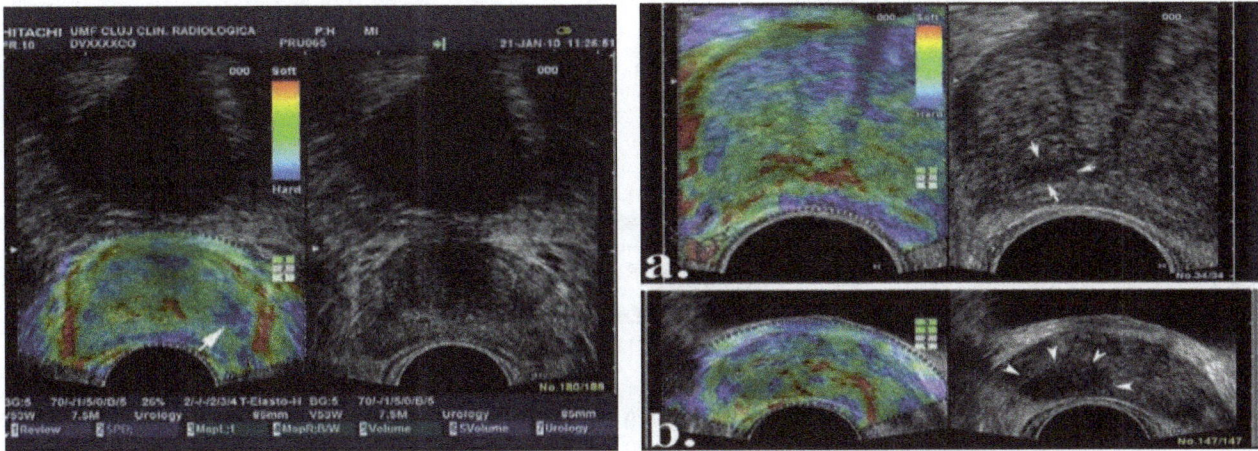

Figure 23. Prostate carcinoma detected using sonoelastography in different areas of the prostate (arrows show tumoral areas; Giurgiu et al., 2011; Dudea et al., 2011).

Table 1. Overall size and coordinates of the spheres.

No.	Sphere diameter size [mm]	X [mm]	Y [mm]	Z [mm]	Achieved
1	5 [mm]	120	20	25	10/10
2	5 [mm]	75	10	25	10/10
3	2 [mm]	25	20	25	9/10
4	2 [mm]	60	25	80	8/10
5	2 [mm]	90	25	80	9/10

Step 1: the Faro Arm has been fixed with respect to the robotic system. Using the laser head (noncontact), the coordinate system of the robot has been determined.

Step 2: using the same laser head, the TRUS probe and biopsy needle have been scanned to determine the tip coordinates and its orientations.

Step 3: data obtained in step 2 have been compared with the CAD model of the robotic system, and an error has been computed.

Step 4: steps 2 and 3 have been performed several times and a database has been created (two sets of results are presented in Table 2).

In real medical scenarios using relevant coordinates for the transperineal prostate biopsy, a set of five consecutive measurements were performed for each set of points. Two different distances were imposed between the insertion points of the biopsy needle and the ultrasound probe to simulate different human anatomies. Based on sonoelastography ultrasound data on a malignant prostate (Giurgiu et al., 2011; Dudea et al., 2011), illustrated below (Fig. 23; the white arrow), two sets of coordinates for the input data were established in the robot coordinate system, considering that the patient is positioned in the gynaecological position aligned to the symmetry line of the robot. In a real scenario, the coordinate transfer

would be achieved by using external markers on the patient and an internal marker placed through the urethra in the center of the prostate.

A surgical prostate biopsy performed manually by a urologist has a statistical accuracy of 9 mm (Kaye et al., 2015). After experimental tests with the presented robotic system, the accuracy of the procedure has been determined between 1 and 2 mm. The results are similar to those presented in Krieger et al. (2005) and Susil et al. (2006). The next step in the robot development is its testing on a human phantom with an elasticity similar to human tissue. For better results, an innovative algorithm for trajectory planning, like the one presented in Girbacia et al. (2017a, b), may be used.

8 Conclusions

The paper presents an innovative parallel robot for transperineal prostate biopsy. The robotic system was designed for both the TRUS probe and biopsy gun guidance with two robotic modules working together and sharing the same coordinate system. An analytical approach has been used to obtain the inverse and forward kinematic models of the structure. The robotic system workspace has been generated using real parameters of the robotic structure. Since the safety of the procedure is of high importance for medical procedures, several constraints have been imposed to avoid collisions be-

Table 2. Coordinates of target points.

Set no.	TRUS module coordinates [mm]						Needle module coordinates [mm]					
	Insertion			Target			Insertion			Target		
	X	Y	Z	X	Y	Z	X	Y	Z	X	Y	Z
1	250	800	420	250	900	430	250	800	460	260	895	452
2	250	800	420	250	880	425	250	800	460	238	875	443

	Measured data, set 1: TRUS module coordinates [mm]							
Mes. no.	Insertion			Error	Target			Error
1	250.68	800.66	420.50	1.07	250.95	899.75	428.87	1.26
2	250.73	800.32	419.30	1.06	249.31	900.51	429.27	1.03
3	249.99	800.50	419.40	0.78	251.05	900.38	430.37	1.31
4	250.14	799.77	419.17	0.87	250.39	898.85	429.04	0.95
5	250.46	800.33	420.74	0.93	250.59	899.63	429.71	1.00

	Measured data, set 1: needle module coordinates [mm]							
Mes. no.	Insertion			Error	Target			Error
1	249.85	800.22	459.04	1.00	259.46	894.45	452.81	1.13
2	249.51	800.95	460.62	1.24	260.15	894.85	451.82	1.15
3	250.03	800.53	460.40	0.66	260.86	894.68	451.30	1.08
4	249.36	799.38	460.86	1.23	259.97	894.42	452.66	1.06
5	249.39	799.04	460.35	1.19	260.37	895.04	452.51	1.09

	Measured data, set 2: TRUS module coordinates [mm]							
Mes. no.	Insertion			Error	Target			Error
1	250.95	800.36	420.68	1.22	249.75	881.16	424.62	0.81
2	249.54	799.34	419.59	0.91	250.34	881.09	425.26	0.85
3	250.41	800.29	419.67	0.60	250.57	880.21	425.31	0.72
4	250.87	800.80	419.01	1.54	248.98	879.71	424.38	1.63
5	250.29	799.09	419.89	0.96	249.62	879.48	424.74	0.99

	Measured data, set 2: needle module coordinates [mm]							
Mes. no.	Insertion			Error	Target			Error
1	249.30	799.04	460.56	1.32	237.99	874.42	442.62	1.12
2	249.11	799.26	460.36	1.21	239.10	874.34	443.76	1.37
3	250.60	799.83	459.78	0.66	237.83	874.69	443.69	0.32
4	250.46	800.65	460.47	0.92	239.02	874.97	443.62	1.30
5	250.60	800.40	460.57	0.92	238.88	874.28	442.72	1.12

tween the robot links. The singularity configurations of the robotic system have been analyzed and solutions for avoiding them have been provided; these are ready to be introduced into the robot control. The experimental model of the structure has been developed and presented. A set of experiments was performed in order to validate the robotic structure. Based on the resulting errors, some improvements will be proposed and consequently applied to the mechanical structure and the control system of the robot, followed by a new set of experimental tests on phantom models. The state of the art in robotic-assisted biopsy includes solutions that combine the motion of the TRUS probe with that of the biopsy gun. In some solutions, only the endorectal probe is auto-

mated, while other solutions present the biopsy gun module as robotic. The majority of robotic solutions are a combination of serial manipulators. Based on its parallel structure (which provides an inherent accuracy and stiffness), BIO-PROS-1 achieved an accuracy of 1–2 mm for the endorectal probe and for the biopsy gun during the simulated medical procedure. Both modules are fully automated; the urologist initializes the robotic structure, inserts the target points for the probe and the biopsy gun, and each step of the procedure has to be validated by the operator. A huge advantage of this robotic system is the development of fusion software between the MRI images and the real-time image provided by the TRUS probe. Prior to the biopsy procedure, an MRI scan of the prostate is performed. The target points on the prostate are identified and marked as reference points and position markers are mounted on the body. When the biopsy procedure starts, the MRI is projected onto the robotic system user interface, and the ultrasound image from the TRUS probe is overlapped (with respect to the same position markers used during the MRI). Using this fusion system, the number of samples is reduced to a minimum (the number of points identified during the MRI).

Competing interests. The authors declare that they have no conflict of interest.

Acknowledgements. This paper was supported by no. 247/2014, code PN-II-PT-PCCA-2013-4-0647 for the project entitled "ROBOCORE: Robotic assisted prostate biopsy, a high accuracy innovative method" and by no. 59/2015, code PN-II-RU-TE-2014-4-0992 for the project entitled "ACCURATE: A multi-purpose needle insertion device for the diagnosis and treatment of cancer." Both were financed by UEFISCDI.

Edited by: Chin-Hsing Kuo

References

ACS: The American Cancer Society's website, available at: http://www.cancer.org/research/cancerfactsstatistics/cancerfactsfigures, last access: January 2016.

Avantgarde Urology's website: available at: http://www.avantgardeurology.com/, last access: January 2016.

Berceanu, C. and Tarnita, D.: Aspects Regarding the Fabrication Process of a New Fully Sensorized Artificial Hand, MODTECH 2010: New face of TMCR, Proceedings of the International Conference ModTech, 123–126, 2010.

Berceanu, C., Tarnita, D., and Filip, D.: About an experimental approach used to determine the kinematics of the human finger, Journal of the Solid State Phenomena, Robotics and Automation Systems, 166–167, 45–50, 2010.

Cheng, W.: Aparatus and method for motorised placement of the needle, Pattent WO 2007085953 A1, Switzerland, 2007.

de Cobelli, O., Terracciano, D., Tagliabue, E., Raimondi, S., Bottero, D., Cioffi, A., Jereczek-Fossa, B., Petralia, G., Cordima, G., Laurino Almeida, G., Lucarelli, G., Buonerba, C., Matei, D. V., Renne, G., Di Lorenzo, G., and Ferro, M.: Predicting Pathological Features at Radical Prostatectomy in Patients with Prostate Cancer Eligible for Active Surveillance by Multiparametric Magnetic Resonance Imaging, PLoS ONE 10, e0139696, https://doi.org/10.1371/journal.pone.0139696, 2015.

Dudea, S. M., Giurgiu, C. R., Dumitriu, D., Chiorean, A., Ciurea, A., Botar-Jid, C., and Coman, I.: Value of ultrasound elastography in the diagnosis and management of prostate carcinoma, Med Ultrason., 13, 45–53, 2011.

Faro: available at: http://www.faro.com/products/metrology/faroarm-measuring-arm/overview, last access: February 2016.

Free Education Network' website: available at: http://www.free-ed.net/sweethaven/MedTech/Surgery02, last access: January 2016.

Gherman, B., Vaida, C., Pisla, D., Plitea, N., Gyurka, B., Lese, D., and Glogoveanu, M.: Singularities and workspace analysis for a parallel robot for minimally invasive surgery, IEEE International Conference on Automation Quality and Testing Robotics (AQTR), https://doi.org/10.1109/AQTR.2010.5520866, 2010.

Gherman, B., Plitea, N., and Pisla, D.: An Innovative Parallel Robotic System for Transperineal Prostate Biopsy, New Trends in Mechanism and Machine Science, 43, 421–429, 2016.

Girbacia, F., Pisla, D., Butnariu, S., Gherman, B., Girbacia, T., Plitea, N.: An Evolutionary Computational Algorithm for Trajectory Planning of an Innovative Parallel Robot for Brachytherapy, New Advances in Mechanisms, Mechanical Transmissions and Robotics, 46, 427–435, https://doi.org/10.1007/978-3-319-45450-4_43, 2017a.

Girbacia, F., Boboc, R., Gherman, B., Girbacia, T., and Pisla, D.: Planning of Needle Insertion for Robotic-assisted Prostate Biopsy in Augmented Reality using RGB-D Camera, New Advances in Mechanisms, Mechanical Transmissions and Robotics, 56, 515–522, https://doi.org/10.1007/978-3-319-49058-8_56, 2017b.

Giurgiu, C. R., Manea, C., Crişan, N., Bungărdean, C., Coman, I., and Dudea, S. M.: Real-time sonoelastography in the diagnosis of prostate cancer, Med Ultrason., 13, 5–9, 2011.

Gosselin, C. and Angeles, J.: Singularity Analysis of Closed-Loop Kinematic Chains, IEEE T. Robot. Autom., 6, 281–290, 1990.

Gosselin, C. M. and Wang, J.: Singularity loci of planar manipulators with revolute actuators, Robotics and Autonomus Systems, 21, 377–398, 1997.

Jemal, A., Siegel, R., Xu, J., Ward, E., Hao, Y., and Thun, M.: Cancer Statistics 2009, CA Cancer J. Clin., 59, 225–249, https://doi.org/10.3322/caac.20006, 2009.

Joshi, S. and Tsai, L.: Jacobian analysis of limited-DOF parallel manipulators, Transactions of the ASME Journal of Mechanical Design, 124, 254–258, 2002.

Kaye, D., Stoianovici, D., and Han, M.: Robotic Ultrasound and Needle Guidance for Prostate Cancer Management: Review of the Contemporary Literature, Curr. Opin. Urol., 24, 75–80, https://doi.org/10.1097/MOU.0000000000000011, 2015.

Krieger, A., Susil, R. C., Menard, C., Coleman, J. A., Fichtinger, G., Atalar, E., and Withcomb, L. L.: Design of a novel MRI compatible manipulator for image guided prostate interventions, IEEE T. Bio-Med. Eng., 52, 306–313, https://doi.org/10.1109/TBME.2004.840497, 2005.

Long, J. A., Hungr, N., Baumann, M., Descotes, J. L., Bolla, M., Giraud, J. Y., Rambeaud, J. J., and Troccaz, J.: Development of a novel robot for transperineal needle based interventions: focal therapy, brachytherapy and prostate biopsies, J. Urol., 188, 1369–1374, https://doi.org/10.1016/j.juro.2012.06.003, 2012.

Merlet, J. P.: Parallel Robots, 2nd Edn., Springer, Dordrecht, Netherlands, 2006.

Ottaviano, E., Rea, P., Errea, P., and Pinto, C.: Design and simulation of a simplified mechanism for Sit-to-Stand assisting devices, Mechanisms and Machine Science, 17, 123–130, https://doi.org/10.1007/978-94-007-7485-8_16, 2014.

Pepe, P. and Aragona, F.: Prostate biopsy :results and advantages of the transperineal approach-twenty year experience of a single center, World J. Urol., 32, 373–377, https://doi.org/10.1007/s00345-013-1108-1, 2014.

Pisla, D., Gherman, B., Tucan, P., Vaida, C., Govor, C., and Plitea, N.: On the kinematics of an Innovative Parallel Robotic System for Transperineal Prostate Biopsy, The 14th IFToMM world Congress, Taipei, Taiwan 14, 438–445, https://doi.org/10.6567/IFToMM.14TH.WC.OS2.042, 2015.

Plitea, N., Pisla, D., Vaida, C., Gherman, B., Tucan, P., Govor, C., and Covaciu, F.: Family of innovative parallel robots for transperineal prostate biopsy, Patent: A/00191/13.03.2015, Oficiul de Stat pentru Inventii si Marci (OSIM), Romania, 2015a.

Plitea, N., Szilaghyi, A., and Pisla, D.: Kinematic analysis of a new 5-DOF modular parallel robot for brachytherapy, Robotics and Computer-Integrated Manufacturing, 31, 70–80, https://doi.org/10.1016/j.rcim.2014.07.005, 2015b.

Podder, T., Buzurovici, I., Huang, K., and Yu, Y.: MIRAB: An Image-Guided Multichannel Robot for Prostate Brachytherapy, Bodine J., 78, S810, https://doi.org/10.1016/j.ijrobp.2010.07.1876, 2010.

Pondman, K. M., Futterer, J. J., Haken, B. T., Schultze Kool, L. J., Witjes, J. A., Hambrock, T., Macura, K. J., and Barentsz, J. O.: MRI-guided biopsy of the prostate: an overview of techniques and a systematic review, Eur. Urol., 54, 517–527, 2008.

Stoianovici, D., Chunwoo, K., and Srimathveeravalli, G.: MRI-Safe Robot for Endorectal Prostate Biopsy, IEEE/ASME T. Mech., 19, 1289–1299, 2014.

Susil, R. C., Menard, C., Krieger, A., Coleman, J. A., Camphausen, K., Choyke, P., Fichtinger, G., Withcomb, L. L., Coleman, C. N., and Atalar, E.: Transrectal Prostate Biopsy and Fiducial Marker Placement in a Standard 1.5T Magnetic Resonance Imaging Scanner, J. Urol., 175, 113–120, https://doi.org/10.1016/S0022-5347(05)00065-0, 2006.

Taneja, S. S., Bjurlin, M. A., and Carter, H. B.: Optimal techniques of prostate biopsy and specimen handling, Am. Urol. Assoc., White Paper: Optimal Techniques of prostate biopsy and specimen handling, AUA guideline, March 2013, 1–29, 2013.

Tarnita, D.: Wearable sensors used for human gait analysis, Rom. J. Morphol. Embryol., 57, 373–382, 2016.

Tarnita, D. and Marghitu, D.: Analysis of a hand arm system, Robot. Cim.-Int. Manuf., 29, 493–501, 2013.

Vaida, C., Pisla, D., Tucan, P., Gherman, B., Govor, C., and Plitea, N.: An innovative parallel robotic structure designed for transperineal prostate biopsy, The 14th IFToMM world Congress, Taipei, Taiwan, https://doi.org/10.6567/IFToMM.14TH.WC.OS2.049, 2015.

Vaida, C., Birlescu, I., Plitea, N., Crisan, N., and Pisla, D.: Design of a Needle Insertion Module for Robotic Assisted Transperineal Prostate Biopsy, MESROB 2016 – 5th International Workshop on Medical and Service Robots, Castle St. Martin, Graz/Austria, in press, 2017.

Walter, D. R. and Husty, M. L.: On implicitization of kinematic constraint equations, Mach. Des. Res., 26, 132–151, 2010.

Zlatanov, D., Bonev, I. A., and Gosselin, C. M.: Constraint singularities of parallel mechanisms, Proceedings of the IEEE International Conference on Robotics and Automation, Washington, D.C., USA, 496–502, 2002.

8

Reduced inertial parameters in system of one degree of freedom obtained by Eksergian's method

Salvador Cardona Foix, Lluïsa Jordi Nebot, and Joan Puig-Ortiz

Mechanical Engineering Department, ETSEIB, Universitat Politècnica de Catalunya, Barcelona, 08028, Spain

Correspondence to: Lluïsa Jordi Nebot (lluisa.jordi@upc.edu)

Abstract. The mechanisms of one degree of freedom can be dynamically analysed by setting out a single differential equation of motion which variable is the generalized coordinate selected as independent. In front of the use of a set of generalized dependent coordinates to describe the system, the method exposed in this work has the advantage of working with a single variable but leads to complex analytical expressions for the coefficients of the differential equation, even in simple mechanisms. The theoretical approach, in this paper, is developed from Eksergian's method and Lagrange's equations. The equation of motion is written by means of a set of parameters – reduced parameters – that characterize the dynamic behaviour of the system. These parameters are function of the independent coordinate chosen and its derivative and can be obtained numerically by direct calculus or by means of a kinetostatic analysis, as is proposed. Two cases of study of the method are presented. The first example shows the study of pedalling a stationary bicycle used in a rehabilitation process. The second one shows the analysis of a single dwell bar mechanism which is driven by an electric motor.

1 Introduction

Analysis of mechanisms from the kinematic point of view is, in general, not difficult to make except in the singular configurations and their environment. There are a number of programs that facilitate it or directly carry out it in lesser or greater extent: Geogebra (Geogebra, 2016) whose main objective is the dynamic geometry, CAD programs including a kinematic analysis module or a number of specific programs (Kurtenbach et al., 2014). The kinetostatic or inverse dynamic analysis, in which the mechanism has as many actuators as degrees of freedom and all movements are imposed, requires extra effort, both conceptual and operational. This fact is particularly difficult when trying to determine the constraint actions in the kinematic pairs of the mechanism. There are also a number of programs that perform kinetostatic analysis, such as PAM – Program of Analysis of Mechanisms – of Mechanical Engineering Department of UPC (Clos and Puig-Ortiz, 2004; Cardona et al., 2006), SAM – Synthesis and Analysis of Mechanisms – by ARTAS Engineering Software (SAM, 2016), WinMecC of Mechanical Engineering Department of Malaga University (WinMecC, 2016) and others. Finally, the direct dynamic analysis, in which the mech-

anism has fewer actuators than degrees of freedom, represents a qualitative leap in difficulty of implementation. To the algebraic manipulation of the above analysis the solution of differential equations must be added. There are different applications that carry out direct dynamic analysis. The use of such applications is not always justified because of their complexity and because they are not trivial to use, even for the simulation of a system of one degree of freedom. Their use may be not recommended or even unfeasible in some cases, such as when the study of the mechanism must be performed in real time, as part of the simulation and control of a production process. In these cases, there should be a low-cost model both computational and in implementation.

In the world of robotics it is usual to use reduced inertial parameters to minimize the time of calculation and simulation. Several authors have proposed methods to determine and to reduce the number of parameters to be used in the inverse dynamics of manipulators (Fogarasy and Smith, 1997; Ebrahimi and Haghi, 2013; Díaz-Rodríguez et al., 2010; Ros et al., 2012; Chen and Beale, 2003; Yoshida et al., 1995) as well as to determine the minimum number of required parameters (Gautier and Khalil, 1988). It has also been investigated the possibility that the parameters vary over time to facilitate manoeuvrability of manipulators in singular configurations (Parsa et al., 2015). With the aim of describing the inertial behaviour of mechanisms, it seems a good idea to use inertial parameters that vary over time.

The dynamic analysis equations of mechanisms are obtained usually by means of Lagrangian formulation (Fogarasy and Smith, 1997), Newtonian formulation and virtual work principle (Wu et al., 2008).

In this paper, similarly to other works of the authors (Cardona et al., 2009; Jordi et al., 2008), the direct dynamic analysis of a mechanism of one degree of freedom by using its reduced inertial parameters and reduced forces and moments is proposed. Both types of parameters are function of the independent generalized coordinate, taken for kinematic description of the mechanism, and may be obtained, for example, using applications that perform the kinetostatic analysis. With these parameters, the direct dynamic analysis leads to a single second order differential equation, equation of motion, which is easily integrated to obtain the time evolution of the generalized coordinate employed. Each constraint action, force or moment, is given by an algebraic expression that includes the reduced parameters associated with such action and the first two derivatives of the coordinate, obtained independently in the process of integration of the equation of motion. In some of the mentioned works, all the reduced inertial parameters are obtained from the Lagrangian formulation. In this study, these parameters, as well as the reduced forces and moments present in the equation of motion, are introduced from the energy theorem or Eksergian's method (Eksergian, 1930; Doughty, 1988). This approach allows the incorporation, in a simple way conceptually and operationally, of motors and passive resistances described by means of the dissipated energy.

The determination of reduced inertial parameters and reduced forces is performed by using the mentioned programs that allow static, kinematic and kinetostatic analysis of planar mechanisms controlled by actuators, angular or linear.

2 Dynamics of a mechanism of one degree of freedom

To obtain the equation of motion of a mechanism of one degree of freedom, the use of energy theorem in differential version or Eksergian's method (Eksergian, 1930) is proposed:

$$\dot{E}_c = P \qquad\qquad (1)$$

Being E_c the kinetic energy of the mechanism and P the sum of the power of the external forces which act on it, and the power of the no-constrain internal forces that make no-null work, as motors or passives resistances, which are not modelled as an explicit function of constrain forces.

For a system of one degree of freedom, kinetic energy E_c can be expressed in terms of the independent generalized co-ordinate q and its time derivative, so that:

$$E_c = \frac{1}{2}m(q)\dot{q}^2 \rightarrow \dot{E}_c = m(q)\ddot{q}\dot{q} + \frac{1}{2}m_q(q)\dot{q}^2\dot{q}$$

Where $m(q)$ is the inertia reduced to the coordinate q and $m_q(q)$ its derivative with respect to the coordinate q.

The total power P of all the forces acting on the mechanism can be expressed in terms of the reduced force $F(q, \dot{q})$ to the independent coordinate as:

$$P = F(q, \dot{q})\dot{q}$$

Thus from Eq. (1), the equation of motion is obtained:

$$m(q)\ddot{q} + \frac{1}{2}m_q(q)\dot{q}^2 = F(q, \dot{q}) \qquad\qquad (2)$$

For the systems of one degree of freedom and the holonomic ones with more than one degree of freedom described by a set of n independent generalized coordinates, the equations of motion can be obtained by means of the ordinary Lagrange equations:

$$\frac{\mathrm{d}}{\mathrm{d}t}\frac{\partial E_c}{\partial \dot{q}_i} - \frac{\partial E_c}{\partial q_i} = F_i^* \quad i = 1\ldots n \qquad\qquad (3)$$

Being F_i^*, the generalized force, associated to the coordinate q_i, of all forces acting on the mechanism.

By developing Eq. (3), for a system of one degree of freedom, an identical equation to Eq. (2) is obtained. This fact shows that the reduced force and generalized force coincide.

When it is possible to use Eksergian's method, the inclusion of not mechanical phenomena, as motors and passives resistances, in the motion equation is conceptually simple. It can be done by means of the power that these phenomena exchange with the system, and that is described with mechanical state variables. In the Lagrange's formulation, these phenomena must be introduced by means of generalized forces associated to the non-conservative forces.

If in the dynamic study of the mechanism is desired to determine a constraint action, the kinematic condition imposed by the constraint is substitutable, conceptually, by an actuator that ensures it – constraint actuator. Figure 1 shows a four linkage bar mechanism activated by an angular actuator T_{act} on the crank OP and how to determine the vertical force F_e at the R joint. This joint is replaced by a slider guide which leaves free the vertical movement, which is now constrained with the actuator. Replacing a constraint condition by an actuator, although conceptually, means that the initial mechanism of one degree of freedom must be studied as a new holonomic mechanism of two degrees of freedom with two independent coordinates q_1 and q_2. The first coordinate q_1 associated with the actual movement of the mechanism and the second q_2 with the movement prevented by the constraint. Thus, the motion equation describing temporal evolution of q_1 can be determined by Eq. (2) obtained with Eksergian's

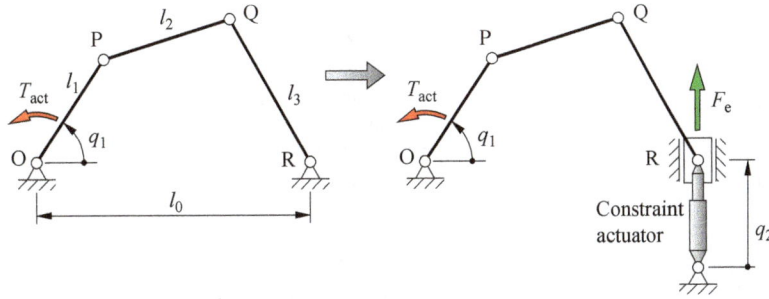

Figure 1. Original mechanism and with a constraint condition replaced by a constraint actuator.

method. The constraint force or moment can be determined with the obtained expressions from Lagrange's method.

For the new system of two degrees of freedom, the kinetic energy is (Jordi et al., 2008; Eksergian, 1930; Cardona and Clos, 2000):

$$E_c = \frac{1}{2}m_1(q_1, q_2)\dot{q}_1^2 + \frac{1}{2}m_2(q_1, q_2)\dot{q}_2^2$$
$$+ m_{12}(q_1, q_2)\dot{q}_1\dot{q}_2 \qquad (4)$$

Applying Lagrange equations to this mechanism, and particularizing for the movement imposed by the constraint actuator $\ddot{q}_2 = 0$ and $\dot{q}_2 = 0$ the algebraic equation including the generalized force F_2^* is obtained. It contains, besides the forces of known formulation, the constraint force or moment desired. In summary, for the example of the original mechanism of Fig. 1 the equation of motion obtained with Eq. (2) is:

$$m_1(q_1)\ddot{q}_1 + \frac{1}{2}m_{1q_1}(q_1)\dot{q}_1^2 = T_{\mathrm{act}} \qquad (5)$$

In general, T_{act} is the reduced force associated to the coordinate q_1 of all the forces different from the constraint forces.

The constraint force obtained with Eq. (3) is:

$$m_{12}(q_1)\ddot{q}_1 + \left(m_{12q_1}(q_1) - \frac{1}{2}m_{1q_2}(q_1)\right)\dot{q}_1^2 = F_2^* \qquad (6)$$

Being $m_{1q_1} = \frac{\partial m_1}{\partial q_1}$; $m_{12q_1} = \frac{\partial m_{12}}{\partial q_1}$; $m_{1q_2} = \frac{\partial m_1}{\partial q_2}$ and $F_2^* = F_e$.

The coefficients of the Eqs. (5) and (6), which are reduced inertial parameters, can be determined from the calculation of the kinetic energy or, as it will be discussed below, using kinetostatic simulation programs as PAM, SAM or Win-MecC.

The kinetostatic analysis allows to determine T_{act} given $q(t)$. Taking the system of one degree of freedom of Fig. 1 as an example, to determine the reduced parameters of the equation of motion the following kinetostatic analyses are performed:

i. $\ddot{q}_1 = 0$ and $\dot{q}_1 \neq 0$; so $m_{1q_1}(q_1) = 2T_{\mathrm{act}}/\dot{q}_1^2$ is obtained.

ii. $\ddot{q}_1 \neq 0$ and $\dot{q}_1 = 0$;
so $m_1(q_1) = \left(T_{\mathrm{act}} - \frac{1}{2}m_{1q_1}(q_1)\dot{q}_1^2\right)/\ddot{q}_1$ is obtained.

The parameters of the Eq. (6) are obtained by the kinetostatic analyses of the same system, so that:

i. $\ddot{q}_1 = 0$ and $\dot{q}_1 \neq 0$,
$$m_{e1_q}(q_1) = \left(m_{12q_1}(q_1) - \frac{1}{2}m_{1q_2}(q_1)\right) = F_2^*/\dot{q}_1^2.$$

ii. $\ddot{q}_1 \neq 0$ and $\dot{q}_1 = 0$, $m_{e1}(q_1) = m_{12}(q_1)$
$$= \left(F_2^* - \left(m_{12q_1}(q_1) - \frac{1}{2}m_{1q_2}(q_1)\right)\dot{q}_1^2\right)/\ddot{q}_1.$$

With programs as PAM, the first analysis is easily implemented using an actuator with a polynomial movement law and imposing the above condition with constant velocity ($\ddot{q}_1 = 0$ and $\dot{q}_1 \neq 0$). This analysis is performed for a set of uniformly distributed instants of time and thus sweep the entire range of values of interest of q_1. So, values of m_{1q_1} and m_{e1_q} are obtained for values of q_1 equispaced.

The necessary conditions of the second analysis ($\ddot{q}_1 \neq 0$ and $\dot{q}_1 = 0$) are impossible to perform with an actuator with a polynomial movement law. So, if you want to get the values of m_1 and m_{e1} for the same values of q_1 of the first analysis, a strategy should be used. This may be the use of an actuator that controls q_1 according to a temporal function with two parts: one polynomial and another one harmonic, so that:

$$q_1(t) = (c_1 + c_2 t) + (c_3 \cos(c_4 t + c_5))$$

where $c_1 \ldots c_5$ are constants which are chosen so that the conditions $\ddot{q}_1 \neq 0$ and $\dot{q}_1 = 0$ occur at points of interest. In short, a continuous function for the actuator is defined and properly sampled provides the required conditions in the desired q_1 configurations.

SAM has a utility that facilitates the realization of the two analyses evaluated at the same instants of time. You can perform a kinetostatic analysis by loading a file in which the temporary values, the value of the coordinate q_1 and its first two derivatives are specified. So, for doing the two analyses two input files have to be created. These files must contain, for the range of values of interest of q_1, the values of this variable equally spaced at time regular intervals. In the first file, these values are associated to a speed $\dot{q}_1 = 1$ and acceleration $\ddot{q}_1 = 0$, and in the second file these values are associated to a speed $\dot{q}_1 = 0$ and acceleration $\ddot{q}_1 = 1$. The values of speed and acceleration have the default units used in the analysis.

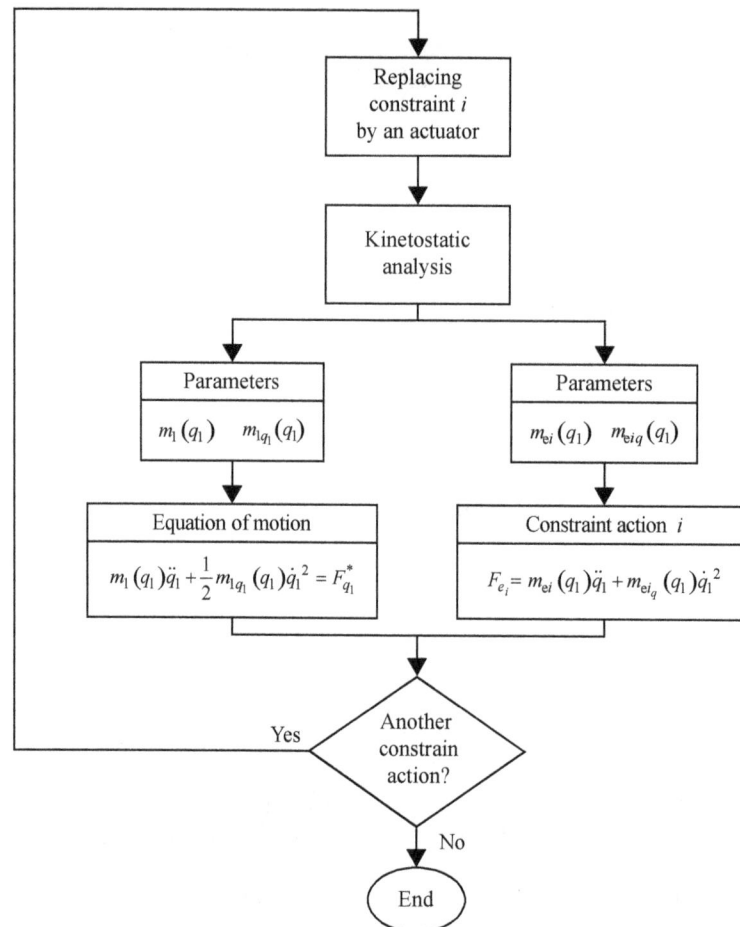

Figure 2. Proposed process for dynamic analysis of a system of one degree of freedom.

The process presented for the determination of the constraint action is extensible to find simultaneously as many actions as desired. In this case, as many constraint actuators as constraint actions to determine must be defined to obtain the corresponding equations. In order not to increase the complexity of the system, as many independent systems of two degrees of freedom as constraint actions to determine must be defined. Each system provides an equation for the constraint action and an equation of motion, that is obviously the same for all systems. Thus, two reduced parameters for the equation of motion and two parameters for each constraint action to determine are obtained. Figure 2 shows a schematic of the process to be followed for determining the constraint actions desired.

3   Cases of study

The procedure described in the previous section is used to study two cases: the pedalling a stationary bicycle used in a rehabilitation process and the analysis of a single dwell bar mechanism which is driven by an electric motor.

3.1   Pedalling a stationary bicycle used in a rehabilitation process of knee damage

The dynamics of the pedalling in a stationary bicycle in a rehabilitation process (Cardona et al., 2009; Jordi et al., 2008; Curià, 2010) is studied. The mechanism of Fig. 3 represents the model for this study. It is assumed that the ankle is fixed to the pedal, so the model has only one degree of freedom. The geometric characteristics and inertial parameters, obtained experimentally by Curià (2010), are shown in Fig. 3a. The translation kinetic energy of the bicycle and the cyclist plus the rotation kinetic energy of the wheels of a conventional bicycle are substituted by the kinetic energy of a flywheel fixed to the pinion. Some aspects as the action of several muscles, that can be modelled as simultaneous actuators, are not taken into account.

The most reasonable kinetostatic analysis to determine reduced parameters is the one that uses an actuator controlling the rotation angle of the pedals φ_p because this coordinate does not have dead-points (Fig. 3b). In this analysis, the variations of potential energy associated to the thigh and the leg are not taken into account.

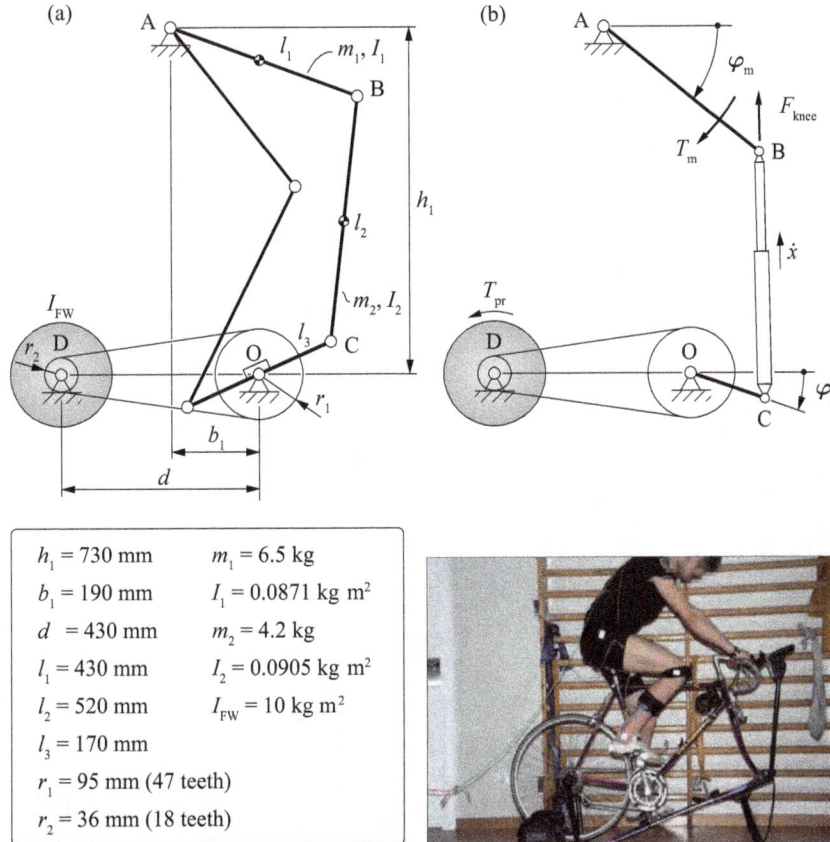

Figure 3. Model for the stationary bicycle and the inferior extremities of the cyclist.

The motion equation of the system of Fig. 3a can be obtained by means of Eksergian's method. The kinetic energy of the system is:

$$E_c = \frac{1}{2}\left(I_1 + m_1\left(\frac{l_1}{2}\right)^2\right)\left(\dot{\varphi}_{m1}^2 + \dot{\varphi}_{m2}^2\right) + \frac{1}{2}I_{FW}\left(\frac{r_1}{r_2}\right)^2\dot{\varphi}_p^2$$

$$+ \frac{1}{2}m_2\left(v_{G21}^2 + v_{G22}^2\right) + \frac{1}{2}I_2\left(\omega_{21}^2 + \omega_{22}^2\right)$$

Where $\dot{\varphi}_{m1}$ and $\dot{\varphi}_{m2}$ are the angular velocities of the two thighs, $\dot{\varphi}_p$ is the angular velocity of the pedals, v_{G21} and v_{G22} are the velocities of the centre of mass of the legs and ω_{21} and ω_{22} are the angular velocities of the legs. Obviously, all these kinematics variables are related by means of the kinematic constrain equations. With this approach, the motion equation can be obtained. If an additional objective of the analysis is to determine some constrain action it is necessary to use Eq. (6) that has been demonstrated with Lagrange's method. It is interesting to remark that the proposed method consists of replacing a constrain by an actuator instead of using Lagrange multipliers method.

The objectives of this analysis are to determine the motion equation and the force in the knee in the leg direction F_{knee} when a torque of passive resistances T_{pr} is acting on the flywheel. So, it is necessary to obtain the generalized forces $F_{\varphi_p}^*$

and F_x^* associated to the rotation of the pedal φ_p and to the extension x of the constraint actuator.

The expressions of generalized forces are obtained, for example, by means of the following virtual movements:

i. $\dot{\varphi}_p^* \neq 0$ and $\dot{x}^* = 0$. This virtual movement is compatible with the constraints of the original system. The relationship between velocities is obtained from the kinematic analysis of the initial mechanism, that is the same as the mechanism with the constrain actuator with $\dot{x} = 0$. The generalized force is:

$$F_{\varphi_p}^* = T_m\frac{\dot{\varphi}_m}{\dot{\varphi}_p}\bigg|_{\dot{x}=0} + T_{pr}\frac{\dot{\varphi}_{flywheel}}{\dot{\varphi}_p}\bigg|_{\dot{x}=0}.$$

ii. $\dot{\varphi}_p^* = 0$ and $\dot{x}^* \neq 0$. The generalized force is:

$$F_x^* = F_{knee} + T_m\frac{\dot{\varphi}_m}{\dot{x}}\bigg|_{\dot{\varphi}_p=0}.$$

In this movement, the mechanism has two degrees of freedom because the restriction $x = l_2$ is not considered. The relationship between velocities is obtained for a movement of the mechanism with the constrain actuator with $\dot{\varphi}_p = 0$.

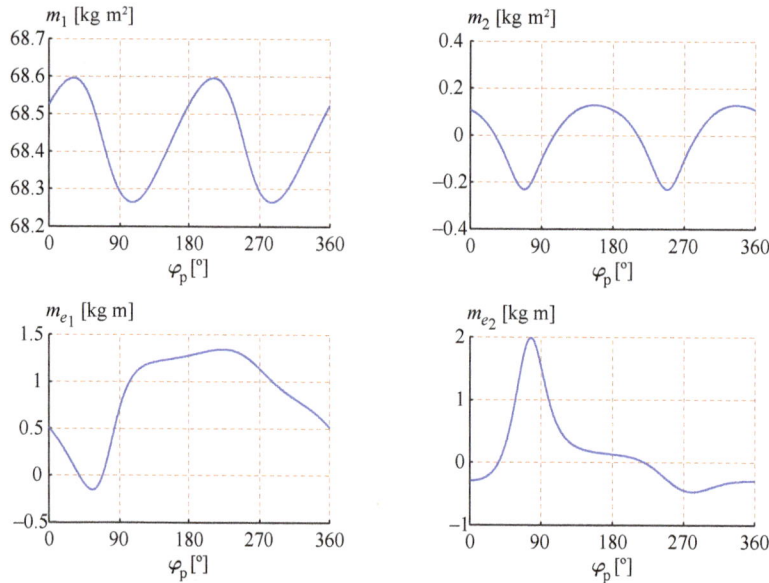

Figure 4. Inertial reduced parameters for pedalling a stationary bicycle.

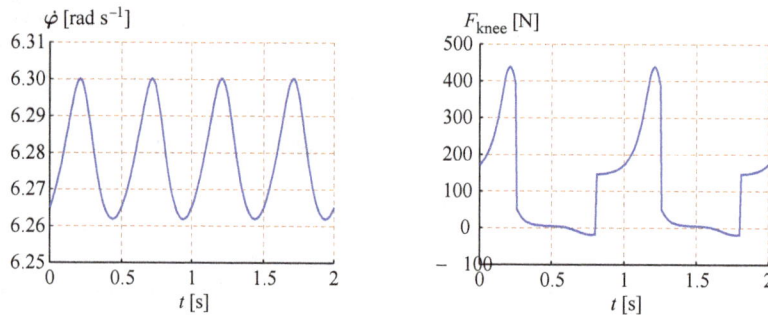

Figure 5. Angular velocity of the pedal and force in the knee in leg direction.

The terms $\frac{\dot{\varphi}_m}{\dot{\varphi}_p}$, $\frac{\dot{\varphi}_{flywheel}}{\dot{\varphi}_p}$ and $\frac{\dot{\varphi}_m}{\dot{x}}$ are obtained directly from the kinematic analysis of the mechanism with the corresponding conditions: only one generalized velocity not null.

In order to finish the analysis, the equations have to be solved; so it is necessary to obtain the values of T_m and T_{pr}. The torque T_{pr}, introduced by the dynamic brake of the stationary bicycle, is assumed as constant and its value has been calculated in order that lost power will be 125 W, when the pedalling rhythm is 1 Hz. The torque T_m, applied to each thigh, is assumed as constant and not null only in the descent phase of the movement of the thigh. Its value has been determined using the described procedure in order to achieve a pedalling stationary regime with the torque T_{pr} previously obtained. The value obtained is $T_m = 69.7\,\mathrm{N\,m}$. The equations to solve are:

$$\begin{cases} m_1\left(\varphi_p\right)\ddot{\varphi}_p + m_2\left(\varphi_p\right)\dot{\varphi}_p^2 = F_{\varphi_p}^* \\ m_{e1}\left(\varphi_p\right)\ddot{\varphi}_p + m_{e2}\left(\varphi_p\right)\dot{\varphi}_p^2 = F_x^* \end{cases} \tag{7}$$

Figure 4 shows the inertial reduced parameters for one revolution of the pedal obtained by means of the exposed procedure of Sect. 2. Calculations have been made with Scilab and simulations with PAM and WinMecC. The results from the simulation have been obtained with intervals of 10° which are sufficient due to the form of the functions. For their use in Scilab, polynomial functions defined with splines of third order have been used.

Figure 5 shows the rotation velocity of the pedal obtained by means of the integration of the equation of movement. Its mean value remains nearly constant because $m_1\left(\varphi_p\right)$ has higher values in front of $m_2\left(\varphi_p\right)$. The figure also shows the force in the knee in the leg direction. The low values of the force correspond when the leg is driven, the high values correspond when the leg is the driving one and the higher value corresponds to the dead-point φ_m.

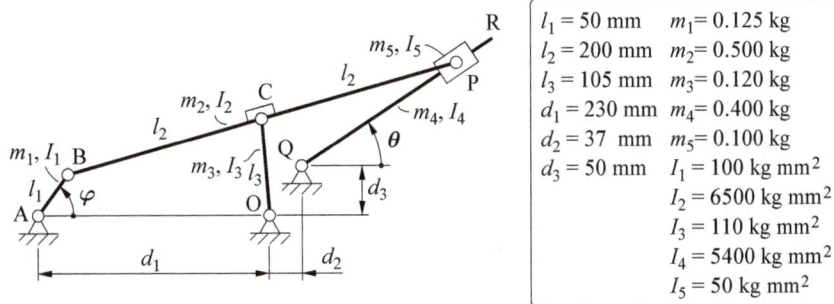

$l_1 = 50$ mm	$m_1 = 0.125$ kg
$l_2 = 200$ mm	$m_2 = 0.500$ kg
$l_3 = 105$ mm	$m_3 = 0.120$ kg
$d_1 = 230$ mm	$m_4 = 0.400$ kg
$d_2 = 37$ mm	$m_5 = 0.100$ kg
$d_3 = 50$ mm	$I_1 = 100$ kg mm^2
	$I_2 = 6500$ kg mm^2
	$I_3 = 110$ kg mm^2
	$I_4 = 5400$ kg mm^2
	$I_5 = 50$ kg mm^2

Figure 6. Sketch of the single dwell bar mechanism used in the case of study.

Figure 7. Prototype of the single dwell bar mechanism.

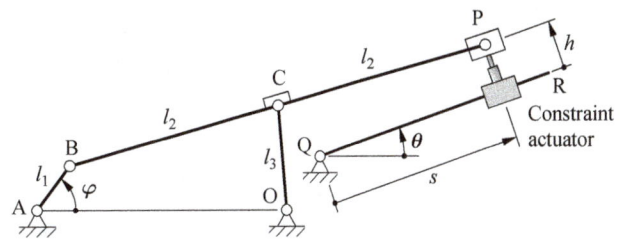

Figure 8. Prototype of the single dwell bar mechanism.

3.2 Single dwell bar mechanism

The procedure described in the Sect. 2 is used to study the dynamics of a single dwell bar mechanism. The mechanism and its characteristics, both geometric and inertial, are shown in Fig. 6 which represents the model for this study. The prototype shown in Fig. 7 is designed, among other functions, to check the temporal movement law of the crank AB, $\varphi(t)$, in order to achieve the prescribed movement law of the rocker QR, $\theta(t)$. A dual axis gyroscope IOG 500 is attached to the rocker to check the accuracy of the result. Even though, in this work, the exposed method in Sect. 2 is used to determine the reduced parameters of this mechanism.

The orientation of the rocker QR is maintained substantially constant for a certain interval of movement of the crank AB. The crank AB is considered balanced so that its centre of inertia coincides with the fixed joint A. The centre of inertia of the connecting-rod BP is at C and the centres of inertia of the rockers OC and QR are at their midpoint. The centre of inertia of the slider coincides with the articulation P.

The motion equation of the system of Fig. 6 can be obtained by means of Eksergian's method. The kinetic energy of the system is the sum of the kinetic energies of the crank AB, the connecting-rod BP, the two rockers OC and QR and the slider. As in the previous example, if some constrain action must be determined it is necessary to use the same procedure of the previous example.

Also in this case, first of all, reduced parameters are determined by kinetostatic simulation programs mentioned and, for a set of time instants, the kinematic and dynamic variables for calculating inertial parameters and reduced forces are obtained. The actuators used in these analyses are an angular actuator that controls the angle φ rotated by the crank AB and a constraint actuator in the prismatic pair as shown in Fig. 8.

From these analyses, the inertial behaviour of the mechanism is obtained. This must be linked with the external forces acting on it, that is the torque of the motor applied in the crank AB T_{motor} and of the passive resistances T_{pr}. In order to take into account these external forces, the generalized forces associated to the coordinates corresponding to the crank angle and the displacement of the constraint actuator must be calculated. The expressions of generalized forces F_i^* are obtained by means of the following virtual movements:

i. $\dot{\varphi}^* \neq 0$ and $\dot{h}^* = 0$. This virtual movement is compatible with the constraints of the original system. For this virtual movement the generalized force F_φ^* is obtained:

$$F_\varphi^* = T_{motor} + T_{pr} \frac{\partial \dot{\theta}}{\partial \dot{\varphi}}.$$

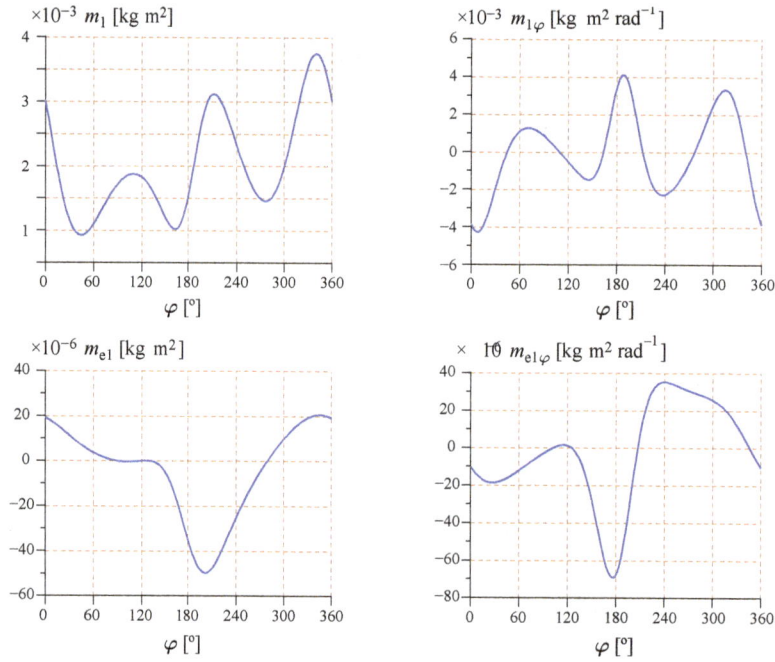

Figure 9. Reduced parameters of the single dwell bar mechanism.

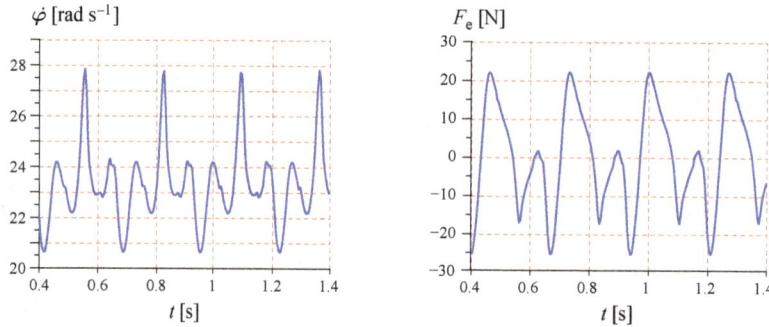

Figure 10. Angular velocity of the crank AB and constraint force in the prismatic pair.

ii. $\dot{\varphi}^* = 0$ and $\dot{h}^* \neq 0$. For this virtual movement the generalized force F_h^* is obtained:

$$F_h^* = T_{pr}\frac{\partial \dot{\theta}}{\partial \dot{h}} + F_e.$$

The terms $\frac{\partial \dot{\theta}}{\partial \dot{\varphi}}$ and $\frac{\partial \dot{\theta}}{\partial \dot{h}}$ are obtained directly from the kinematic analysis of the mechanism.

The differential equation of movement for the coordinate φ that must be integrated is:

$$m_1(\varphi)\ddot{\varphi} + m_{1\varphi}(\varphi)\dot{\varphi}^2 = F_\varphi^* \tag{8}$$

The equation for finding the constraint force in the prismatic pair, from the values of φ obtained by means of the integration of Eq. (8) and its derivatives, is:

$$m_{e1}(\varphi)\ddot{\varphi} + m_{e1\varphi}(\varphi)\dot{\varphi}^2 = F_h^* \tag{9}$$

The motor used to determine the reduced parameters is a DC motor with permanent magnets. The operation of this type of motors can be described by the following equations:

$$U = Ri + Li + K\dot{\varphi}_{\text{motor}}$$

$$T_{\text{motor}} = Ki$$

where U is the tension of the armature, i is the intensity that flows through it, R and L are the resistance and inductance in terminals of the armature, K is the constant of the torque, T_{motor} is the generated torque and $\dot{\varphi}_{\text{motor}}$ is the angular velocity of the motor.

The motor chosen in this application is a model Dunkermotoren GR 63×55 powered 12 V with the following characteristics: $R = 0.6\,\Omega$, $L = 1.5\,\text{mH}$, $K = 64\,\text{mN m A}^{-1}$, $I_{\text{mot}} = 75\,\text{kg mm}^2$. In order to adjust the velocity of rotation of the motor to that necessary in the entrance shaft of the mechanism, the use of a gear reducer is proposed. The chosen one is PLG 52 of the same brand with a transmission ratio $\tau_{\text{red}} = 0.125$ and an inertia momentum reduced to the entrance shaft $I_{\text{red}} = 20\,\text{kg}\,\text{mm}^2$.

The passive resistances have been considered concentrated in bar QR and are modelled as:

$$T_{\text{rp}} = -T_0 \text{sign}(\theta) - c\dot{\theta}$$

where $T_0 = 0.1\,\text{N}\,\text{m}$ and $c = 0.03\,\text{N}\,\text{m}\,(\text{rad}\,\text{s}^{-1})^{-1}$.

Figure 9 shows the inertial reduced parameters for one revolution of the crank AB obtained by means of the exposed procedure of Sect. 2. Calculations have been made with Scilab and simulations with PAM and SAM. All the results agree among them.

Figure 10 shows the rotation velocity of the crank obtained by means of the integration of Eq. (8) and the constraint force in the prismatic pair obtained with Eq. (9). Rotation velocity and force are periodic. Although the value of motor torque is almost constant the considerable variation of the rotation velocity is consequence of the great variation of the inertial parameters.

4 Conclusions

The proposed approach for doing the dynamic analysis of one degree of freedom mechanisms (not necessary planar) based on the use of reduced parameters leads to a simple and efficient procedure for their study.

This approach enlarges the possibilities and utilities of easy use programs that cannot do direct dynamic analysis. In particular, the inertial reduced parameters have been obtained with programs PAM, SAM and WinMecC and the reduced parameters for calculating the generalized forces with PAM.

This approach is useful when the study of the mechanism must be performed in real time, as part of the simulation and control of a production process. A comparison between the time used to integrate two cycles of stationary regime of the example 1 with a commercial software and with the proposed method has been performed. The computation time on a PC with Windows7 operating system and 64-bit processor Intel Core i7 have been 658 ms with commercial software and 68 ms with the proposed method. The results support the statement that this method can be used in real-time simulations more efficiently.

The inertial reduced parameters are obtained as a table of values. So, the integration algorithm for the equation of movement must access to them by means of an interpolation function.

The use of Eksergian's method allows the incorporation, in a simple way conceptually and operationally, of motors and passive resistances described by means of the exchanged energy, instead of doing it by means of non-conservative forces.

The exposed procedure is easily extensible to conservative forces, that only depend on position. The generalized forces associated to those can be obtained by kinetostatic analysis. Forces that depend on velocity must be dealt directly in the equation of movement. This type of forces can also be reduced to the coordinate, but the associated generalized force is, then, function of position and velocity and its determination by means of a kinetostatic analysis is not operative.

Competing interests. The authors declare that they have no conflict of interest.

Edited by: A. Müller

References

Cardona, S. and Clos, D.: Teoria de Màquines, UPC, Barcelona, 2000.

Cardona, S., Clos, D., Jordi, L., and Puig-Ortiz, J.: Curs d'autoaprenentatge de simulació de mecanismes, Universitat Politècnica de Catalunya, Barcelona, 2006.

Cardona, S., Jordi, L., and Puig-Ortiz, J.: Utilización de fuerzas y parámetros reducidos para el estudio dinámico de mecanismos de un grado de libertad, Congreso Iberoamericano de Ingeniería Mecánica, 17–20 November 2009, Las Palmas de Gran Canaria, Spain, 1503–1510, 2009.

Chen, K. and Beale, D. G.: A linear approach for experimental dynamic parameter estimation of planar mechanisms, Multibody Syst. Dyn., 9, 165–184, doi:10.1023/A:1022552113353, 2003.

Clos, D. and Puig-Ortiz, J.: PAM, un programa de análisis de mecanismos planos de n grados de libertad enfocado a la docencia universitaria, An. Ing. Mec., 15, 757–765, 2004.

Curià, E.: Una modificació de la pedalada. Estudi mecànic i biomecànic. Repercussió en la limitació de la flexió del genoll, PhD Thesis, Universitat de Lleida, Spain, 2010.

Díaz-Rodríguez, M., Mata, V., Valera, A., and Page, A.: A methodology for dynamic parameters identification of 3-DOF parallel robots in terms of relevant parameters, Mech. Mach. Theory, 45, 1337–1356, doi:10.1016/j.mechmachtheory.2010.04.007, 2010.

Doughty, S.: Mechanics of Machines, John Wiley & Sons, USA, 1988.

Ebrahimi, S. and Haghi, A.: Characterization of the contribution of inertial parameters to the dynamics of multibody systems, Multibody Syst. Dyn., 30, 449–460, doi:10.1007/s11044-013-9355-x, 2013.

Eksergian, R.: Dynamical Analysis of Machines, J. Franklin I., 209, 21–36, doi:10.1016/S0016-0032(30)90992-1, 1930.

Fogarasy, A. A. and Smith, M. R.: A unified tensor approach to the analysis of mechanical systems, P. I. Mech. Eng. C–J. Mec., 211, 313–322, doi:10.1243/0954406971522079, 1997.

Gautier, M. and Khalil, W.: A direct determination of minimum inertial parameters of robots, 27th IEEE Conference on Decision and Control, 7–9 December 1988, Austin, Texas, 1682–1687, doi:10.1109/CDC.1988.194738, 1988.

Geogebra: available at: http://www.geogebra.org, last access: 25 October 2016.

Jordi, L., Zayas, E. E., and Cardona, S.: Reduced parameters for the dynamic study of a system of one degree of freedom, 12th International Research/Expert Conference "Trends in the Development of Machinery and Associated Technology", 26–30 August 2008, Istanbul, Turkey, 1049–1052, 2008.

Kurtenbach, S., Prause, I., Weigel, C., and Corves, B.: Comparison of Geometry Software for the Analysis in Mechanism Theory, in: New Trends in Educational Activity in the Field of Mechanism and Machine Theory, Springer International Publishing, Switzerland, 193–201, 2014.

Parsa, S. S., Boudreau, R., and Carretero, J. A.: Reconfigurable mass parameters to cross direct kinematic singularities in parallel manipulators, Mech. Mach. Theory, 85, 53–63, doi:10.1016/j.mechmachtheory.2014.10.008, 2015.

Ros, J., Iriarte, X., and Mata, V.: 3D inertia transfer concept and symbolic determination of the base inertial parameters, Mech. Mach. Theory, 49, 284–297, doi:10.1016/j.mechmachtheory.2011.09.006, 2012.

SAM (Mechanism Design Software, Kinematics, Simulation, Optimization): available at: http://www.artas.nl, last access: 25 October 2016.

WinMecC (Programa de análisis cinemático y dinámico de mecanismos planos): available at: http://winmecc.uma.es, last access: 25 October 2016.

Wu, J., Wang, J., Wang, L., and Shao, H.: Dimensional synthesis and dynamic manipulability of a planar two-degree-of-freedom parallel manipulator, P. I. Mech. Eng. C–J Mec., 222, 1061–1069, doi:10.1243/09544062JMES830, 2008.

Yoshida, K., Mayeda, H., and Ono, T.: Base parameters for manipulators with a planar parallelogram link mechanism, Adv. Robotics, 10, 105–137, doi:10.1163/156855396X00147, 1995.

9

Automated local line rolling forming and simplified deformation simulation method for complex curvature plate of ships

Yao Zhao[1,2], Changcheng Hu[1], Hongbao Dong[1], and Hua Yuan[1]

[1]School of Naval Architecture and Ocean Engineering, Huazhong University of Science and Technology, Wuhan, 430074, P. R. China
[2]Collaborative Innovation Center for Advanced Ship and Deep-Sea Exploration (CISSE), Shanghai, 200240, P. R. China

Correspondence to: Yao Zhao (yzhaozzz@hust.edu.cn)

Abstract. Local line rolling forming is a common forming approach for the complex curvature plate of ships. However, the processing mode based on artificial experience is still applied at present, because it is difficult to integrally determine relational data for the forming shape, processing path, and process parameters used to drive automation equipment. Numerical simulation is currently the major approach for generating such complex relational data. Therefore, a highly precise and effective numerical computation method becomes crucial in the development of the automated local line rolling forming system for producing complex curvature plates used in ships. In this study, a three-dimensional elastoplastic finite element method was first employed to perform numerical computations for local line rolling forming, and the corresponding deformation and strain distribution features were acquired. In addition, according to the characteristics of strain distributions, a simplified deformation simulation method, based on the deformation obtained by applying strain was presented. Compared to the results of the three-dimensional elastoplastic finite element method, this simplified deformation simulation method was verified to provide high computational accuracy, and this could result in a substantial reduction in calculation time. Thus, the application of the simplified deformation simulation method was further explored in the case of multiple rolling loading paths. Moreover, it was also utilized to calculate the local line rolling forming for the typical complex curvature plate of ships. Research findings indicated that the simplified deformation simulation method was an effective tool for rapidly obtaining relationships between the forming shape, processing path, and process parameters.

1 Introduction

At present, forming of the complex curvature plate for ships can be classified into mechanical cold forming and thermal forming from the viewpoint of the loading method. From a process viewpoint, it is divided into local line loading forming and monolithic loading forming. In detail, the line loading here does not signify that the loading comes into play on an identical line of the plate, but makes the point load exert continuous actions on a certain line in order. As far as the current shipbuilding enterprises are concerned, line heating and mechanical cold press forming are often used for manufac-

turing these curvature plates. In line heating, the shrinkage strain generated during heating and cooling in some parts of the metal plate causes deformation of the plate. With regard to such a line heating forming approach, Ueda et al. (1991, 1993, 1994a, b) predicted line heating forming paths based on the inherent strain calculation. And Nguyen et al. (2009) used artificial neural networks to carry out similar studies. Based on theoretical studies, automation equipment adopting the line heating forming approach has been successfully developed so that the processing efficiency for hull plates can be effectively improved (Yoshihiko et al., 2011). Nevertheless, owing to the characteristics of the line heating forming approach, such as small thermal strains incurred by line heating, the shape obtained through line heating forming is still under restrictions to a certain extent for thicker plates and regions with a larger curvature (Yoshihiko et al., 2011). Correspondingly, the cold mechanical press forming utilizes mechanical devices to apply a bending load to realize plate deformation. Compared to line heating forming, it has a higher efficiency in processing thick plates with a large curvature. Therefore, it is clear that the cold mechanical press forming has its own advantages. However, it is rather difficult to apply the mold forming method, a commonly used cold mechanical press forming method, for the automated molding of complex curvature plates in batches, with considerable differences in shapes and sizes. If variable mold molding is used, Hwang et al. (2010) performed studies on springback control and deviation compensation for multi-indenter molding and developed a multi-indenter molding device. Shim et al. (2011) studied, the mechanism of complex curvature plate molding based on an adjustable punch increment. In addition, they also investigated the use of a multi-roller forming device to realize complex curvature plate forming. Another representative mechanical cold press forming method is the die-free local forming method. In this method, complex curvature plate forming is ultimately realized by constituting a local small-range plastic processing area on the premise that the processing plate enters such an area in diverse orders and from various parts. However, the processing path determination and forming shape control in the above mechanical cold pressing process during contour machining of complex curvature planking for ships still mainly depend on artificial experience.

Clearly, the die-free local forming method mentioned above is very suitable for shipbuilding applications. The local line rolling forming is one of the die-free local forming methods commonly used (see Fig. 2) and has been extensively applied in shipyards. The corresponding basic approach can be described as follows. A pair of concave-convex wheels is mounted onto the upper and lower surfaces of the processing plate to drive the relative up and down motions of the upper and lower wheels. Therefore, local pressure on the processing plate can be realized. Meanwhile, the upper convex wheel or the lower concave wheel drives the motion of the processing plate that further passes between the upper and lower wheels through rotation to realize local bending of the plate. Based on the motions of the concave-convex wheels on the processing plate along with paths in diverse directions, complex curvatures of various shapes can be generated for the plate. From the perspectives of the upper and lower concave-convex wheels as well as the processing plate, the sphere of loading actions from those wheels is relatively narrower if compared with the length-width dimension of the plate. Moreover, their loading actions can be deemed as a process that approximates rolling line mechanical rolling, which is referred to as the local line rolling forming method here. Similar to the line heating plate forming method, it also involves

gradual forming; the overall shape is realized based on the control over local deformation. Such a forming method requires small equipment dimensions and low costs. In this machining process, the processing path and loading force of the roller should be primarily determined according to operator experience. Finally, forming of the target complex curvature plate is completed. In fact, the processing mode based on artificial experience is in line with special knowledge systems. Theoretically speaking, through rational path planning and loading force settings, forming of complex curvature plates with diverse sophisticated shapes can be realized. If an inherent mapping relationship expression involving the forming shape and processing parameters, such as loading path and loading force, that can be accepted by a computer system is developed, a rolling scheme can be accurately formulated. Then, by virtue of the corresponding processing equipment, automation of local line rolling forming can be realized.

During manual operation, the operator determines the preliminary loading magnitude and path according to his/her experience as well by estimating the unloading elastic resilience of the plate in the first place. Second, comparisons between the processing plate after rolling forming and the target shape are carried out by using the cardboards. Third, according to his/her experience, both the rolling feed and rolling path are modified to obtain the target shape by repeating comparisons and modifications. The plate is deformed into the target shape with complex curvature from its initial form, which is mainly caused by uneven in-plane and bending strains inside the plate (Liu and Yao, 2005). After obtaining the in-plane and bending strain distributions required to form the target shape, a processing technique able to exert such strains can be found. These are the general considerations for plate forming processes. Till date, relevant studies in this regard have emphasized the following three issues:

i. an application path of evenly dispersed in-plane or bending strain incurred by the in-plane and bending strain distributions of continuous and uneven target shapes;

ii. finding the corresponding relation between process parameters and in-plane or bending strains;

iii. obtaining a successive correction method for the applied strain approximate to the target shape.

These issues are the focus of this study. With the objective of resolving these problems, experiential knowledge stored in the human brain should be integrally transformed into driving data for automation equipment; plate forming as a sophisticated nonlinear problem should be solved completely using theoretical analysis or experimental measures. However, it is almost impossible to execute these approaches. Numerical calculations with the help of a computer become an inevitable choice. It is feasible to compute different loading paths and loading forces through computer-based numerical calculations to acquire the corresponding relation between them and the in-plane and bending strains as well as deformation shapes. Further, the numerical computation for the local line rolling forming process involves complicated factors such as contact variations, moving loading, unloading resilience, material type and geometric nonlinearity, if the three-dimensional elastoplastic finite element method is adopted. The time and capacity required for computation are very high In practice, it is also very difficult to complete formation of massive driving data supporting automated forming system for the complex curvature plates of ships. Based on the above discussions, to constitute such driving data, a calculation method with high precision and efficiency should be found, and this is an inevitable link on the entire development cycle.

In this paper, direct at the above local line rolling forming method and for obtaining a highly precise and efficient calculation method, the three-dimensional elastoplastic finite element method is studied for the single local line rolling by taking moving loading, contact point variation, elastoplastic deformation, and unloading resilience into consideration first. Therefore, detailed strain and deformation fields are obtained. In addition, after analyzing the distribution characteristics of the corresponding in-plane and bending strains, a simplified deformation simulation approach acquired for deformation is put forward in allusion to strain application. Then, based on the three-dimensional elastoplastic finite element method, deformations obtained by one- and two-local line rolling are compared relative to their computational accuracies and time. On this basis, the possibility of employing this method to determine the typical complex curvature plate shape in multi-local line rolling is examined. Moreover, with the objective of obtaining relationships between the simplified deformation simulation method and deformation shape, the possibility of utilizing such a method to rapidly conduct a large number of deformation simulation calculations based on the local line rolling forming is verified under conditions of diverse loading paths and forces.

2   Overall scheme design of local line rolling forming

The core of the automated local line rolling forming of the complex curvature plate for ships is the formation of driving data. Concrete formation procedures of such driving data are shown in Fig. 1; the connotations of the key points (i) to (iii) mentioned above, corresponding to the formation framework of driving data, are also contained.

Focusing on the key points (i) to (iii) described above, the relationship between the automated local line rolling forming method for the complex curvature plates of ships and deformation simulation calculation method is demonstrated.

Key point (i) pertains to the determination of the local line rolling forming loading path. Here, plate forming mainly concerns both plate deformation and the strain distribution closely related to it (that is the geometrical relationship of plate). Therefore, the large deformation elastic finite element method can be adopted to implement mechanical calculations for the target shape to further obtain deformation and strain distributions. Generally, for convenience, all strain components acquired through calculations are synthesized into the principal strain. Then, according to the definitions of in-plane and bending strains, the in-plane and bending strain distributions represented by the principal strain can be obtained. From the above process, it was found that the acquired strain distributions specific to complex curvature shapes were not only continuous, but also uneven. With regard to the practical process in local line rolling forming, the line-to-line disperse loading is applied to the processing plate in reality. Obviously, the continuity of strain cannot be guaranteed. Seen from another perspective, during practical processing, it is very hard to realize constant changes in the press amount in the process of one line loading owing to the related equipment. In other words, the unevenness of strain cannot be satisfied as well. Therefore, direct at the practically continuous and uneven strain distributions on the target shape plate, a multi-line dispersion loading scheme with invariant strain levels can be generated approximately for a single loading path by formulating a strain accepting/rejecting principle or by integration. In addition, studies in this regard are still under way (Keisuke et al., 2012; Park et al., 2016a, b). Considering that this is not the focus of this study, it will not be described in detail. However, the approximation degree of such a strain-loading path acquired in this manner or the rationality of the accepting/rejecting principle mentioned above should be explored and verified by simulations and calculations.

Key point (ii) states that the corresponding relation between the strain and process parameters should be determined. As far as the local line rolling forming loading is concerned, the principal process parameter is the press amount loaded by the upper/lower wheel. Under practical processing, as described above, the load is exerted in the form of a line and intervals exist between lines. Moreover, for a single loading line, its press amount usually remains unchanged. As a result, the corresponding relation between the strain and processing parameters can be acquired using the three-dimensional elastoplastic finite element method under the condition of a single loading line with diverse press amounts. Despite such a calculation conducted for a single loading line, it is difficult to fully determine the relationship corresponding to a single strain component regardless of process parameter changes, because such parameters have a rather broad variation range. The relationship between the technological parameters and proportion occupied by a certain strain component can be determined. Relational data for technological parameters provided with a small matching difference in the loading paths of key point (i) are presently still obtained through numerical calculations (Liu and Yao, 2005; Shi et al., 2013). However, if their acquisition is com-

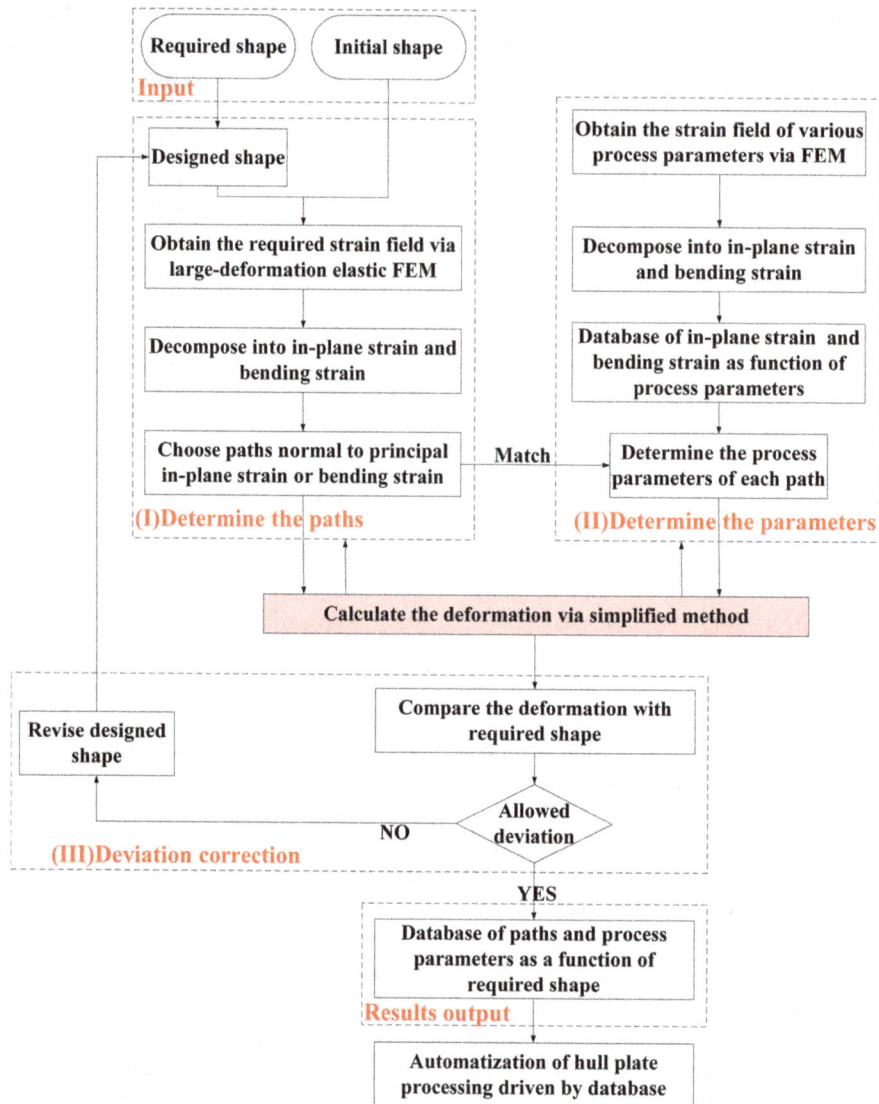

Figure 1. Formation procedure for the automated local line rolling forming driving data of complex curvature plates for ships.

pletely dependent on the three-dimensional elastoplastic finite element method, considerable computational efforts will be required.

For key point (iii), through repeated iterative correction computations for the target shape, the error between (i) and (ii) can be eliminated to satisfy the accuracy requirement of processing driving data. As such an error that cannot be erased easily, it may be confirmed that the application of (ii) into (i) fails to meet the accuracy demand of the target shape. Here, relying on the results of key points (i) and (ii), the difference between the target shape and the numerical forming outcome can be discovered by numerical calculation. Then, this difference is adopted to modify the target shape, so that the designed shape can be obtained. Subsequently, the strain distribution and the corresponding loading path and process parameters are all obtained, followed by the implementation

of new numerical forming calculations and accuracy verification. Based on these iterative loop calculations, the required precision for the target shape can be satisfied. In particular, if the numerical forming calculations are performed using the three-dimensional elastoplastic finite element method, it is difficult to imagine how significant the computational efforts and how low the computational efficiency will be.

Figure 1 indicates not only the relation between the automated local line rolling forming driving data and the key points, but also the relevance of a highly precise and efficient computational method. The simplified deformation simulation method proposed in this study plays an important role throughout the forming process.

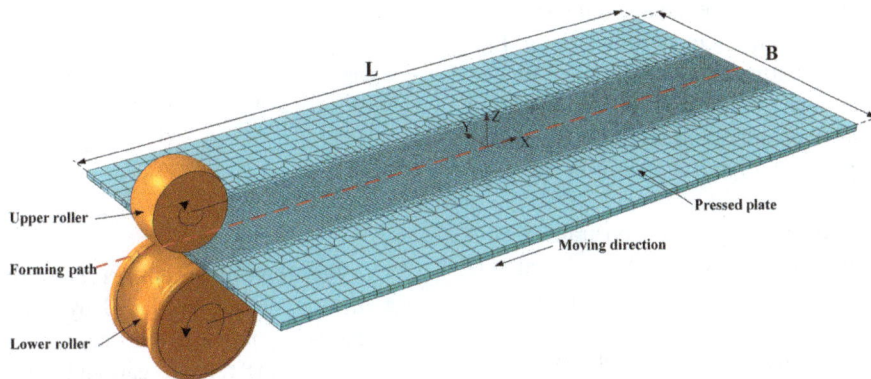

Figure 2. Finite element model and mesh generation for local line rolling forming.

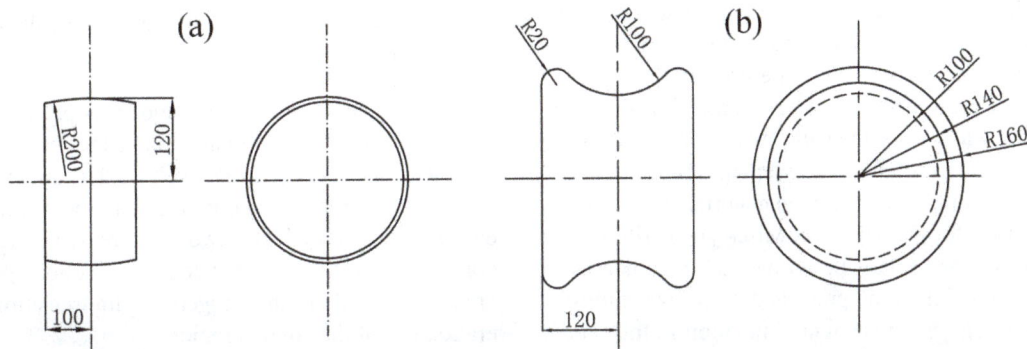

Figure 3. Shape and dimensions of upper/lower roller (unit: mm). **(a)** Upper Roller. **(b)** Lower Roller.

3   Model and results of elastoplastic finite element analysis

To understand the mechanical process of local line rolling forming in a comprehensive manner and lay a foundation for the presentation of a simplified deformation simulation method, three-dimensional elastoplastic finite element simulations and result analyses were first carried out for local line rolling. As the local line rolling forming is a discrete line loading performed on the pressed plate, it is assumed that a certain distance exists between lines. In other words, mechanical quantity variations caused by one loading line have little influence on another loading line. For this study, a specific three-dimensional elastoplastic finite element method simulation was carried out direct at the local line rolling forming under the action of a single loading line in the first place.

As for local line rolling forming, there exist multiple nonlinear problems such as large deformations, material elastoplasticity, and contact point variation. Therefore, the commercial finite element software ABAQUS was chosen to simulate the three-dimensional elastoplastic finite element calculation. In line with the finite element simulation calculation for plastic forming, ABAQUS/Explicit results were imported into ABAQUS/Standard for springback analyses

following calculations based on ABAQUS/explicit (You et al., 2014). The reason is that ABAQUS/Explicit provides a higher computational efficiency than other simulation calculation methods for the forming process. However, it is not suitable for the springback process simulation (Shaohui et al., 2012). In this study, the relative position between the roller and processed plate keeps changing during rolling forming so that the processed plate moves in and out of the rolling area. Furthermore, this process is always accompanied by the forming and springback phenomena. Considering that ABAQUS/Explicit is not suitable for springback calculations, ABAQUS/Standard is employed in this study to perform the numerical simulation of rolling forming.

The geometrical shapes and dimensions of the rollers are shown in Fig. 3. The dimensions of the plate are 2000 mm × 1000 mm × 20 mm. All of the rollers in the model are assumed rigid and are modeled as analytical rigid surfaces, which do not need to be meshed. The finite element mesh for the plate is S4R, which is a four-node, doubly curved, quadrilateral shell element; it includes large rotations, transverse shear deformations, and finite membrane strains. The S4R mesh is formulated for large strains and deformations. To increase the computation speed, the isoperimetric S4R elements were evaluated using reduced integration with one integration point per element, and effective hourglass control techniques were used to avoid the spurious deformation modes during finite element simulations. Figure 2 shows the finite element meshing of the plate model; the finite element model total contains 29 200 elements and 37 067 nodes. The initial elemental dimensions in the plastic area near the loading path are 10 mm × 10 mm in the plate plane.

Contact between the roller and plate is constrained by a penalty function method. In other words, the external surface of the roller is set as the master surface, while upper and lower surfaces are its slave surfaces. Moreover, interactions exist between the roller and plate in the case where the normal pressure during forming is positive; in comparison, separated rollers fail to interact with each other when the normal pressure is zero or negative. Under the condition that there are interactions between two contact surfaces, the corresponding contact status is defined as finite sliding.

The material parameters of the plate used in the numerical simulation are presented in Table 1. As mentioned above, it is not necessary to endow the roller with material properties because it makes use of an analytic rigid model. At the time of numerical simulation, the plate material is considered isotropic, and it conforms to the Von Mises yield criterion.

Finite element simulation is based on load increment calculation, with each simulation process divided into multiple load increments. The finite element simulation in this study involves the following loading steps:

- Step 1: assemble the components (plate and rollers) according to their geometrical relationship (Fig. 1), and set the contact surface and contact properties between the processed plate and rollers.

- Step 2: apply an artificially set z-axial negative minimum displacement load to the upper roller, with the lower roller location unchanged, and carry out the numerical simulation by load increment computation, to establish contact between the rollers and the processed plate.

- Step 3: the upper roller applies an appropriate displacement load to the plate and causes local deformation of the plate.

- Step 4: apply an angular displacement load to the lower roller with the location of the upper and lower rollers unchanged. The contact friction leads to a change in the relative location of the plate and rollers, and the plate gradually enters the rolling zone. In this process, the plate keeps entering and exiting the rolling zone, accompanied by forming and springback. After the completion of rolling on the entire forming path, the upper and lower rollers are removed gradually away from the plate. The contact action between the rollers and plate gradually decreases in this process, and the plate springs back as well, until it is completely separated from the

Table 1. Material parameters.

Young's modulus (E)	Poisson's ratio (ν)	Yield stress (σ_s)	Constitutive relation	Density (ρ)
210 GPa	0.3	290 MPa	$\sigma = 395 \times \varepsilon^{0.05}$	7800 kg m^{-3}

rollers. During the entire process of line rolling forming, the springback produced by unloading exists, so it is necessary to predict mechanical springback. In the simulation, the relative gap and contact pressure between the rollers and the plate is used to determine whether they are in contact or separated. After the static load is removed, it is assumed that the roller is separated from the pressed plate, and the resilience calculation is carried out.

The established finite element model is presented in Fig. 2. The directions of the short and long edges of the plate are defined as x- and y-axes respectively, and its thickness direction corresponds to the z-axis. In addition, the origin of the axes coordinates is on the center of the pressed plate. Within the contact region of the upper and lower rollers, grid elements are appropriately refined, together with transitional grid generation out of the contact region.

Geometric and material parameters of the pressed plate adopted for the associated calculation model are presented in Table 1. A power hardened and incompressible material is adopted, and the stress–strain curve is derived from previous experimental results (Wu, 2014).

The geometrical shapes and dimensions of the rollers are shown in Fig. 3. The rollers apply a press amount rolling load equal to 0.2 times the plate thickness onto the pressed plate.

To verify the applicability of this method, a three-dimensional elastoplastic finite element calculation is performed for comparison with the experimental results in Shaohui's results (Shaohui et al., 2012), wherein the above-mentioned similar computational features were applied. The comparison results are presented in Table 2, where 2a is the distance between the support rolls, d_Z is the forming depth, Rb is the radius of curvature of the workpiece boundary, (Rb)$_{EXP}$ refers to the experimental results and (Rb)$_{CAL}$ refers to the FEM results. The excellent correlation between the numerical and experimental results is revealed in Table 2, demonstrating the feasibility of the three-dimensional elastoplastic finite element method.

In accordance with the objective of the simplified deformation simulation method mentioned above, the aim is to accurately simulate the deformation shapes of the pressed plate after rolling loading. In other words, determining the transient variation in mechanical quantities in the process of rolling loading as well as the calculation of other non-geometrical mechanical quantities are not the main concerns. From the perspective of geometrical relationships, deforma-

Table 2. Comparison of experimental and FEM results (Yoon et al., 2003).

No.	2a (mm)	d_z (mm)	Rb (mm)		
			$(Rb)_{EXP}$	$(Rb)_{CAL}$	$(Rb)_{EXP}/(Rb)_{CAL}$
1	40	1.2	142.4	149.3	0.95
2	45	1.2	191.8	198.5	0.96
3	40	1.4	125.4	134.2	0.93
4	45	1.4	167.0	172.7	0.97

tion in its ultimate state after loading and unloading shows a one-to-one correspondence with strain at that time. Therefore, the strain distribution for the entire plate should be found, which is directly associated with the determination of plate shape.

Rolling forming is a large deformation process; the relationships between the corresponding deformation and the strain can be expressed as follows.

$$\begin{cases} \varepsilon_x = \dfrac{\partial u}{\partial x} + \dfrac{1}{2}\left(\dfrac{\partial w}{\partial x}\right)^2 - z\dfrac{\partial^2 w}{\partial x^2} \\ \varepsilon_y = \dfrac{\partial v}{\partial x} + \dfrac{1}{2}\left(\dfrac{\partial w}{\partial y}\right)^2 - z\dfrac{\partial^2 w}{\partial y^2} \\ \gamma_{xy} = \dfrac{\partial u}{\partial y} + \dfrac{\partial v}{\partial x} + \dfrac{\partial w}{\partial x}\dfrac{\partial w}{\partial y} - 2z\dfrac{\partial^2 w}{\partial x \partial y} \end{cases} \tag{1}$$

where ε and γ as well as their subscripts refer to the axial strain and shearing strain in one direction, respectively. Considering the thinness characteristics of the shell plate, $\varepsilon_z = \gamma_{yz} = \gamma_{zx} = 0$. As described above, when rolling depression loading and unloading are completed, its strain element should correspond to the strain, just as the deformation given in Eq. (1) corresponds to the ultimate residual deformation of the pressed plate. For the convenience of describing the strain, the strain components obtained through calculations are transformed into a principal strain expression. Hence, based on the above strain, the corresponding principal strain and direction can be expressed in the following equation set.

$$\begin{cases} \varepsilon_1 = \dfrac{\varepsilon_x + \varepsilon_y}{2} + \dfrac{1}{2}\sqrt{\left(\varepsilon_x - \varepsilon_y\right)^2 + \gamma_{xy}^2} \\ \varepsilon_2 = \dfrac{\varepsilon_x + \varepsilon_y}{2} - \dfrac{1}{2}\sqrt{\left(\varepsilon_x - \varepsilon_y\right)^2 + \gamma_{xy}^2} \\ \alpha = -\dfrac{1}{2}\arctan\left(\dfrac{\gamma_{xy}}{\varepsilon_x - \varepsilon_y}\right) \end{cases} \tag{2}$$

where ε_1 and ε_2 are the first and the second principal strains respectively; α refers to the direction angle of the principal strain. Considering the relationship between pressed plate forming and the in-plane/bending strain, the distributions of principal strains are decomposed into in-plane and bending strains. Then, ε_i^m and ε_i^b (i=1,2) are respectively used to represent the in-plane and bending strain components of the principal strain, which can be obtained based on the following equation set.

$$\begin{cases} \varepsilon_i^m = \dfrac{1}{h}\displaystyle\int_{-h/2}^{h/2} \varepsilon_i \, \mathrm{d}z \\ \varepsilon_i^b = \dfrac{2}{h^2}\displaystyle\int_{-h/2}^{h/2} z\left(\varepsilon_i - \varepsilon_i^m\right) \mathrm{d}z \end{cases} \tag{3}$$

where h stands for the plate thickness. The in-plane and bending strain direction for principal strain is still denoted by α from Eq. (2).

The three-dimensional elastoplastic finite element calculation results are shown in Fig. 4. Specifically, the overall deformation W and the curvature K as well as their subscripts denoting direction are separately represented by results of the sections from a–a to f–f along the directions of plate x and y. According to Fig. 4a–d, the local line rolling loading is able to generate deformations and curvatures in two directions simultaneously, and the curvature in the x direction is small; Based on the existing press amount, the maximum deflection ratio and curvature ratio on directions of y and x are approximately 5 and 100. Further, it can be seen from Fig. 4a and b that the regions where deformation in the y direction takes place are basically centralized between two contacts points of the lower concave wheel, that is the rolling depression part. From the results for sections a–a to d–d of the pressed plate perpendicular to the y direction of the rolling line, it is clear that the b–b and c–c sections near the middle part of the pressed plate have fundamentally similar deformation sizes and curvatures. There is a greater difference however between the a–a and d–d sections. In the existing condition of local line rolling loading, except smaller scopes on both loading terminals of the pressed plate, its deformation and curvature can stay stable for a certain duration. Thus, that with the increase in plate length can be inferred, such a feature will become increasingly prominent.

If the strain of the pressed plate is transformed according to Eqs. (2) and (3), the corresponding results expressed in principal in-plane and bending strains can be seen in Fig. 5a and b. Figure 5c shows the distribution of maximum principal strain. The orthogonal line segments of one point stand for the first and the second principal strain; the length of an arrow signifies the size of such a strain, and its direction denotes the direction of strain. Two arrows extending outward indicate a positive value; else, the value is negative. In Fig. 5a and b, changes in the principal in-plane and bending strains along the x and y coordinates of the pressed plate are shown. For the in-plane strain, two outward arrows refer to tension strain, while two inward arrows refer to compression strain. In contrast, with regard to the bending strain, the direction of arrows indicates the bending direction of the plate; two outward arrows refer to upward bending, while two inward arrows refer to downward bending.

As seen in Fig. 5c, all strains are concentrated within the region near the rolling line, approximately in the range of the

(a)

(b)

(c)

(d)

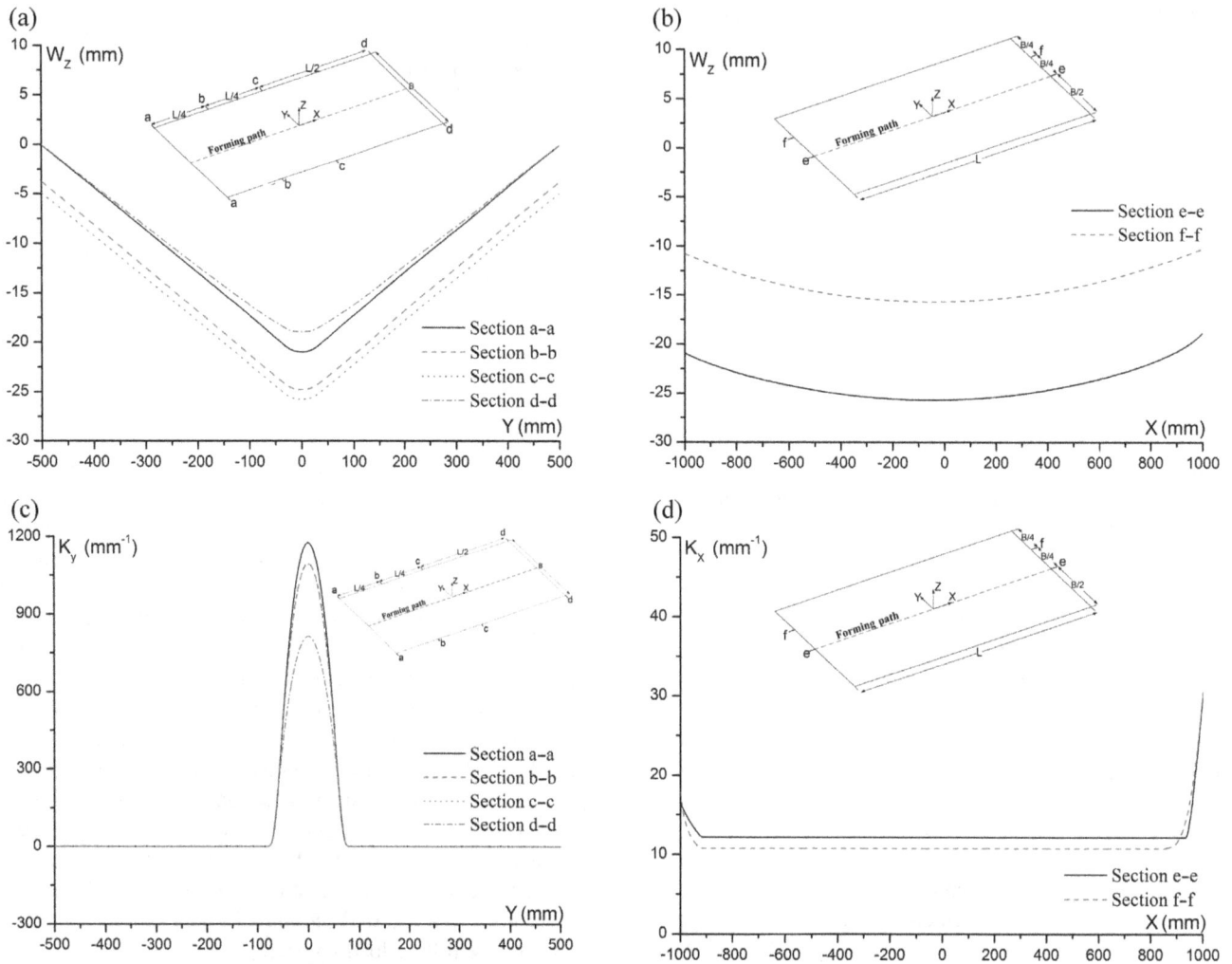

Figure 4. Results of deformation and curvature ($\delta = 0.2\,h$). (**a**) Deformation in y direction. (**b**) Deformation in x direction. (**c**) Curvature in x direction. (**d**) Curvature in y direction.

lower roller wheel width, which is approximately 180 mm. Deviating from such a region, the strain distribution is almost zero. Furthermore, inside this region, they mainly are together near the center line of rolling depression; in comparison with the midpoint value, that of strain near the center line of rolling depression decreases and rapidly approaches zero. In addition, not only are the maximum in-plane and bending strains of the first principal strain larger than those of the second principal strain, but its direction is perpendicular to the direction of the loading line. In addition to the rising and descending of the starting and ending points of the strain of the rolling loading path owing to the edge effect, the distribution of strains along the rolling loading path almost remains unchanged. Both in-plane and bending strains have the above features. As for the in-plane strain, they are mainly tensile strains vertical to the loading path, while elements of bending strain are upward bending strains vertical to such paths. Their maximum values are 0.003 and 0.024, respec-

tively. Clearly, the bending strain is one order of magnitude greater than the in-plane strain, indicating that the bending strain is a major component of deformation depending on the extent of pressing.

On the premise that a press amount of 0.2 times the plate thickness remains for rolling loading, the in-plane and bending strains are determined on different unloading positions of the pressed plate. They are from one end of this plate to 1/4, 1/2, and 3/4 of its length. Figure 6 shows the variation in distribution of the in-plane and bending strains on the center line of the pressed plate. As seen in the figure, irrespective of the location of the termination point of the loading path, the in-plane and bending strains are both at a steady stage throughout the rolling path, despite that they go up and down within a very small scope from the initiating terminal and of loading termination (unloading). Their values are certain and fundamentally coincide with each other.

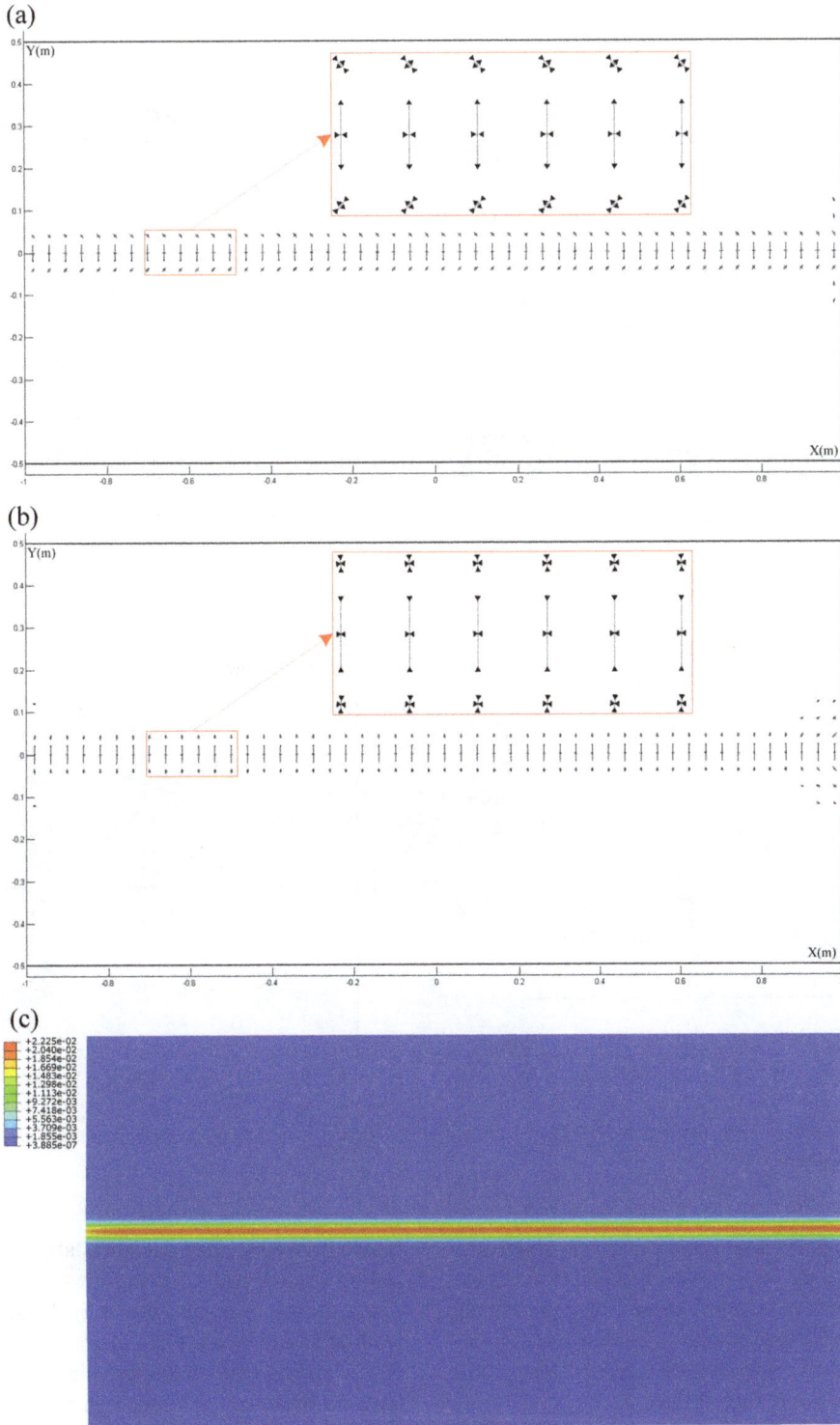

Figure 5. Results of strain distributions. **(a)** Distribution of in-plane strain (maximum value = 0.003). **(b)** Distribution of bending strain (maximum value = 0.024). **(c)** Distribution of maximum principal strain.

Figure 7 shows the calculation results obtained by changing the initial press amount. According to it, the maximum in-plane principal strain and the maximum bending princi-

pal strain both increase with the press amount. However, the change laws of the corresponding strain values remain the same. In other words, within a very small area of loading

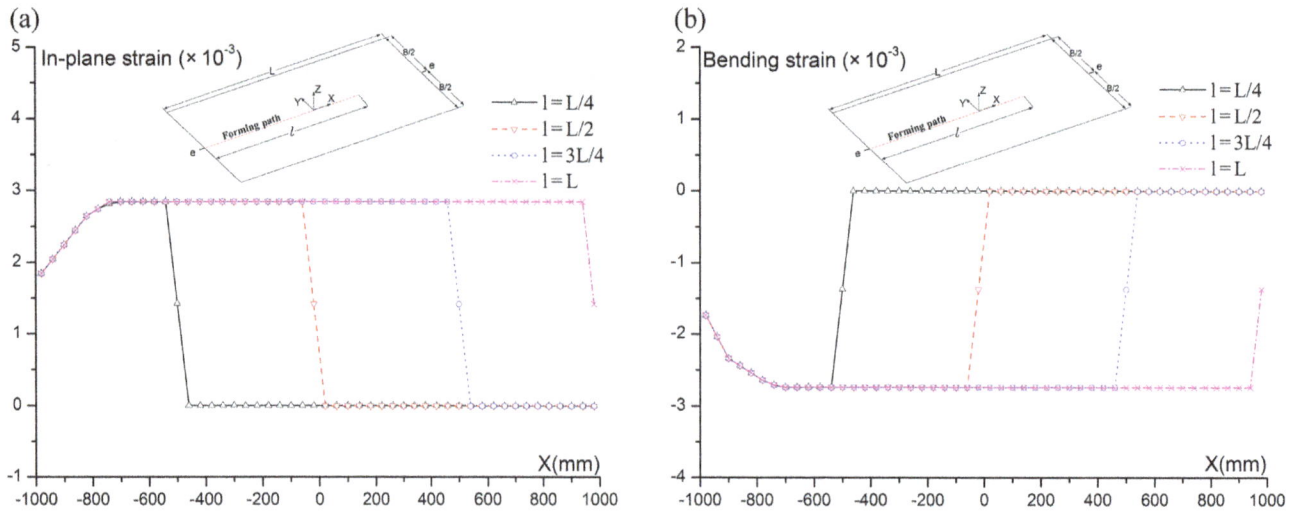

Figure 6. Relationship between length of loading path and midline strain variation of the rolling line of pressed plate. **(a)** In-plane strain distribution. **(b)** Bending strain distribution.

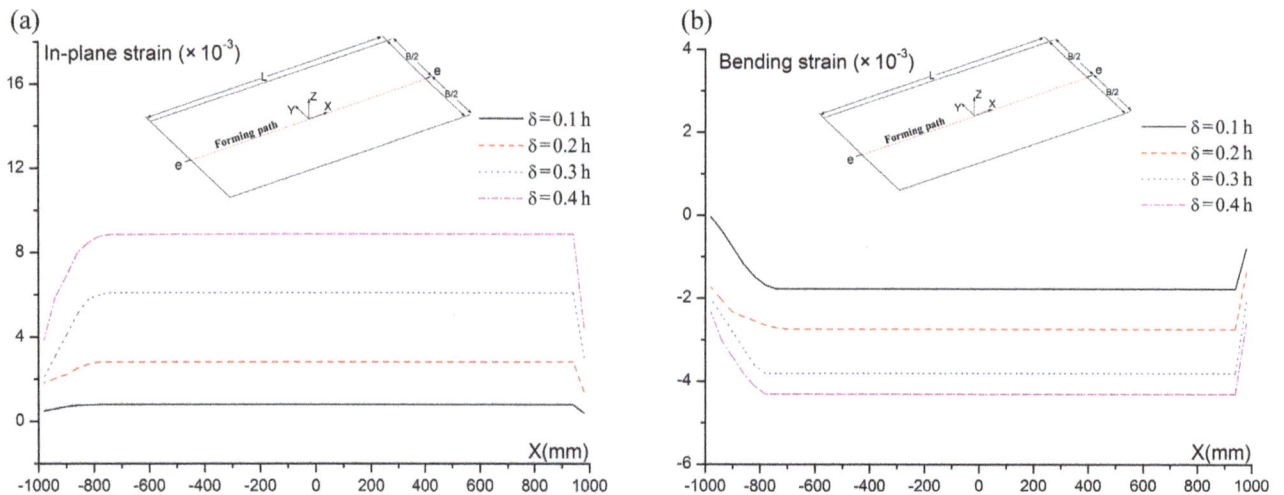

Figure 7. Relationship between press amount and midline strain variation of the rolling line of pressed plate. **(a)** In-plane strain distribution. **(b)** Bending strain distribution.

away from the initiating terminal and within a very small area of loading termination, the in-plane and bending strains both maintain definite values throughout the rolling path. As the press amount increases, the lengths of such two variation areas may increase slightly. However, they are still very short when compared to the entire rolling line.

Figure 8 shows the calculation results of the diverse thicknesses of the plate. Clearly, as the thickness of the plate rises, the distributions of the relevant strains also change. In addition, the maximum in-plane principal strain decreases while the maximum bending principal strain increases. Meanwhile, the lengths of the strain variation regions at the initiation and termination of loading increase slightly despite the corresponding variations being very small. As for in-plane and bending strains, they remain stable on a majority of the rolling loading paths.

According to results shown in Figs. 6–8, as long as the loading length of the plate is not excessively short, changes in the loading path length, press amount, and plate thickness have no significant influence on the lengths of the strain increase and decrease segments at the initiation and termination of loading. In in most cases, in practical manufacturing processes, the loading path can be very long. Therefore, it can be considered that in-plane and bending strains basically remain invariant for most of the entire rolling loading path.

(a)

(b)

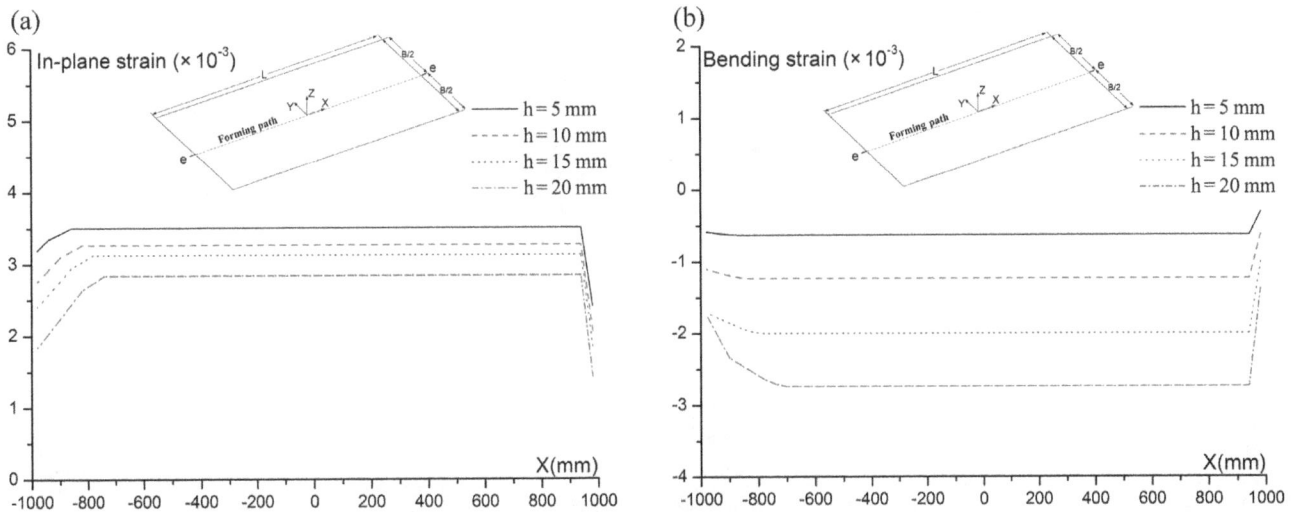

Figure 8. Relationship between plate thickness and midline strain variation of the rolling line of pressed plate. **(a)** In-plane strain distribution. **(b)** Bending strain distribution.

4   Simplified deformation simulation method

The above three-dimensional elastoplastic finite element calculation results demonstrate the basic strain distribution characteristics of local line rolling. First, the strain is mainly concentrated near the center line of rolling. Second, within a very short distance from loading initiation and termination, the strain increases or decreases; it remains at a constant value when it is not affected by the length of loading path, press amount, and the thickness of the plate. In addition, in-plane and bending strains follow the same distribution law.

Depending on the above characteristics, the method mentioned below can be adopted to replace the three-dimensional elastoplastic finite element for performing the relevant calculations. The calculation based on the three-dimensional elastoplastic finite element is performed first for single short-range rolling line loading under diverse press amounts. Here, the short range refers to the minimum distance to reach a stable strain stage described above. According to the previous results, it is known that the loading path is generally able to attain a stable area of strain through approximately 10 times plate thickness from the beginning. In fact, during such a calculation, the corresponding relationship between press amount and strain is established simultaneously. That is, the calculation depends on the relationship between the pressed plate deformation and some process parameters. Subsequently, the variations in strain at the beginning and end of the loading line should be neglected, so as to extract all strain components of a stable strain region, which are determined by the three-dimensional elastoplastic finite element calculation. Furthermore, these components approximately serve as strain values on the complete loading path. These strains are applied onto the pressed plate by means of a linear elastic calculation based on the mode of initial strain. There-

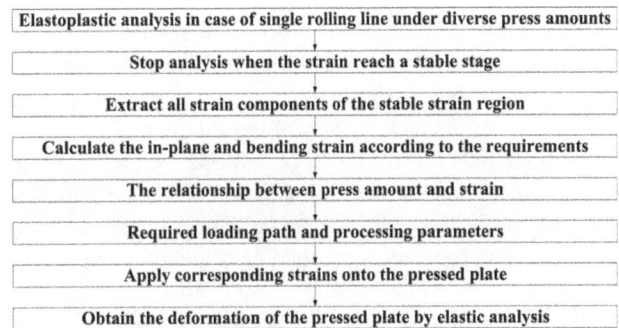

Figure 9. Simplified deformation simulation method.

fore, deformations of the pressed plate under different press amounts or various lengths/directions of loading paths can be obtained. The advantage of such a computational method is the substantial reduction in the three-dimensional elastoplastic finite element calculations. The extraction of effective strains, centralized within a narrow area of the pressed plate requires little workload. While the secondary calculation is an elastic calculation that applies a strain with low approximation, the original features are essentially maintained. The computational method described above is referred to as the simplified deformation simulation method, and the corresponding computational flow is summarized in Fig. 9.

Depending on the simplified deformation computation method, the large deformation elasto-plastic finite element method is adopted to establish the correspondence between strain distribution and processing parameters. On this basis, calculation of elasticity can be carried out to acquire the ultimate deformation result. Compared to the large deformation elasto-plastic finite element method, the simplified deformation computation method omits iterative computations

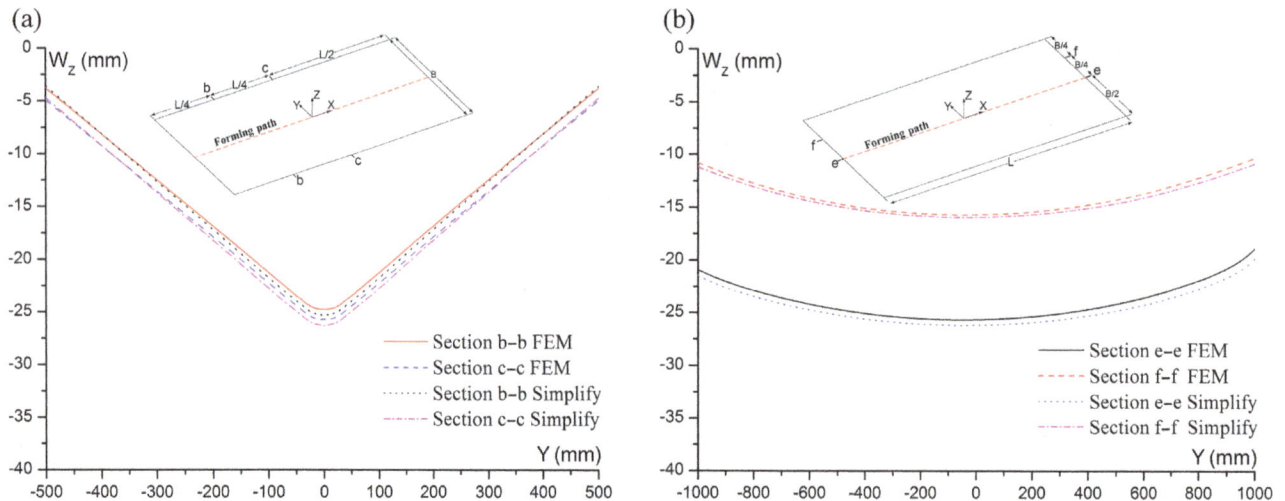

Figure 10. Comparison of deformation between elastoplastic FEM and simplified deformation simulation method. **(a)** Results from lateral sections. **(b)** Results from longitudinal sections.

required by the nonlinear finite element method during forming and springback simulations. Instead, it directly exerts initial strain distribution on the plate model to obtain the final deformation outcome by means of large deformation elasticity analysis. Specifically, the large deformation elasto-plastic finite element method can be used to follow variations in the stress strain and displacement field for the entire forming process; in contrast, the simplified deformation computation method focuses on deformation results after forming. Evidently, such a difference is the fundamental reason why the former requires a long computing time and a large amount of computation, while the latter requires a short computing time and a small amount of computation.

4.1 Comparison between elastoplastic FEM and simplified method

In order to test the computational accuracy and efficiency of the simplified deformation simulation method, based on the premise that calculation procedures of this method presented in Fig. 8 are followed, the comparison calculation is carried out under conditions of pressed plate and rolling loading; these conditions are identical to those adopted by the three-dimensional elastoplastic finite element calculation in the above section. The computational grid of the simplified deformation simulation method is not required to be consistent with that of the three-dimensional elastoplastic finite element calculation. As the simplified deformation simulation method falls into the elastic calculation category, the requirement for the element grid is rather low. For the convenience of mesh generation and adopting an identical element grid calculation model to determine multiple rolling loading paths, homogeneous element grids that are divided in line with appropriate density can be selected to perform

calculation based on the simplified deformation simulation method. Thus, it is conceivable that grids cannot be divided in accordance with Fig. 2 under the condition of multiple rolling loading paths as far as the three-dimensional elastoplastic finite element calculation is concerned. Consequently, full-plate grid subdivision must be conducted for the pressed plate with the aim of adapting rolling loading paths to different directions and positions.

First, three-dimensional elastoplastic finite element calculation is carried out. The calculation ends when the length of the corresponding loading path is approximately 10 times the plate thickness. After all strain components are extracted near the single rolling line, they are applied to every element of an identical loading path of the pressed plate as initial strains to perform elasticity calculations. In this way, the deformation of such a pressed plate can be acquired. The calculation results are compared with deformations of the vertical and horizontal cross sections, which are obtained by the three-dimensional elastoplastic finite element calculation, and with the deformations of sections b–b, c–c, e–e and f–f obtained by the three-dimensional elastoplastic finite element calculation. The related locations of sections for such comparisons are shown in Fig. 10. It can be seen that along the x- and y-axes of the pressed plate, the deformation results achieved by the simplified deformation simulation method perfectly coincide with those obtained using the three-dimensional elastoplastic finite element calculation.

The calculation time comparisons between the two methods are presented in Table 3; it is clear that the calculation time required by the simplified deformation simulation method is 1/100 of that required by the three-dimensional elastoplastic finite element method. The full-plate uniform rectangular grid partition with 200 and 100 elements on the x and y-axes, respectively, is considered in the simplified de-

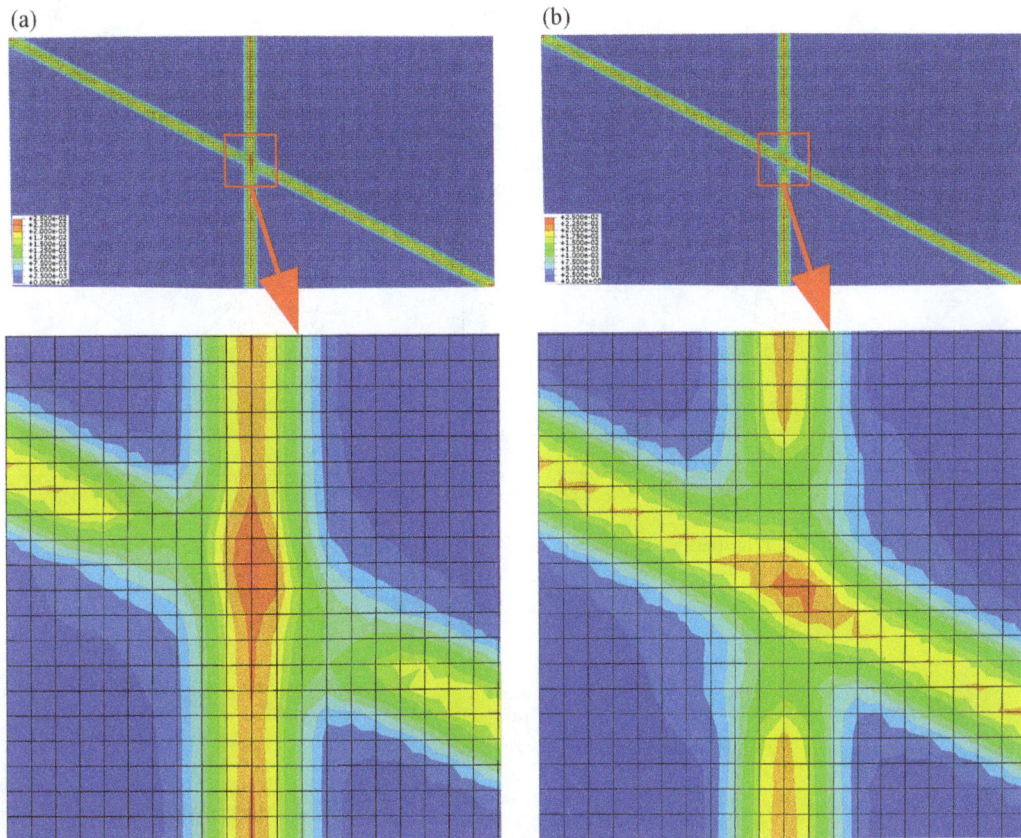

Figure 11. Maximum principal strain distribution of different loading sequences. **(a)** Loading scheme 1. **(b)** Loading scheme 2.

Table 3. Comparison for elastoplastic FEM and simplified deformation simulation method.

Method	Number of elements	Number of nodes	Incremental steps	Time
Elastoplastic FEM	29 200	37 067	2001	960 min
Simplified method	80 000	101 505	1	10 min

formation simulation method. In other words, the numbers of elements and nodes are both greater than those of the three-dimensional elastoplastic finite element method. Despite such facts, the calculation time taken is far less than that required by the three-dimensional elastoplastic finite element method.

4.2 Application of the simplified deformation simulation method

In reality, local line rolling forming is the process of multi-channel rolling processing in diverse directions. Without loss of generality, two rolling loading paths in diverse directions are selected for this study to perform the three-dimensional elastoplastic finite element method; moreover, the simplified deformation simulation method is also adopted to carry out

comparison calculations, so as to validate the applicability of such a simplified deformation simulation method under the condition of multiple loading lines. According to the preceding calculation results, deformation or plastic zone is focused within a narrow band under the action of rolling loading. Approximately, it lies between the widths of lower rollers. Therefore, if the width of the rollers do not overlap with each other for twice rolling, the loading sequence may not have an enormous impact on the deformation results of the pressed plate. However, in order to investigate effects of the loading sequence, calculations are also conducted for various loading sequences. By taking the existing calculation results given in the previous section, direction y from the long edge midpoint of the pressed plate and one of its diagonals are selected as the two loading paths. They have a point of intersection; the associated calculation conditions are the same as those given in Tables 1 and 3. The press amount of roller is taken to be 0.2 times that of the plate thickness. Here, the loading along the y direction from the long edge midpoint of the pressed plate is referred to as Loading Scheme 1; while, that along the diagonal is known as Loading Scheme 2. Figure 11 shows the dependency of strain distributions on loading sequence. According to this figure, the two loading schemes only show evident differences at the intersection of loading lines as far as plastic strain distributions are concerned.

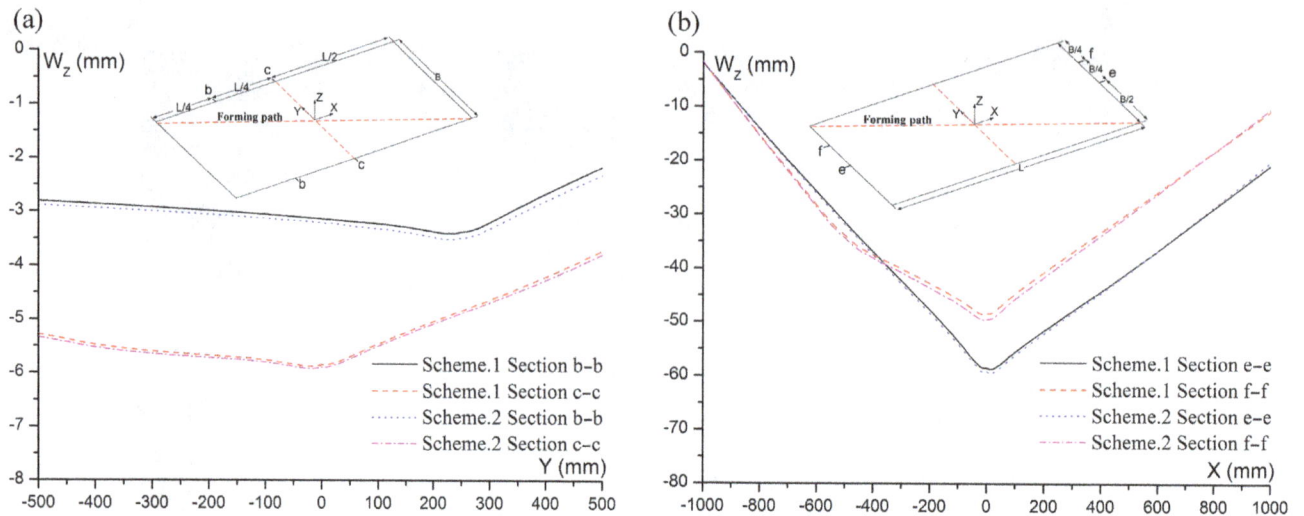

Figure 12. Results of deformation of different loading sequences. **(a)** Results from lateral sections. **(b)** Results from longitudinal sections.

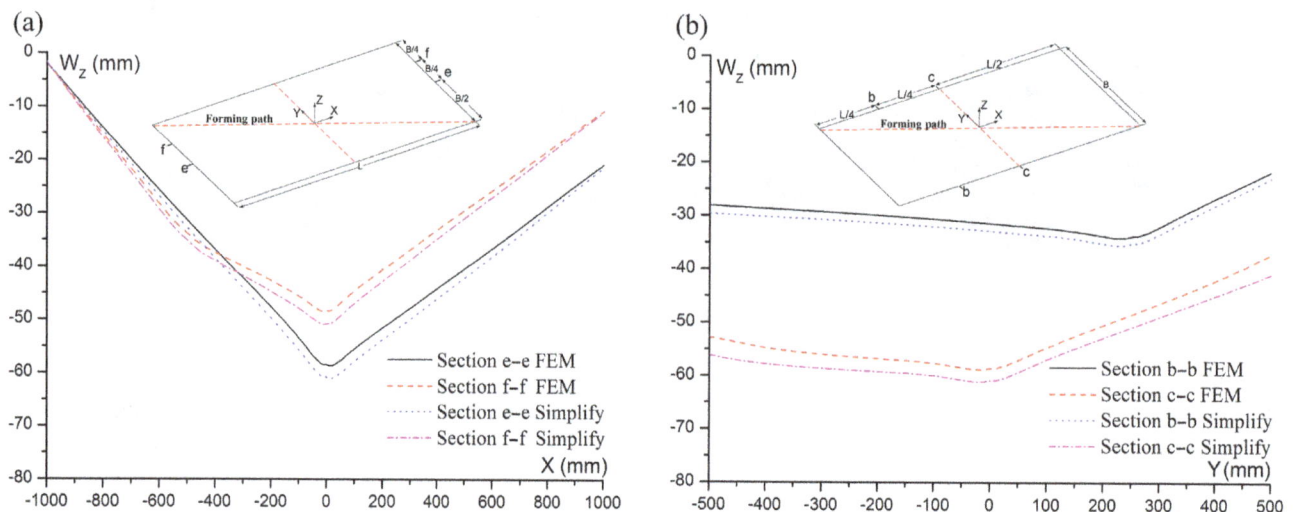

Figure 13. Comparison of deformation between elastoplastic FEM and simplified deformation simulation method (two lines). **(a)** Results from longitudinal sections. **(b)** Results from lateral sections.

In addition, the corresponding variation range is very small and its size is only approximately 100×100 mm (element size $= 10 \times 10$ mm). From the perspective of area, it occupies approximately 1/200 of the total area of the plate. In terms of length, it is approximately 1/20 of the shorter loading line. Figure 12 shows the deformation comparison results for such two schemes of loading sequence. They indicate that almost no differences lie in their deformations. With the increase in press amount, such an involved area may expand appropriately. However, its suffering area is mainly concentrated within a very small area around the intersection. Clearly, such a characteristic has no dramatic changes. In other words, such a variation has no significant influence on the entire deformation of the pressed plate.

The calculation results of the three-dimensional elastoplastic finite element method and simplified deformation simulation method are shown in Fig. 13. The calculation results indicate that for two loading lines, a preferable coincidence exists between the calculation results of the simplified deformation simulation method and those of the three-dimensional elastoplastic finite element method in terms of deformations. Deformation differences caused by the compliance of plastic deformation sequences near the intersection of loading lines can be neglected. Therefore, the simplified deformation simulation method is applicable to calculations of multiple loading lines. Concerning the calculation time comparisons of two loading lines, please see Table 4. According to this table, it is clear that the simplified deformation simulation method is more computational efficient.

(a)

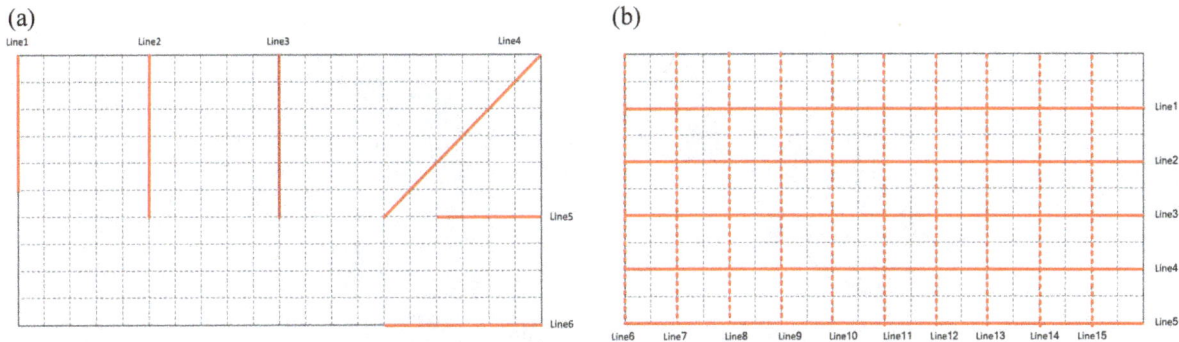

(b)

Figure 14. Schematic diagram for two different rolling line loading paths. **(a)** Path 1. **(b)** Path 2.

Table 4. Comparison for elastoplastic FEM and simplified deformation simulation method (two lines).

Method	Number of elements	Number of nodes	Incremental steps	Time
Elastoplastic FEM	29 200	37 067	3400	4000 min
Simplified method	80 000	101 505	1	15 min

In accordance with the above results, after the practical plate forming loading scheme has been acquired from key points (i) and (ii) in the actual calculation, simplified deformation simulation computing can be adopted during verification computation according to all loading paths and the relevant loading processes so as to complete strain application on the corresponding elements at a time, provided that the widths of loading paths do not overlap with each other. Then, an elasticity calculation is able to achieve deformation of the pressed plate. As an application of the simplified deformation simulation method, the possibility of adopting this method to realize complex curvature shape calculations through multiple loading paths is explored.

In this study, the formation of a sailed plate and a saddle plate is taken as an example to define process parameters related to such forming. The loading path obtained by the above method is shown in Fig. 1, which can be divided into three steps.

- In Step 1, the large deformation elastic finite element method is used to obtain the required strain field distribution according to the target shape desired. As the length and width of planking are both larger than its thickness, the strain distribution is decomposed into the in-plane strain and the bending strain.

- In Step 2, the processing path is determined in accordance with distributions of in-plane and bending strains. In-plane and bending strain distributions can be utilized to determine the processing path of local line rolling because local line rolling forming mainly leads to bending and in-plane tensile strains perpendicular to such a path.

In addition, the workpiece surface for loading can be also confirmed based on bending strain symbols here.

- In Step 3, the processing parameters are determined according to the processing path. As both the shape and dimension of the roller remain unchanged in this study, this step principally aims at making the corresponding loading force clear.

Owing to symmetry, only 1/4 of the pressed plate is shown in this figure. Concerning the loading Path 1 for the pressed plate, 20 loading lines are chosen in total, and all loading lines loading on an identical surface of the pressed plate. For loading Path 2, 28 loading lines are adopted. While 19 of those loading lines act on one surface of the pressed plate (denoted by the dotted lines in Fig. 14b), another 9 act on the other surface of the pressed plate (signified by the full lines in Fig. 14b).

After determination of the loading paths, the next step is to determine the press amount. It is done as follows:

- Step 1: in terms of all the determinated loading paths, calculating the average value of in-plane principal strain or bending principle strain along loading path and calculating integrals to get total in-plane or bending strain.

- Step 2: using elastoplastic finite element method to simulate the local line rolling forming to get the strain distribution on pressed plate. Transferring strain into the forms of in-plane strain and bending strain by Eqs. (2) and (3) and calculating integrals along plastic zone to get total in-plane strain and bending total strain.

- Step 3: according to Step 2 calculating a series of values under conditions of different press amounts, getting the database of relationships bewteen press amount and total in-plane strain or total bending strain.

- Step 4: matching the total in-plane strain or total bending strain with data in database of Step 3, selecting the press amount which total strain is closest to the total strain of Step 1.

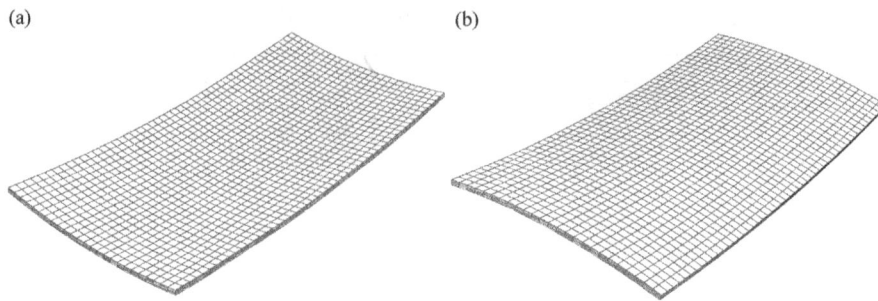

Figure 15. Deformation results for diverse loading paths based on the simplified deformation simulation method. **(a)** Result for Path 1 of pressed plate. **(b)** Result for Path 2 of pressed plate.

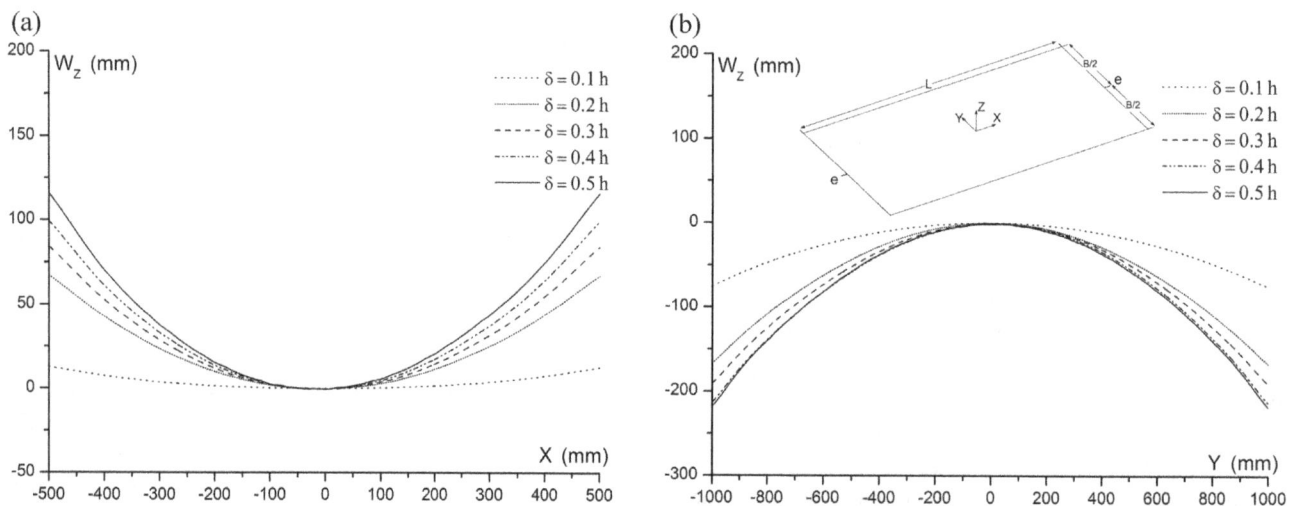

Figure 16. Impacts of roller press amount on saddle plate forming. **(a)** Results from lateral section. **(b)** Results from longitudinal section.

– Step 5: setting the press amount in database of Step 4 as the press amount of this loading path and processing all loading paths one by one.

The total calculation time consumed is approximately 40 min for the calculations of two loading paths. As for the overall calculation results of deformations, please refer to Fig. 15. Based on this figure, the deformation results obtained by Path 1 are exhibited as the pillow shape integrally; however, the deformation results are saddle-shaped as a whole, as far as Path 2 is concerned. As two different complex curvature shapes, they are common hull plates.

After Path 2 shown in Fig. 15b is selected for use, the press amount is set as 0.1, 0.3, 0.4 or 0.5 times the plate thickness to carry out the corresponding calculations in conditions identical to those for a press amount equal to 0.2 times the plate thickness. As observed from the calculation results, the overall deformation shape still remains saddle-shaped. Deflections for the center lines (section c–c and section e–e) of edges along the x- and y-axes are shown in Fig. 16. As seen in the figure, as the press amount increases, the deflections of the saddle shape along the x and y directions also increase.

During practical forming loading, Eqs. (2) and (3) are usually adopted to form in-plane and bending strain distribution diagrams with the aim of more conveniently searching for path rules constituted by strain distributions in the case that key point (i) is implemented. In addition, in-plane or bending strain can be respectively applied on the pressed plate in line with a definite principle. Such analysis with diverse strain components can be applied by using the simplified deformation simulation method; on the other hand, in-plane and bending strains can also be separately exerted. Meanwhile, owing to the features of elastic calculation, if the rolling loading path calculations or calculation of other strain components applied is carried out on any plates with an initial shape, the relevant calculation process is only a repetition of the above contents. This indicates that it is feasible for the simplified deformation simulation method to be used to calculate complex curvature forming.

With regard to the actual automated local line rolling forming process, the generation of related driving data is jointly constituted by key points (i)–(iii) described above. As approximation exists among loading paths, the loading force magnitude, material property, process parameter matching, etc., the generation process for ultimate driving data of the automated processing equipment must be inherent with plenty of iterative numerical simulation calculations. The previous calculation results show that the simplified deformation simulation method, which is a calculation method with powerful functions, high accuracy, and efficiency, is able to play such a critical role.

5 Conclusion

This study focuses on a local line rolling forming method commonly used in complex curvature plate forming in the manufacture of ships. This study determines key technologies that are adopted to form automation equipment driving data for such a forming processing method. Additionally, a three-dimensional elastoplastic finite element calculation is employed to investigate the impacts of loading by virtue of such an approach on distributions of strains generated by the pressed plate. Based on the three-dimensional elastoplastic finite element calculation, a high-precision, high-efficiency simplified deformation simulation method is further proposed to support these studies on key technologies from the perspective of strain distribution laws. The relevant conclusions can be drawn as follows:

1. Regarding the local line rolling forming method used in complex curvature plate forming in the manufacture of a ship, the strain is mainly distributed within the width between the two supports of the lower roller. Specifically, the maximum principal strain appears at the center of the width. Along the direction of roller loading, strain values generated essentially remain unchanged, except at a very small area near the initiating terminal and loading termination. Moreover, when the length of the loading line extends beyond approximately 10 times the plate thickness, such a law is under a very minor impact of the loading force, loading line length, and plate thickness.

2. Strain distribution near the intersection between loading lines conforms to the corresponding loading sequence. In contrast to the length of the pressed plate or general rolling line, the involved area or length is very small or short. Therefore, they fail to exert significant influences on the overall deformation of the pressed plate. It can be estimated that the loading sequence dependency of this contour machining method can be neglected in the case that the widths of the loading lines do not mutually overlap.

3. The simplified deformation simulation method is able to preferably reflect geometrical characteristics of such a contour machining method and substantially reduce the calculation time compared with the three-dimensional elastoplastic finite element method. In addition, the workload required by strain extraction is also low. As

for multiple loading paths, the strain distributions of the corresponding technological parameters can be applied to all elements of the associated paths during processing before calculations. Subsequently, only one calculation of elasticity should be implemented. This is a simple computational method. In terms of precision, its calculation results are similar to those of the three-dimensional elastoplastic finite element calculation. Nevertheless, its computational efficiency is enormously superior to the latter.

4. The simplified deformation simulation method that adopts multiple loading paths can be used to determine the shape of the plate constituted by complex curvature.

Competing interests. The authors declare that they have no conflict of interest.

Acknowledgements. This work was supported by grant No. 2012DFR80390 from the Ministry of Science and Technology of the People's Republic of China. The authors wish to thank Kirill Rozhdestvensky of the State Marine Technical University of St. Petersburg for useful academic and constructive advices during the project.

Edited by: Z. Gronostajski

References

Hwang, S. Y., Lee, J. H., Yang, Y. S, and Yoo, M. J.: Springback adjustment for multi-point forming of thick plates in shipbuilding, Comput.-Aid. Design, 42, 1001–1012, 2010.

Keisuke, K., Nakamura, T., Aoyama, H., Matsushita, N., and Ushimaru, A.: Basic study on laser forming method for curved surfaces, in: 14th International Conference on Precision Engineering, ICPE 2012, Hyogo, Japan, 995–1000, 2012.

Liu, C. and Yao, Y. L.: FEM-Based Process Design for Laser Forming of Doubly Curved Shapes, J. Manufact. Process., 7, 109–121, 2005.

Nguyen, T. T., Yang, Y. S., Bae, K. Y., and Choi, S. N.: Prediction of deformations of steel plate by artificial neural network in forming process with induction heating, J. Mech. Sci. Technol., 23, 1211–1221, 2009.

Park, J., Kim, D., Hyun, C., and Shin, J.: Thermal forming automation system for curved hull plates in shipbuilding, Analysis and design, Int. J. Comput. Integr. Manufact., 29, 287–297, 2016a.

Park, J., Kim, D., Mun, S., Kwon, K., and Lee, J.: Automated thermal forming of curved plates in shipbuilding: system development and validation, Int. J. Comput. Integr. Manufact., 29, 1128–1145, 2016b.

Shaohui, W., Zhongyi, C., Mingzhe, L., and Yingwu, L.: Numerical simulation on the local stress and local deformation in multipoint stretch forming process, I. J. Adv. Manufact. Technol., 9, 901–911, 2012.

Shi, Y. J., Yi, P., and Hu, J.: Effects of process parameters on forming accuracy for the case of a laser formed metal plate, Lasers Eng., 26, 295–310, 2013.

Shim, D. S., Yang, D. Y., Lee, D. J., and Han, M. S.: Investigation into the process design for the manufacture of doubly curved plates in the incremental roll forming process and its experimental verification, in: 10th International Conference on Technology of Plasticity, ICTP 2011, Aachen, Germany, 373–378, 2011.

Ueda, Y., Murakawa, H., Rashwan, A. M., Okumoto, Y., and Kamichika, R.: Development of computer-aided process planning system for plate bending by line heating (report 1): Relation between the Final Form of Plate and the Inherent Strain, Trans. JWRI, 20, 275–285, 1991.

Ueda, Y., Murakawa, H., Rashwan, A. M., Kamichika, R., Ishiyama, M., and Ogawa, J.: Development of computer-aided process planning system for plate bending by line heating (Report 4): Decision Making on Heating Conditions, Location and Direction, Trans. JWRI, 22, 305–313, 1993.

Ueda, Y., Murakawa, H., Mohamed, A., Okumoto, Y., and Kamichika, R.: Development of computer-aided process planning system for plate bending by line heating (report 2). Practice for plate bending in shipyard viewed from aspect of inherent strain, Trans. Jwri, 21, 123–133, 1994a.

Ueda, Y., Murakawa, H., Mohamed, A. M., Neki, I., Kamichika, R., Ishiyama, M., and Ogawa, J.: Development of computer-aided process planning system for plate bending by line heating (report 3): Relation between heating condition and deformation, Trans. Jwri, 22, 145–156, 1994b.

Wu, Y.: Analysis on Springback of High Strength Steel Plate with Experiments and Finite Element Simulation, MS Thesis, Huazhong University of Science & Technology, Huazhong, 2014.

Yoon, S. J. and Yang, D. Y.: Development of a Highly Flexible Incremental Roll Forming Process for the Manufacture of a Doubly Curved Sheet Metal, CIRP Ann. Manufact. Technol., 52, 201–204, 2003.

Yoshihiko, T., Morinobu, I., and Hiroyuki, S.: "IHIMU-α" a fully automated steel plate bending system for shipbuilding, J. IHI Technol., 51, 24–29, 2011.

You, W., Mingzhe, L., Daming, W., and Anyuan, W.: Modeling and numerical simulation of multi-gripper flexible stretch forming process, Int. J. Adv. Manufact. Technol., 73, 279–288, 2014.

10

Elasto-kinematics design of an innovative composite material suspension system

Shuang Xu, Alessandro Ferraris, Andrea Giancarlo Airale, and Massimiliana Carello

Mechanical and Aerospace Engineering Department, Politecnico di Torino, C.so Duca degli Abruzzi 24 Turin, 10129, Italy

Correspondence to: Shuang Xu (shuang.xu@polito.it)

Abstract. In this paper, a lightweight suspension system for small urban personal transportation vehicle is presented. A CFRP (Carbon fiber reinforce polymer) beam spring has been used to efficiently integrate the functions of suspension control arm and anti-roll bar. Composites materials were chosen to tailor the required behavior of the beam spring and to reduce the weight. Furthermore, larger space for engine compartment has been provided thanks to the compact arrangement of beam suspension components. This suspension could be installed on electric/hybrid vehicles and conventional automobiles.

1 Introduction

The suspension system is very important in an automobile, since it directly affects the handling performance and the ride comfort. All the driving/braking force and lateral force during cornering are transferred to the car body from the ground though the suspension system.

Weight reduction is an important task in the current trend of automobile development. Aluminum alloys were commonly used to substitute steel, which can obtain a weight reduction up to 30 % (Fuganti and Cupitò, 2000). However, using aluminum for structural components such as coil springs and anti-roll bars can be very difficult for effectively reducing weight while keeping the same reliability. By introducing composite materials into lightweight design[1], engineers also have the possibility to improve vehicle dynamic performance. The usage of CFRP has been usually associated with high-end racing cars for building their body shell and chassis. For example, Formula 1 cars have achieved around 80 % of its components made with composite material (Kulshreshtha, 2002).

Although not widely used, composite leaf springs can be found in automotive sector and in these cases the composite leaf spring has the same performance in terms of component stiffness with an increase of durability as much by five times and weight reduction of 65 % (Wood, 2014). The composite leaf spring presented in this paper, is a highly deformable "beam component" made with CFRP integrating the function of spring, anti-roll bar and control arm. The proposed design saves space, system weight and complexity.

Due to the architecture chosen, the CFRP works as a complaint mechanism (CM; Hao et al., 2016), which has certain stiffness under loading during in working condition. In this research, the CFRP beam spring sustains the load to support the vehicle, while the deformation is far beyond the linear range. For this reasons particular attention has been given to manufacturing process and the simulation modeling.

2 Suspension topology

A front wheel driven (FWD[2]) vehicle has been chosen as a case study, which is designed to use McPherson strut suspension on the front axis and SLA suspension on the rear axis. That is the most competitive combination for space saving of power train and improved lateral performance of the vehicle

The conventional McPherson suspension is well known for its simple design, compact space as well as the large shock absorber and the heavy "banana shape" lower control

---

[1] More than 50 % mass can be reduced using composite material (Beardmore, 1986).

[2] FWD: Front wheel drive, in this case, the engine is also mounted on the front axis.

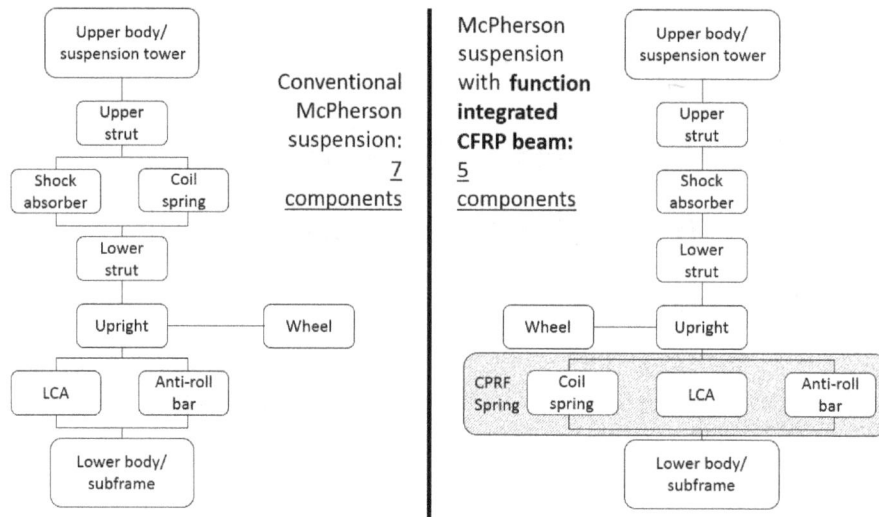

Figure 1. Topology comparison. (Upper/lower body and wheel are not count for suspension components).

arm. Since no upper control arm is present, the lateral space is larger compared to SLA suspension, which is the main reason for front mounted engine arrangement. It usually has the following components:

- Wheel hub
- Upright
- Lower control arm*
- Shock absorber (strut)
- Coil spring*
- Antiroll bar*
- Tie rod

In the purposed architecture, components marked with "*" are made redundant by integrating its function and substituted by a single CFRP beam spring as a shaded block shown in Fig. 1, the rear SLA suspension topology is similar but with an additional aluminum upper control arm.

Benefits of using a beam suspension solution are that additional space can be available with no spring (and its seat) presented and the strut can be placed closer to the wheel assembly. This can also provide better handling with a small scrub radius at relative high speed (Milliken and Milliken, 1995).

The rear suspension cannot have the same level of space reduction, but the weight is reduced by component and function integration.

2.1 Beam spring functionality

The most common application of CFRP is weight reduction of the chassis frame and body panels while keeping the same durability. In this application, CFRP is used to create a deformable beam spring to substitute the suspension lower control arm, which is virtually divided into three sections by the bushing mounting point on the left and right (Fig. 2).

During the parallel wheel travel (wheels of the same axis on both side moves in the same direction vertically), the beam deforms like a bow, functioning like a normal coil spring. When one wheel is moving vertically in the opposite direction of the other wheel (known as the opposite wheel travel) the beam deforms into an "S" shape, working like an anti-roll bar.

To achieve the desired vehicle dynamics performance, the stiffness of the beam under parallel wheel travel and opposite wheel travel are achieved by the correct number of the ply and the stacking sequence that is used during manufacturing of CFRP beam spring.

2.2 Metallic components

Using components such as upper control arm of rear SLA suspension and some other pieces made of metal (mostly aluminum alloy) has several reasons. First of all, to simplify the manufacturing process for non-definite components. Being a prototype vehicle, it is important to keep some design space and make it possible to implement some modifications of the components. Carbon fiber components are optimal when considering weight reduction, but once they have been built, it is very difficult to modify.

Besides, metallic component design procedure is relatively simple; for carbon reinforced plastics, instead, the static and dynamic loading design is difficult due to its anisotropic structure. A precise finite element analysis for such components can be very complex and unreliable. For these reasons, using the metallic components in prototyping phase is more reasonable.

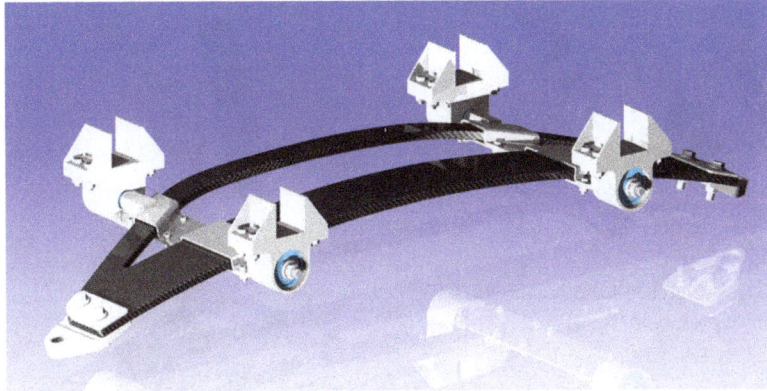

Figure 2. Unloaded front CFRP beam.

Figure 3. XAM (left), XAM 2.0 (right).

Due to this, nearly all the other components apart from the CFRP beam have been made with machined aluminum alloy. For future mass production, some of those components can be replaced with stamped aluminum sheets or hollow casted components with optimized structure design, so further weight reduction could be achieved (European Aluminum Association, 2013).

3 Suspension modeling method

The entire development of the prototype virtual model has been performed using MSC ADAMS/Car and Altair Radioss. The multi-body dynamic model has been made in ADAMS/Car in order to achieve a good precision on multibody simulation and modeling. A reference suspension for setting performance targets has been set-up with conventional suspensions (with coil springs) as well.

Altair Radioss has been used to correlate the material properties and the design parameters. To verify the stiffness with respect to design target after the prototyping of the beam spring, a simple test bench has been used.

The modeling of the CFRP beam suspension has been divided into 4 steps:

– Reference suspension modeling

– Rough prediction for beam spring characteristic using "Non-linear" beam tool in ADAMS

– Finite Element Method simulation and correlation with physical experiments

– "Non-linear" beam calibration with reference to the FEM results

Since it is complicated to start with CFRP beams, a "reference suspension model" with conventional suspensions has been defined in the beginning. MSC ADAMS is the most commonly used MBD simulation environment in the industry, two previous award-winning vehicles (Fig. 3) with conventional suspensions are designed successfully in the past. The experiment data on the test track is well correlated with the simulation results (Carello et al., 2014, 2012).

The reference suspension has been defined to have ±70 mm wheel travel, and an understeering behavior has been chosen to ensure safety. Consequently, the front suspension has a −0.4° of camber angle variation at maximum stroke and a toe-out of 0.7° when the coil spring is fully compressed.

Due to the rear suspension design for beam spring, no significant toe variation has been set. The camber variation for the reference rear suspension has been set to −2.1°.

Beam suspension from geometric and kinematic point of view is identical for the two suspension models (reference and beam suspension models) as shown in Fig. 4.

Figure 4. MBD full vehicle model, reference-handling model (left), BEAM Spring suspension vehicle prototype (right).

3.1 Non-linear CFRP beam

In this particular project, the beam has to be mounted transversely rather than longitudinally. The study shows in order to maximize the limit for elastic strain energy storage, the beam should be designed to have a tapered shape for vertical loading along the length (Yu and Kim, 1988). Still, a design with constant thickness along the length has been chosen to balance cost and performance, also because the space required for mounting the tapered beam is not available, and it is very difficult to prototype.

The beams have been designed to have two different shapes for front and rear axle. The beam on the front is designed to have an isosceles trapezium shape to have higher longitudinal stiffness considering driving axle may sustain heavier load on the longitudinal direction during acceleration. Another reason is that the mounting system became more reliable for constraining the beam movements during driving condition as shown in Fig. 5 for front suspension beam and its mounting design. The CFRP beam has the shape of a bow to have a higher constructional strength and curved to have correct preload.

At the end of both sides, two plates are mounted by thread fasteners, making a "sandwich structure". The beam is drilled through to let the thread fastener pass. On the lower plate, there is another hole left for the spherical joint, which is further connected with upright.

The bushing housings have been milled from a block of 7075 aluminum alloy. To prevent potential damage on the fiber beam (due to high-localized stresses) a layer of elastomer has been inserted between the beam and the steel plate.

On the rear axle, the geometry of the beam is different from the front and its shape is similar to a simple rectangular plate with curvature. The curved shape is to ensure the necessary preload when the vehicle is assembled and to permit the right vertical wheel travel same as front.

Comparing with the front McPherson suspension (shock absorber mounted between the upright and the top mount on the chassis), the rear SLA suspension shock absorber is usually mounted on the lower suspension control arm.

In Fig. 6, on the metal plate for mounting the upright, apart from the two coaxial hole (drilled to fix the upright), there is

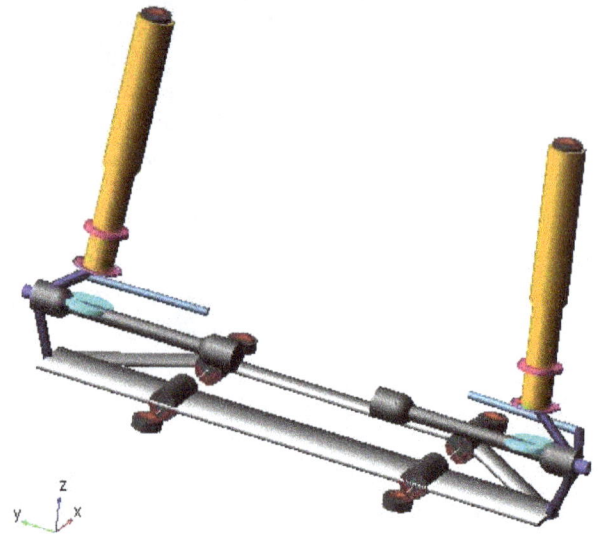

Figure 5. Front beam suspension in ADAMS/Car.

another small hole drilled on the vertical surface, for mounting the shock absorber, because drilling any more holes on the CFRP beam is not recommended for its reliability. The rear suspension is chosen to use a "H arm" topology[3] to eliminate the need for another linkage for toe control. The side effect is that the toe variation during suspension stroke is limited.

To recreate the performance of the beam suspension, the tool "non-linear beam" in ADAMS/Car is used, which is shown in Fig. 4. The non-linear beam is defined as a flexible body made with several connected deformable segments, which is an ideal component to present the behavior of beam spring in multi-body dynamics model.

3.2 Parametrization of beam spring

Suspension geometry has been defined using the reference model to reach the performance target. To reach the same performance after substitution with the beam, several tuning and modification of the model may be necessary. As the beam is relatively deformable compared to the rigid control arm in conventional solution, kinematic performance for beam suspension has some variance using the same geometry from reference model.

Since CFRP beam springs have been used to substitute the lower control arm, the hardpoints of the reference suspension need to be recreated in the CFRP beam spring. Other characteristics such as the length, thickness, cross-section shape and material properties used are very important for calculating the stiffness, damping and weight during simulation with the "non-linear beam" tool for CFRP beam.

---

[3]Such as the rear suspension of Gunnell (2008) Ford Thunderbird (1992–1996).

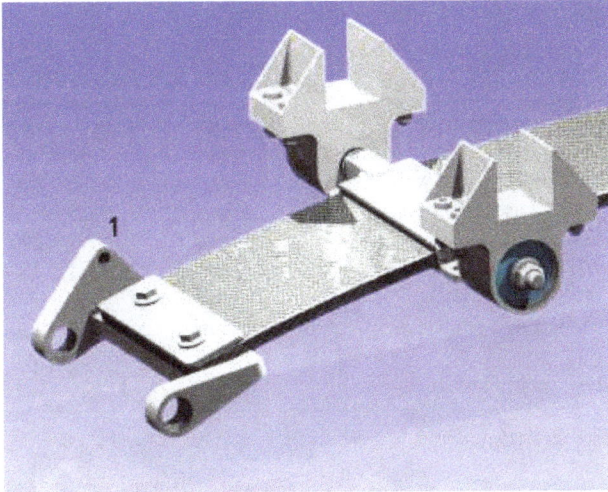

Figure 6. Drilled hole for mounting the shock absorber on rear beam (1).

To have the expected stiffness for the beam according to the performance design is very difficult, and for the FEM analysis, an initial geometry for the beam is required. In order to integrate the functionality of spring and anti-roll bar together in the single beam spring, it is difficult to obtain the desired result by manual "trial and error", considering high time consuming computational FEM simulations.

MBD simulation with DoE has been applied to obtain the correct design parameters, and the original MBD model was improved as the DoE needs parametrized model.

The beam spring has been modeled with simple parameters considering the cross section width and thickness (beam spring length is fixed by the overall design of the track width). During the development, it was intended to avoid cross section variation along the lateral direction. Variable thickness of the beam may lead to stress concentration (which increases the probability of delamination) during the deformation and production complexity will be increased as well.

It needs to be pointed out that in ADAMS/Car system, the geometry of the beam cannot be effectively controlled by the "parameter variable"[4], and so it is necessary to modify the model in code level.

Once the material properties[5] of CFRP are determined through experimental test, the only aspect could be modified in the design process is the moment of inertia (the cross-section). The non-linear beam component in MBD model reads the moment of inertia calculated by the parameter variables using simple formulas, which can be modified to create DoE iterations.

In order to perform the calculation for moment of inertia automatically during DoE iterations, additional scripts are necessary.

Front suspension has a large slot in the center of the beam component (giving space for electric motor and other components), which is modeled as two separate beam attached with each other at the beam ending (as shown in Fig. 2). The rear beam is also divided into two separate part to achieve H-arm structure on rear suspension. In such configuration, the beam components in the model are named as:

- Front suspension front beam (beam_front_nrl/nrr/nrs[6]_lca_front)

- Front suspension rear beam (beam_front_nrl/nrr/nrs_lca_rear)

- Rear suspension front beam (beam_rear_nrl/nrr/nrs[7]_lca_front)

- Rear suspension rear beam (beam_rear_nrl/nrr/nrs_lca_rear)

3.3 DoE script preparation

There are 3 different DoE scripts for this model:

- Generate parameter variables

- Modify beam component force calculation formula

- Create target measurement reading function

Nonlinear beam elements in a single beam are sharing the same geometry properties[8] (thickness and width). Even though user can modify the thickness and width by click on different beam segment (beam elements), the beam will act as a single component just using the last input value user configured in any individual beam segment.

When writing the script, it is mandatory to use "Adams/view Command", which is based on concept of "object-variable" and every variable belongs to its parent component. To make the variable accessible to DoE environment, users can choose any segment of a beam component, and reconfigure the variable of "height" (stands for thickness) and "width" into "parameter variables".

---

[4]The geometry can be controlled normally in the environment of template, but the simulation can only be done in the normal environment.

[5]The properties used in MBD model include: elastic modulus, shear modulus, Poisson ratio and density. Stacking sequence of CFRP beam spring is designed in FEM model.

[6]nrl: nonlinear beam left; nrr: nonlinear beam right; nrs: nonlinear beam single.

[7]nrl: nonlinear beam left; nrr: nonlinear beam right; nrs: nonlinear beam single.

[8]In Adams environment, properties of an object are called "VARIABLE", more like the "dependent variable" in general mathematic formula, its value is determined by other input values. "Parameter variables" are user-defined additional "parameters", more like the "independent variables", which give input values to the system.

It needs to be pointed out that "variables" in MBD model cannot be modified DoE iteration, but "parameter variables" can be accessed easily by user using graphic interface or command line same as DoE functions. Eight individual parameter variable were created in total (two for each beam component, four beams in total). Changing the variables into parameter variables makes the beam lose the original source of properties, the force calculation is no longer valid. The force calculating functions need to be manually redirected to the new parameter variables that have been modified.

The functions of beam component forces are:

- Ixx

- Iyy

- Izz

- Y_shear_area_ratio

- Z_shear_area_ratio

- Youngs_modulus

- Shear_modulus

- Area_of_cross_section

- Damping_ratio

The above functions can be modified using Adams/view Command.

The final step for the script preparation is to create "DoE design objective" using the function builder. For this research, the beam spring stiffness under different loading cases (parallel wheel travel/opposite wheel travel) is needed and is calculated using simple "force-displacement" function.

Another function is defined as an additional indicator for beam spring stiffness: the stiffness linearity. As the experiment results show the CFRP beam spring has a certain non-perfect linear stiffness behavior. It tends to increase its stiffness when the displacement is increasing. The minimal value of stiffness appears when the beam is not loaded, similar to a uniform cubic function passing through origin. DoE shows the results only when an objective function is defined, so there is no way to observe the non-linearity without a function. The non-linearity function is very simple in this case, the function calculates the ratio ($r_{nl} = k_{fs}/k_{hs}$) between the mean stiffness within $1/2$ the total wheel displacement (k_{hs}: ± 35 mm) and the stiffness at the maximum stroke (k_{fs}: ± 70 mm).

DoE simulation has to use only one simulation script for every iteration. In this case, the script need to perform parallel wheel travel and opposite wheel travel simulation in a single simulation run. The DoE script is written to perform first parallel wheel travel, then return the wheel back to its original position, at last the suspension will be asked to perform an opposite wheel travel.

Table 1. General set-up of vehicle model.

Property	Unit	Value
Loaded weight at SD A*	kg	785
Weight repartition (front/rear)	–	45/55
Maximum loaded weight	kg	870
Wheel base/ front track/rear track	mm	1912/1170/1170
Ground clearance at SD A	mm	140
H_{cg} at SD A	mm	350
Suspension stroke	mm	± 70
Minimum ground clearance at SD A	mm	120
Tire	–	95/80R16
Powertrain layout	–	FF

* One driver (75 kg) and one luggage (10 kg) are included.

The design objective functions are written as "time dependent" functions. For example, wheel center force (F_{CW}) for calculating the k_{hs} is read in four different moments using Adams/View function "VALAT (array,array,REAL)[9]":

- FCW at -35 mm for parallel wheel travel[10]

- FCW at $+35$ mm for parallel wheel travel

- FCW at -35 mm for opposite wheel travel

- FCW at $+35$ mm for opposite wheel travel

The other functions are written in the same way as for k_{fs} and r_{nl}.

4 Simulation results

The research is processed through reference model study to the beam suspension mode; the general set-up of the vehicle model is shown in Table 1. The kinematic characteristics and dynamic behavior has been defined through the simulation trails referred with specific performance targets for riding comfort and safety

DoE simulation is performed only to identify the geometry size of beam components in this article. The kinematic simulation is geometry-dependent, while tuning is based on suspension performance targets and CAD design space. Dynamic simulation is more component-dependent, as the stiffness and damping of elastic components can significantly change the dynamic response.

---

[9]The function VALAT is a linear interpolation for arrays, the measurement results are arrays for example.

[10]Adams/View function is written as: VALAT (ANALYSIS.left_hub_forces.TIME,ANALYSIS.left_hub_forces.normal, TIME). "TIME" is the real number for the moment when wheel hub reaches -35 mm wheel travel.

Main effects for response

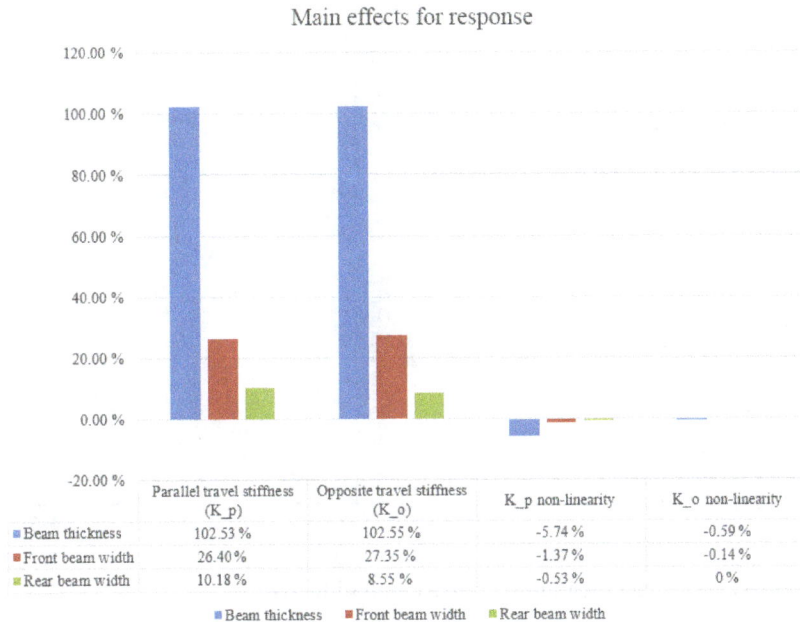

	Parallel travel stiffness (K_p)	Opposite travel stiffness (K_o)	K_p non-linearity	K_o non-linearity
■ Beam thickness	102.53 %	102.55 %	-5.74 %	-0.59 %
■ Front beam width	26.40 %	27.35 %	-1.37 %	-0.14 %
■ Rear beam width	10.18 %	8.55 %	-0.53 %	0 %

■ Beam thickness   ■ Front beam width   ■ Rear beam width

Figure 7. Effect for response.

4.1 DoE simulation for suspension system

The beam suspension system is designed using DoE with respect to the reference suspension model (conventional McPherson and SLA) performance target. The stiffness for front/rear beam spring are designed considering the first resonance frequency of the vehicle and vehicle designed load. The anti-roll stiffness, which is known as opposite travel stiffness in this paper is designed with respect to minimize the maximum rolling angle under maximum lateral acceleration and rolling resonance during dynamic maneuver.

During the DoE simulation, the beam stiffness in front suspension for both front and rear beam component in the model are bounded together to have the same value. In the manufacturing phase, the beam for each suspension subsystem (front/rear suspension) should be cut and formed with a single piece of stacked carbon fiber laminate.

The parameter variables have been set within the range considering the available space inside the motor compartment using uniform distribution.

– Beam thickness: $10\,mm \pm 2\,mm$

– Front beam width: $90\,mm \pm 20\,mm$

– Rear beam width: $40\,mm \pm 10\,mm$

The simulation has been performed with Latin Hyper cubic method with 200 iterations, calculating time: 4 h. The effects for response is shown in Fig. 7.

"Beam thickness" influence is most significant as shown in Fig. 7, "beam width" is less significant for its smaller contribution in moment of inertia under suspension-loaded condition, and wider beam cannot be used in the application con-

sidering the available space. The "parallel wheel travel nonlinearity" shows an inverse correlation with the beam thickness, the thicker the beam, the non-linearity decreases in the beam spring, which means the spring stiffness is more constant over the whole deformation range (for spring stiffness). It is very interesting to point out that the beam thickness has negligible effect on the opposite travel stiffness linearity (linearity of anti-roll stiffness), it is not clear if the mounting structure used on the beam is the cause of this phenomenon. Unfortunately, it is not possible to investigate more on this phenomenon in MBD solution, which is the main concern of this article.

In the end, the front suspension beam is designed as 100 mm wide on the front, 40 mm wide on the rear and 9.75 mm thick with a parallel/opposite stiffness nonlinearity of 1.151/1.048. It is the best compromise between manufacturing cost and stiffness requirements. The stiffness comparison is available in Fig. 8.

4.2 Suspension sub-system kinematic simulation

The kinematic characteristics are focused on the following aspects and the comparison between reference suspension and beam suspension of the simulation results are shown:

– Camber variation

– Toe variation

– Caster

– Beam spring stiffness

– Quasi-static constant radius cornering

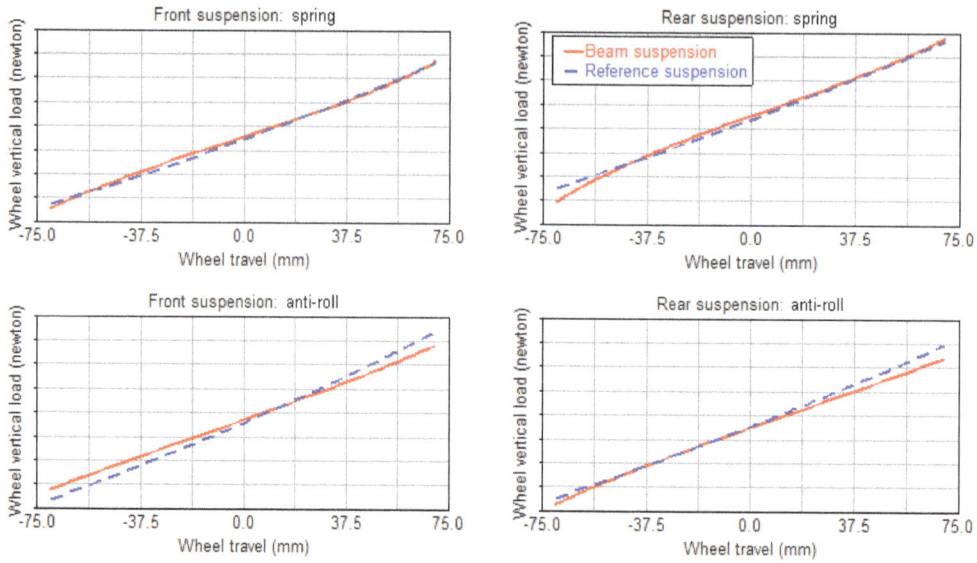

Figure 8. Beam spring stiffness tuning (compression stiffness and anti-roll stiffness).

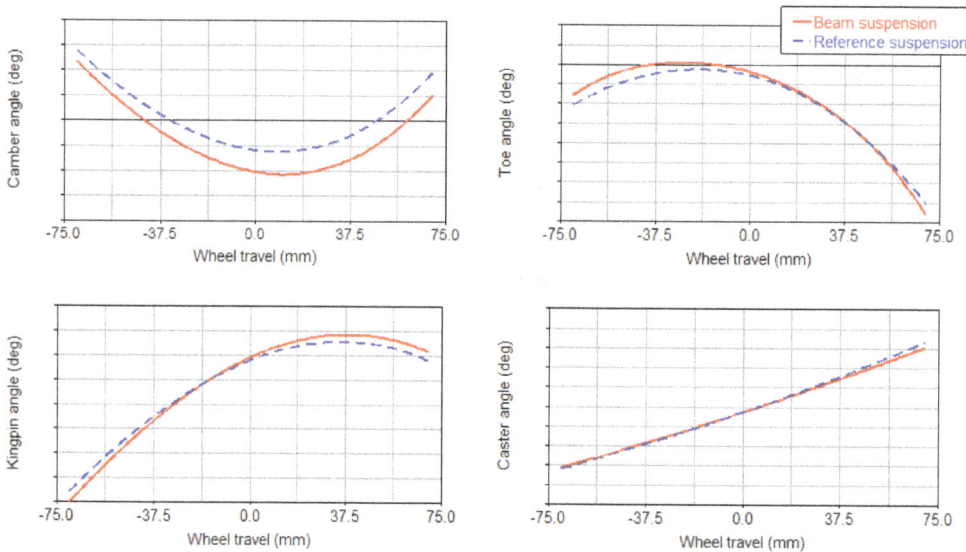

Figure 9. Front suspension kinematics.

Simulations have been done for both reference model and beam model, using the common parallel wheel travel on front and rear suspension. According to the common experience of chassis design for passenger cars (Gunnell, 2008), the expected performance has the following values in Table 2.

After the reference model has been developed, the beam model is set-up with all the coordinates of hardpoints from reference model, afterwards the tuning of beam model has been done to fit the performance of the reference model. The comparisons of results are shown in the Fig. 9 for the front suspension and Fig. 10 for rear suspension.

As it can be seen, the kinematic characteristics are perfectly reproduced for the beam suspension, such as camber, kingpin, toe-in/out and caster angle variations[11].

Beam spring stiffness has been carefully tuned with respect to FEM analysis. The correlation for the stiffness has been compromised[12] for the reason of manufacturing cost.

---

[11]Correlation factor is more than 99 % between the reference curve and beam suspension result, using "Pearson correlation method" (Pearson, 1985)

[12]In order to keep the riding height under different loading condition respect to the reference model, the beam spring stiffness has been set as the primary target, while the anti-roll stiffness is compromised.

Table 2. Performance target* for beam suspension.

Front suspension	Unit	Range		
Static Camber angle	deg	−0.80	to	−0.30
Anti-Dive Angle	deg	5.00	to	7.00
Anti-lift Angle	deg	−1.00	to	−0.20
Ride Steer at ±10 mm bump	deg $100\,\text{mm}^{-1}$	−1.60	to	−0.40
Ride Camber at ±10 mm bump	deg $100\,\text{mm}^{-1}$	−2.00	to	−1.00
Rear suspension	Unit	Range		
Static Camber Angle	deg	−1.00	to	0
Anti-Dive Angle	deg	5.00	to	7.00
Ride Steer at ±10 mm bump	deg $100\,\text{mm}^{-1}$	0.40	to	1.60
Ride Camber at ±10 mm bump	deg $100\,\text{mm}^{-1}$	−3.00	to	−1.50

* The reference suspension has been designed respect to these requirements.

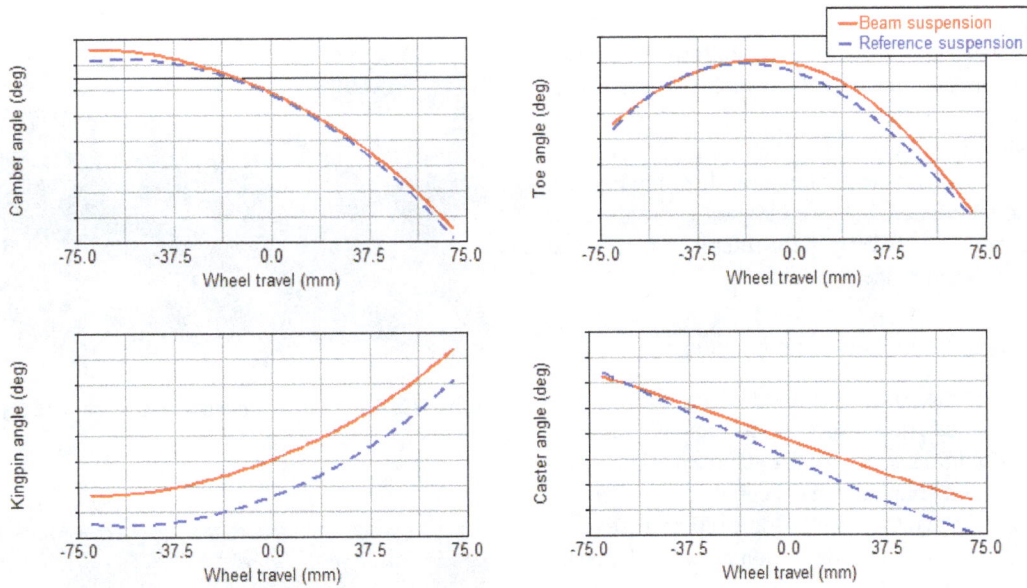

Figure 10. Rear suspension kinematics.

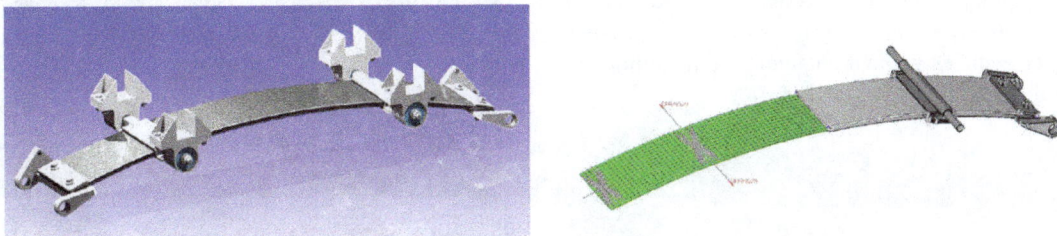

Figure 11. CAD model (left) and FEM model (right), the rear suspension has H arm layout and the mounting points are located near the beam geometry centre using bushings.

Table 3. Swept-sine steering simulation result comparison.

Yaw rate – Steering angle[a]	Unit	Reference model	Beam model
Steady Gain	s^{-1}	0.40	0.43
Gain at 0.5 Hz	s^{-1}	0.43	0.52
Delay at 0.5 Hz	s	0.029	0.029
DF[b]	–	1.20	1.23
Roll angle – Lateral acceleration	**Unit**	**Reference model**	**Beam model**
Steady Gain	$\deg g^{-1}$	4.01	4.76
Gain at 0.5 Hz	$\deg g^{-1}$	4.06	4.06
DF	–	1.74	1.05
Slide slip angle – Steering angle	**Unit**	**Reference model**	**Beam model**
Steady Gain	–	0.044	0.064
Gain at 0.5 Hz	–	0.047	0.073
Delay at 0.5 Hz	s	0.81	0.80
DF	–	1.07	1.03

[a] INPUT – OUTPUT; [b] Damping factor = Gain(max)/Gain(steady).

The beam has been designed to have uniform thickness. Complex lamination will increase the complexity and uncertainty during manufacturing.

The rear suspension has been designed using the "H arm", the toe angle cannot be efficiently tuned as shown in Fig. 10.

The skid pad simulation[13] showed a good correlation for under-steering ratio, $37.19\ \deg g^{-1}$ for reference model and $36.26\ \deg g^{-1}$ for beam model.

4.3   Dynamic simulation-swept sine steering

The swept sine steering is very important for understanding the full vehicle steering response on frequency domain (Fisher, 2011). The initial speed for the vehicle has been chosen as $80\ km\,h^{-1}$ and simulation has been performed covering the frequency range 0.05 to 5 Hz. Some representative results are shown in Table 3, the correlation between the beam suspension and reference suspension is good.

At this moment, it is clear that the beam spring suspension proposed can reproduce with a good agreement the performance of the reference suspension (conventional type) in terms of global handling. Which means it is possible to substitute the conventional suspension (coil spring and metallic control arms) with CFRP beam spring suspension for standard passenger vehicles without changing the handling performance of the existing ones.

Figure 12. Closed mould filled with Carbon fibre ply and resin.

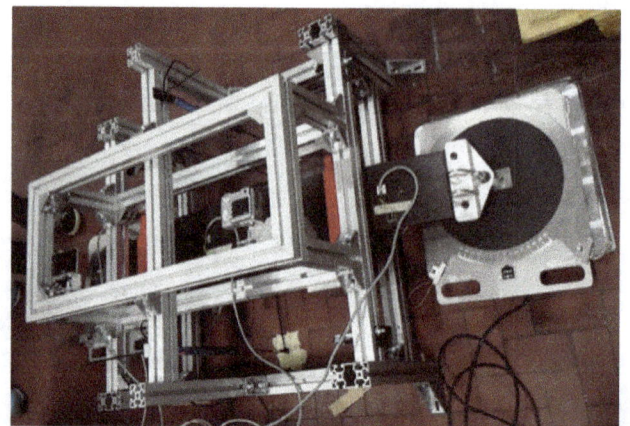

Figure 13. CFRP beam stiffness test bench.

---

[13] Skid pad test with a radius of 40 m and simulate the vehicle from steady until reaching 1 g lateral acceleration.

Table 4. Improvements respect to conventional aluminum made suspension system.

Improvements	Unit	Value	Difference
Front suspension weight reduction	kg	−10.2	−29.1 %
Rear suspension weight reduction	kg	−10.7	−26.8 %
Wheel to strut distance at front suspension	mm	−17	−34 %
Wheel to strut distance at rear suspension	mm	−22	−44 %
Strut mounting height at front suspension	mm	−11	−7.4 %
Front spring seat plate	–	Removed, no spring needed	−100 %
Rear spring seat plate	–	Removed, no spring needed	−100 %

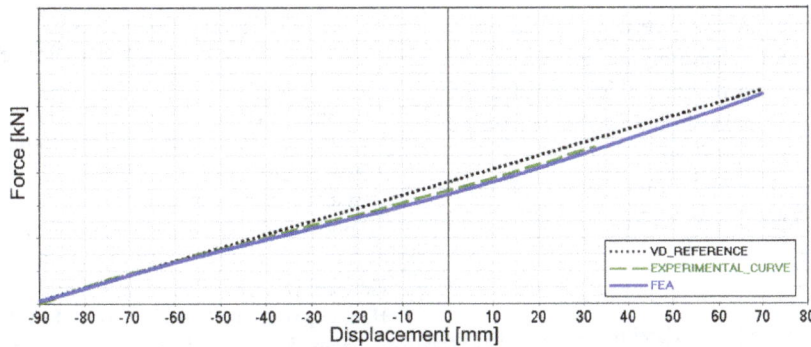

Figure 14. Comparison between FEM analysis result, test bench experimental data and Vehicle Dynamic design target.

5 Building the Beam suspension

Using the approved vehicle-dynamics simulation results, the characteristics of the beam components were approximately defined and the first component CAD was created.

FEM analysis has been done to obtain the optimum ply sequence (55 layers) and orientation (Fig. 11).

The beam has been produced completely manually, by using a special vacuum bag technology and autoclave process (Uddin, 2013) to ensure the correct polymerization without any defects in the melt resin. The mold is floating with the carbon fiber to give the better surface finishing and ensure the uniform pressure on the whole piece (shown in Fig. 12).

After cleaning the flanges and drilling for addition components (joint housing), the component is mounted on the chassis prototype.

A simple test bench for evaluating the beam spring stiffness has been made (Fig. 13), with five strain gauges mounted on the spring and weight loaded on the bushing mountings. In Fig. 14, the final Force-Displacement correlation for the beam spring is shown. The displacement starts from −90 mm to generate correct preload at normal loaded position. The wheel travel will be limited with the rebound-stop integrated in the shock absorber, in order to guarantee the negative wheel travel won't exceed the lower limit (−70 mm).

The experimental data shows that the beam spring has successfully manufactured with respect to the design target for the beam suspension model.

6 Conclusion

In this research, an innovative construction of suspension system has been developed. Using the CFRP beam, the function of lower control arm, coil spring and anti-roll bar has been integrated into a single lightweight composite component. The solution provides new possibilities for saving space and weight reduction (comparison data with conventional vehicle on the market is presented in Table 4).

For front suspension tower, the spring plate has been removed since there is no coil spring present, which benefits the frontal vehicle part design in styling smoother curved surface. When using McPherson suspension on the front, the coil spring is usually placed over the tire to make the strut close to the wheel (decrease scrub radius while keep the kingpin more vertical), the mounting point at suspension tower is very high to fit the coil spring preloaded length. Using CFRP beam suspension can completely solve this issue, the suspension tower can be as low as the shock absorber stroke with additional space for bump stop, the mounting point is lowered more than 10 mm in this case.

Weight reduction is the main benefit for using CFRP beam suspension. More than 25 % weight is reduced[14] compared to the conventional suspension arm, coil spring and antiroll bar with their mountings.

---

[14]The suspension beam with mounting weighs less than 5 kg for one axle.

MBD model has been developed with reference to the performance target. FEM analysis using exact material property was done to predict the component behavior and to design the CAD details for production.

For the future development, the study will be focused on the riding comfort, considering the vibration filtering, the noise proofing and impact smoothing while driving on uneven surface conditions. Also complex lamination design will be studied to improve the correlation with respect to the conventional suspension. Modeling compliant modules or joints (Li and Hao, 2017) in the mounting system for beam spring will be furtherly studied.

The DoE method provide an effective way to find the aspect that is more significantly changing the behavior of the beam spring. The phenomenon of beam thickness effecting non-linearity of anti-roll stiffness should be studied further more.

Production process design will be done later on considering mass production method like RTM and compression molding.

Author contributions. Massimiliana Carello carried out the CFRP beam concept and designed the suspension. Andrea Airale performed the composite material characterization, analysis and production process study. Shuang Xu made the MBD model and completed the suspension fine-tuning. Alessandro Ferraris is in charge of DoE model setting and results analysis.

Competing interests. The authors declare that they have no conflict of interest.

Acknowledgements. This work was supported by the laboratory of Innovative Electric and Hybrid Vehicle DIMEAS – Politecnico di Torino cooperating with BEOND Spin Off (http://www.beondrive.it) and FCA Group.

Edited by: G. Hao

References

Beardmore, P.: Composite structures for automobiles, Compos. Struct., 5, 163–176, 1986.

Carello, M., Filippo, N., and D'Ippolito, R.: Performance optimization for the XAM hybrid electric vehicle prototype, SAE International Conference, 24–26 April 2012, doi:10.4271/2012-01-0773, Detroit, 2012.

Carello, M., Airale, A., Ferraris, A., and Messana, A.: XAM 2.0: from student competition to professional challenge, Computer-Aided Design and Applications, Taylor & Francis, 11, 61–67, doi:10.1080/16864360.2014.914412, 2014.

European Aluminum Association: The Aluminum Automotive Manual: Application – Car body – Body structures, ©European Aluminium Av. de Broqueville, Brussels, 2013.

Fisher, D.: Handbook of driving simulation for engineering, medicine, and psychology, CRC Press, Boca Raton, 2011.

Fuganti, A. and Cupitò, G.: Thixoforming of aluminum alloy for weight saving of a suspension steering knuckle, Metallurgical science and technology, Vol. 18, available at: http://www.gruppofrattura.it/ors/index.php/MST/article/viewFile/1058/1011 (last access: 20 February 2017), 2000.

Gunnell, J.: Standard catalog of Ford, 4th Edn., KP, Iola, Wis., 2008.

Hao, G., Yu, J., and Li, H.: A Brief Review on Nonlinear Modelling Methods and Applications of Compliant Mechanisms, Frontiers of Mechanical Engineering, 11, 119–128, 2016.

Heissing, B.: Chassis handbook fundamentals, driving dynamics, components, mechatronics, perspectives, Vieweg Teubner, Wiesbaden, 2011.

Kulshreshtha, A.: Handbook of polymer blends and composites, 1st Edn., RAPRA Technology, Shrewsbury, 2002.

Li, H. and Hao, G.: Constraint-force-based approach of modelling compliant mechanisms: Principle and application, Precis. Eng., 47, 158–181, 2017.

Milliken, W. and Milliken, D.: Racecar vehicle dynamics, SAE International, Warrendale, PA, USA, 1995.

Pearson, K.: Notes on regression and inheritance in the case of two parents, P. Roy. Soc. Lond., 58, 240–242, 1895.

Uddin, N.: Developments in fiber-reinforced polymer (FRP) composites for civil engineering, Woodhead Publishing Limited, Cambridge, UK, 2013.

Wood, K.: Composite leaf springs: Saving weight in production suspension systems: Composites World, available at: http://www.compositesworld.com/articles/composite-leaf-springs-saving-weight-in-production-suspension-systems (last access: February 2017), 2014.

Yu, W. J. and Kim, H. C.: Double tapered FRP beam for automotive suspension leaf spring, Compos. Struct., 9, 279–300, 1988.

11

Energy calculation model of an outgoing conveyor with application of a transfer chute with the damping plate

Vieroslav Molnár[1], **Gabriel Fedorko**[1], **Nikoleta Husáková**[1], **Ján Král' Jr.**[2], **and Mirosław Ferdynus**[3]

[1]Technical University of Kosice, Letna 9, 042 00 Kosice, Slovak Republic
[2]Faculty of Mechanical Engineering, Technical University of Kosice, Letna 9, 042 00 Kosice, Slovak Republic
[3]Faculty of Mechanical Engineering, Lublin University of Technology, Nadbystrzycka 36,
 20-616 Lublin, Poland

Correspondence to: Vieroslav Molnár (vieroslav.molnar@tuke.sk)

Abstract. The energy efficiency of transport systems consisting of several belt conveyors is significantly affected by re-direction. The proper sizing of several conveyor belts using deflector plates can significantly affect their efficiency. At present, there are no uniform rules (models) which specify the methodology and procedures for their design. This paper brings proposals of design of optimal parameters for energy-efficient operation of the transport system consisting of belt conveyors based on the new analytical simulation models. Recommendations for the practical application of transport systems at the transfer point have been designed according to optimization. The results are analysed in detail in three phases of shifting by means of a physical approach with the support of computing methods and simulation experiments with the transfer model. We can state that the direction and orientation of material impact have a direct influence on the conveyor's energy intensity. Thus, the inevitable condition for operation of arbitrary type of belt conveyor is to pay greater attention to the construction of the transfer model, particularly the intensity of the energy of the outgoing conveyor.

1 Introduction

Material handling is an important sector of industry (Zhang and Xia, 2011). Belt conveyors are being employed to form the most important parts of material handling systems because of their high efficiency of transportation (Alspaugh, 2004). Belt conveyors are, in most cases, the most cost-effective solution for handling bulk material mass flows over short and medium conveying distances. The belt is a key component of these conveyors, and its dynamic characteristics determine the working performance to a great extent (Hou and Meng, 2008). Belt conveyors are basic intra-plant transport machines, especially in the mining industry. Due to their numerous advantages, belt conveyors are also used in other industries, such as in natural resource processing, smelting, cement and lime production, pulp and paper production, sea and river ports, civil engineering, agriculture, sugar factories, and power plants. The reason for this is that belt conveyors are simple in construction, flexible in trans-

port system configuration, and versatile in use, and they also may be used to transport goods over considerable distances (Mazurkiewicz, 2008). Wensrich and Wheeler researched the method of optimization for buckling occurring on a conveyor belt (Wensrich and Wheeler, 2004). During the tumbling and movement process, motion resistance occurs, which has been described in detail (Spaans, 1999). The knowledge obtained is necessary to create the energy calculation model. The model for energy calculation of the belt conveyor is much needed for the optimization of its operating efficiency. There are two categories of models in the literature: one relying on resistance force calculation and the other on energy conversion through a compensation length. Zhang and Xia (2009) proposed a model which evolved by the interlinkage of the two categories. This model is characterized by two compensation length variables. Evidence is provided in comparative studies in terms of better accuracy and applicability.

One of the most important areas of bulk solids handling is the efficient flow of materials at transfer points within the system. Bulk material transfer points are found in a wide range of industries, including mining, mineral processing, chemical processing, thermal power plants, and many other areas that deal with bulk solids (Huque and McLean, 2002). Scott and Choules (1993) suggest the importance of the application of damping plates or impact boards as being the most important factor regarding their wear and tear. The simple use of damping plates and their effectiveness in terms of functionality, simple maintenance, and maximum lifetime of the transfer point is a widely discussed topic. The authors suppose that the angle under which the material impacts the damping plate can significantly affect its wear. Their article presents experiments carried out in a mining enterprise in which the transport performance of 40 000 t per day resulted in the excessive wear of damping plates. The results of the experiments indicated that modification of the damping plate shape to a certain angle of curving resulted in lower wear by impact of a high abradant. Roberts (2003) presented criteria for the selection of the most appropriate chute geometry to minimize chute wear and belt wear at the feed point. He dealt with the determination of optimum chute profiles to achieve the specified performance criteria. Benjamin et al. (2010) analysed the problem of transfer chutes as well as the design concept, maintenance, and operation of the transport system, with practical examples in broad terms. Scott (1992) analysed conveyor transfer chute design, modern concepts in belt conveying, and handling of bulk solids. Roberts (unpublished data) described how the relevant flow properties of bulk solids are measured and applied to chute design. Chute flow patterns and the application of chute flow dynamics to the determination of the most appropriate chute profiles to achieve optimum flow are described. Wensrich (2003) researched a small part of chute design, i.e. the choice of the curve or profile that the chute follows. Overall, this choice relies on minimizing wear in the chute/product and impact on the belt. Seifried et al. (2005) presented a method to evaluate the coefficient of restitution for multiple impacts between material bodies.

The most widespread applied method examining interaction of material flow particles during the re-direction by the help of transfer chutes is the discrete element method (DEM), and tools are developed according to this simulation. Simulation experiments on the basis of this method are presented by most of the authors in order to optimize and design elements of transfer chutes. Bertrand et al. (2005) discussed the current state of the art in the modelling of granular flows in mixing processes; the authors then focus on the DEM, which has recently proven worthy of interest for the mixing of granular materials. There are numerous methods available with which to analyse particle flow through a conveyor transfer, including continuum-based analytical methods, the DEM, and experimental analysis. Research is presented by Gröger and Katterfeld (2007), Kessler and Prenner (2009), Grima and Wypych (2010, 2011), and Dewicki (2003) regarding the interaction caused by the contact of particles of material flow

during impacts on the walls of transfer chutes or damping plates. Results of simulation experiments indicate that the shape, material, and construction of transfer chutes affect the structure and deformation of spherical particles of the material as a result of their mutual contact. Minkin (2012) reports about the application of the DEM for an analysis of the "rock box" transfer stations, which are widely used in the mining industry. David and Kruses described input parameters obtained by simulation with current application of the DEM (for example, static or dynamic loads of material, material density, adhesion and cohesion of the particles of material, restitution coefficient, viscosity). Di Renzo and Di Maio (2004) examined the evolution of the forces, velocities, and displacements during the collision, emphasizing the importance of correctly accounting for non-linearity in the contact model and micro-slip effects. Chandramohan and Powell (2005) researched preliminary numerical simulations using the standard viscous damping model are performed using the DEM and a comparison between experimental and numerical results. Bierwisch et al. (2009) investigated the rapid granular flow from a moving container and angle of repose formation by means of numerical simulations using the DEM and experiments. McIlvenna and Mossad (2003) reported an investigation into a continuum model based on kinetic theory as an alternative to the DEM for basic transfer chute analysis. Software for the fluent analysis has been used to perform two-dimensional modelling of chutes, and the result is compared to traditional design methods. Hastie and Wypych (2010) detailed their findings regarding three methods for granular cohesionless materials. The experimental investigations were performed on a conveyor transfer research facility located at the University of Wollongong, using high-speed video to capture the flow and subsequently analysing this with Image Pro Plus. Donohue et al. (2010) modelled granular flow through a constant radius chute using computational fluid dynamics (CFD) in which a multiphase simulation is considered. This approach provides valuable information about the flow of dust, and the results are confirmed by experimental observations. Jensen et al. (1999) presented an enhanced DEM for the numerical modelling of particulate media. This method models a particle of general shape by combining several smaller particles of simpler shape, such as a circle, into clusters that act as a single larger particle. The clusters more accurately model the geometry-dependent behaviour of the particles, such as particle interlock and resistance to rolling. The method is implemented within the framework of an existing DEM program without the introduction of new contact or force algorithms. Zhang and Xia (2011) researched the model-based optimization approach to improve the efficiency of belt conveyors at the operational level.

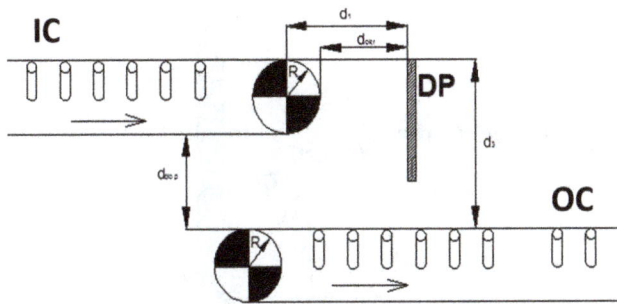

Figure 1. Simplified model of material re-direction. DP – damping plate; IC – incoming conveyor; OC – outgoing conveyor.

2   Energy calculation model

The energy calculation model of the outgoing conveyor with application of a transfer chute with the damping plate was determined by a basic model of transport system consisting of

1. two belt conveyors with the belt (transbelt) with a one-pulley drive suitable for transport of granular and bulk material,

2. a firmly and perpendicularly gripped damping plate of metal material aligned to the x axis of the incoming conveyor belt (Fig. 1).

When we examine kinematic and dynamic effects on the basis of theoretical analysis, we consider the basic model of the conveying systems with re-direction of one particle of granular material, not with material flow of particles, or pieces. Incoming and outgoing conveyors have the same direction and orientation of movement, and the drive pulley has the same diameter.

2.1   Input parameters of the basic model of re-direction

For the calculation of kinematic and dynamic effects of the basic re-direction model, the following technical and selected parameters have been used:

1. speed of conveyor belts v_0,

2. radius of pulleys R for rubber–textile conveyor belts,

3. horizontal distance of the damping plate from the incoming conveyor d_{okr},

4. vertical distance among conveyors d_3,

5. mass of particles for granular bulk material m,

6. coefficient of restitution ε.

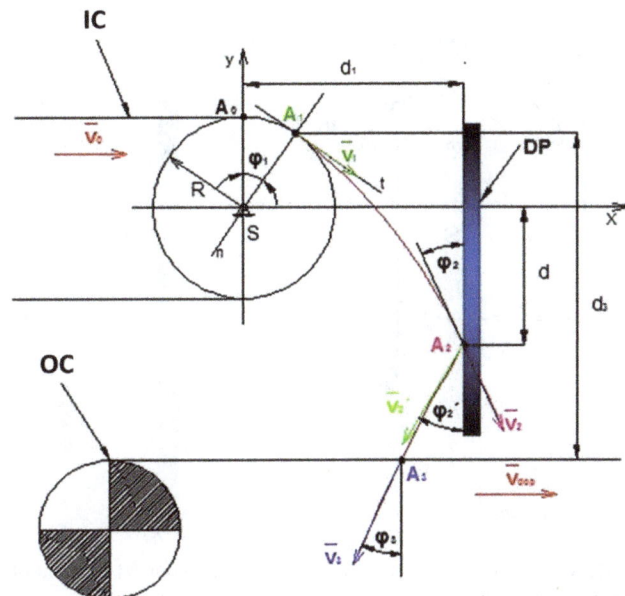

Figure 2. Trajectory of material particle movement during three phases. DP – damping plate; IC – incoming conveyor; OC – outgoing conveyor.

2.2   The goal of the kinematic and dynamic analysis

The analysis carried out was intended to research the effect of mechanical energy of a material particle on the energy demand of an outgoing conveyor during three phases (Fig. 2).

The first phase was aimed at the separation of the particle from the drive pulley of the incoming conveyor (IC) at point A_1 and the movement of the free material particle for impact on the damping plate (DP) from point A_1 to point A_2. It was assumed at the interval of movement A_0A_1 that the adhesion among bulk material and conveyor belt is such that there is no relative movement among the particle of bulk material and conveyor belt; therefore,

$$\overline{v_1} = \overline{v_0}, \tag{1}$$

where $\overline{v_0}$ is the speed of the conveyor belt [m s^{-1}] and $\overline{v_1}$ is the speed of the particle for bulk material [m s^{-1}].

In the interval A_1A_2, the trajectory of the material particle (theoretical depiction of movement kinematics) is at an angle with input values v_1, φ_1. During this movement, if we neglect air resistance, the mass point is affected only by the force of gravity $\overline{G}$ (Fig. 3).

The second phase focuses on the material particle's impact on the DP and reflection → A_2. At the moment of the material particle meeting with a fixed damping plate, the phenomenon of impact, in terms of mechanics, occurs. In the moment of impact on the mass point, the impact force $\overline{F_R}$ occurs. The impact force $\overline{F_R}$ is the result of surface load which occurs on the contact surface after the impact of the material particle with the damping plate (for our case, point A_2).

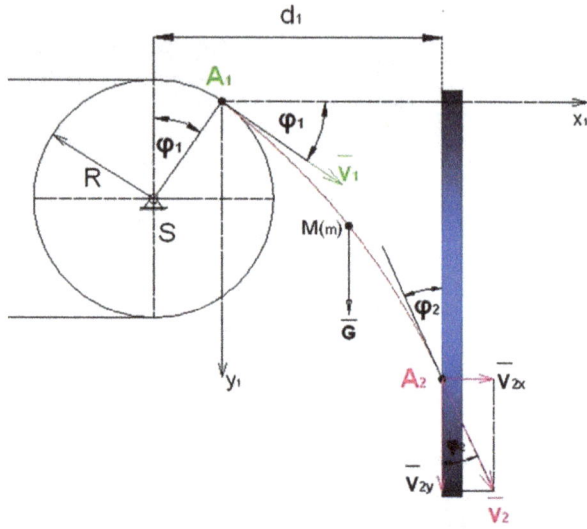

Figure 3. Movement of material particle from point A1 to point of impact A2.

Its carrier is a normal line which faces in a perpendicular direction towards the damping plate (Fig. 4). Resulting from Fig. 4, the change in material particle momentum is only at the direction of the x axis, so the material particle has an imperfect elastic central impact. Because the position of the damping plate is at rest, based on the theory of direct central impact, their common speed at the end of both phases is zero. In the direction of the y axis, the material particle performs an imperfectly elastic direct central stroke and the position of damping plate is unchanged. The velocity component of the material particle in the direction of the y axis stays unchanged.

Furthermore, the dynamics of the impact are affected by different elastic material properties of the mass point and damping plate, the effect of which is expressed as the coefficient of restitution ε. To calculate the input parameters of the model of rebound, the values for the coefficient of restitution are selected in the range $0 < \varepsilon < 1$.

The third phase describes the impact of the material particle on the outgoing conveyor (OC) $\rightarrow$ A$_3$. Figure 5 presents the movement of the mass point from position A$_2$ to position A$_3$ as well as the resolution of the velocity components at the point of the material particle's impact on the OC as a projection at an angle. To calculate the distance of the impact of material particle x_3 on the OC from the damping plate, value d_2 was used as defined in the equation in Fig. 5.

These phases of shifting were analysed as three separate tasks in which material particle movement analysis was oriented to the determination of input values with the following defined equations:

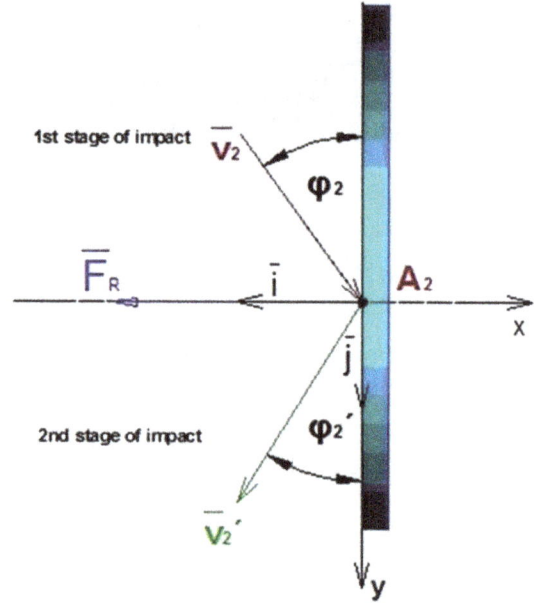

Figure 4. Impact force at the moment of the material particle impacting the damping plate and immediately after the impact.

– The angle of bulk material particle separation from the drive pulley of IC φ_1 according to CEMA (2007):

$$\cos\phi_1 = \frac{v_1^2}{g \cdot R} \Rightarrow ar\cos\phi_1 = \frac{v_1^2}{g \cdot R}[°]. \quad (2)$$

– Coordinates of bulk material particle impact $[x_2, y_2]$ at time t_2:

$$x_2 = v_1 \cdot t_2 \cdot \cos\phi_1 \, [\mathrm{m}], \quad (3)$$

$$y_2 = g \cdot \frac{t^2}{2} + v_1 \cdot t_2 \cdot \sin\phi_1 \, [\mathrm{m}], \quad (4)$$

$$t_2 = \frac{d_1 - R \cdot \sin\phi_1}{v_1 \cdot \cos\phi_1} \, [\mathrm{s}]. \quad (5)$$

– The size of bulk material velocity at the time of impact on the DP v_2:

$$v_2 = \sqrt{v_{2x}^2 + v_{2y}^2} \, [\mathrm{m\,s^{-1}}]. \quad (6)$$

– The angle of velocity direction v_2, or the angle φ_2, under which the bulk material particle impacts the DP:

$$\tan\phi_2 = \frac{v_{2x}}{v_{2y}} \Rightarrow \phi_2 = \arctan\frac{v_{2x}}{v_{2y}}[°]. \quad (7)$$

– The angle of bulk material reflection from the DP $\varphi_{2'}$:

$$\tan\phi_{2'} = \varepsilon \cdot tg\phi_2 \Rightarrow \phi_{2'} = \arctan(\varepsilon \cdot \tan\phi_2)[°]. \quad (8)$$

– The size of bulk material velocity $v_{2'}$ after reflection from DP:

$$v_2' = v_2\sqrt{\sin^2\phi_2 + \varepsilon^2 \cdot \cos^2\phi_2}[\mathrm{m\,s^{-1}}]. \quad (9)$$

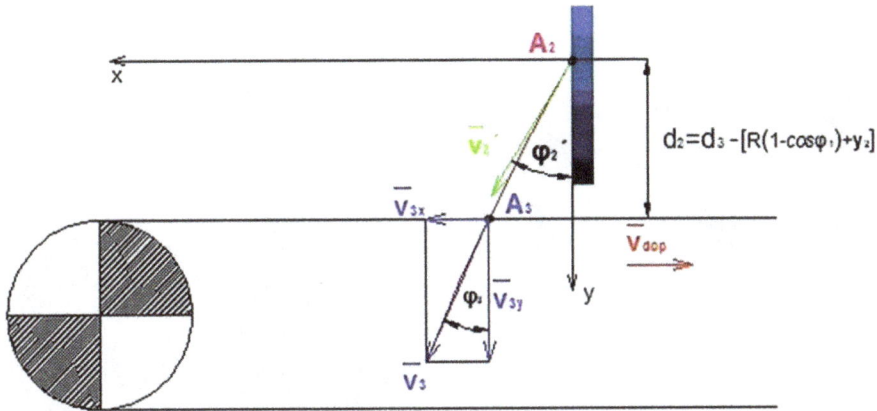

Figure 5. Decomposition of velocity components at the point of the material particle's impact on the OC.

– Coordinates of bulk material particle impact $[x_3, y_3]$ at time t_3:

$$x_3 = v_{2'} \cdot t_3 \cdot \sin\phi_{2'} [m], \qquad (10)$$

$$y_3 = g \cdot \frac{t_3^2}{2} + v_{2'} \cdot t_3 \cdot \cos\phi_{2'} [m], \qquad (11)$$

$$t_3 = \frac{-b \pm \sqrt{b^2 - 4 \cdot a \cdot c}}{2a} [s], \qquad (12)$$

where $a = \frac{g}{2}$, $b = v_{2'} \cdot \cos\phi_{2'}$, and $c = [R \cdot (1 - \cos\phi_1) + y_2] - d_3$.

– The size of velocity of bulk material particle at the time of impact on the belt of OC v_3:

$$v_3 = \sqrt{v_{3x}^2 + v_{3y}^2} [m\,s^{-1}]. \qquad (13)$$

– The change in kinetic energy E_k of the material particle after the impact on the OC:

$$\Delta E_k = \frac{1}{2} \cdot m \cdot \left(v_{3x}^2 + v_{dop}^2\right) [J]. \qquad (14)$$

– Instantaneous engine power of the outgoing conveyor P:

$$P = \frac{m}{2 \cdot t} \cdot \left(v_{3x}^2 + v_{dop}^2\right) [W]. \qquad (15)$$

The angle under which the material impacts the damping plate markedly affects its wear, as also mentioned in Scott and Choules (1993). The correct setup of the transfer chute minimizes wear in the chute/product and impact on the belt of the outgoing conveyor.

2.3 Comparison of energy calculation model for the velocity $v_0 = (2\,m\,s^{-1}; 4\,m\,s^{-1})$

The energy calculation model of the outgoing conveyor compares the change in kinetic energy of a material particle after

hitting the OC with dependence on the radius of a drive pulley by doubling of the conveyor velocity. The change in conveyor velocity has an equally strong influence on the perpendicular and vertical distance of the material particle's impact on the damping plate and OC as well as the change in the material particle's velocity during the three phases of shifting, with perceptible differences in the values of impact angles, rebounds, and impacts.

For the calculation of the investigated kinematic and dynamic effects by interaction of material with damping plate, or material with the conveyor belt, the following technical and selected parameters of the transfer model were used:

1. normalized velocities of conveyor belts;

2. diameters of pulleys for rubber–textile conveyor belts with allowable stress in tensile force of up to 30 % and with a polyamide insert according the manufacturer's recommendations;

3. horizontal distance of the damping plate from the incoming conveyor d_1, which is equal to the sum of the values of the interval $d_{okr} = <0.5\,m \div 1\,m>$ and values of the radius of the drive pulley R;

4. consistently chosen vertical distance among conveyors d_{dop}, with the distance d_3 determined as

$$d_3 = R \cdot (1 - \cos\phi_1) + y_2 + d_2 [m], \qquad (16)$$

where R is the radius of the drive pulley of IC [m], ϕ_1 is the angle of material particle separation from the drive pulley [°], and y_2 is the y coordinate of the material particle from the place of separation on the upward-carrying pulley to the place of impact on the damping plate [m];

5. weight of bulk material particle $m = 7.95\,g - 8\,g$;

6. coefficient of restitution $\varepsilon = 0.5$.

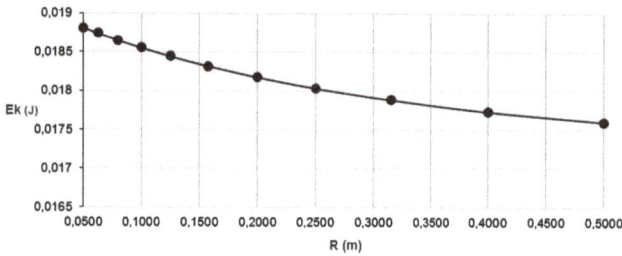

Figure 6. Dependence of kinetic energy E_k and radius R of the drive pulley at velocity $v_0 = 2\,\mathrm{m\,s}^{-1}$.

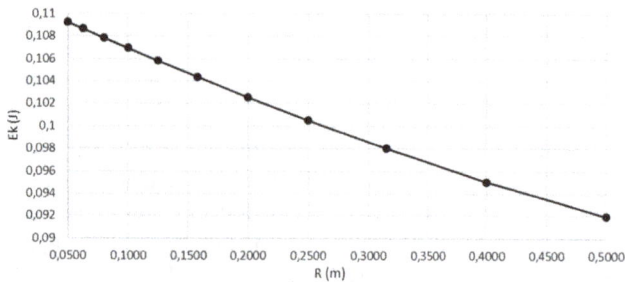

Figure 7. Dependence of kinetic energy E_k and radius R of the drive pulley at velocity $v_0 = 4\,\mathrm{m\,s}^{-1}$.

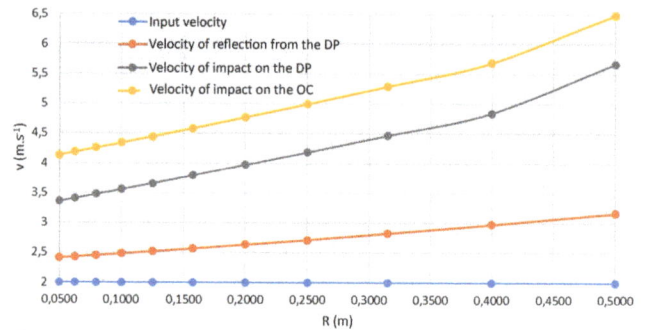

Figure 8. Graph of the velocity change for the material particle at $v_1 = 2\,\mathrm{m\,s}^{-1}$.

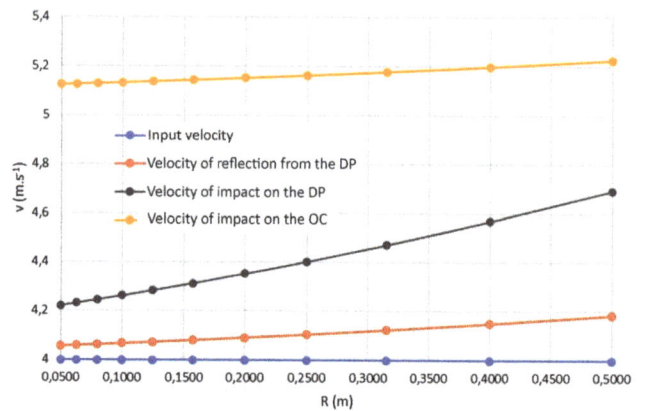

Figure 9. Graph of the velocity change for the material particle at $v_1 = 4\,\mathrm{m\,s}^{-1}$.

Figures 6 and 7 present a comparison of kinetic energy changes for the material particle after impact on the OC. These figures indicate that the increasing radius of the IC results in a decrease in the velocity component for material particle v_{3x}, thereby also causing a decrease in the kinetic energy of material particle E_k after the impact.

Figure 7 presents the increase in the kinetic energy of material particle E_k after the impact by double velocity v_0.

In the first phase of re-direction, the rate of distance x_2 at the point of the material particle impacting the DP depends on the point of separation of material particle from the drive pulley of IC, or the angle φ_1. With regard to the distance of DP d_1, the distance x_2 can not be more than d_1. When the angle $\varphi_1 = 0°$, $x_2 = d_1$. The distance y_2 gradually decreases with velocity v_1 increasing, and it increases with regard to the larger radius of the drive pulley R. This is also affected, besides the velocity v_1 and radius R, by the horizontal distance of the DP d_{okr}; therefore, for practical reasons, a smaller distance d_{okr} between IC and DP is convenient.

In the second phase of re-direction, distance x_3 increases with the change in velocity v_1; however, it decreases with the increase in radius of the drive pulley R. With regard to the vertical distance from the point of the material particle impacting the DP up to the point where the material particle lands on the OC $d_2 = 0.5\,\mathrm{m}$, the distance y_3 stays constant for both velocities v_1 and radiuses of drive pulleys R, i.e. $d_2 = y_2$.

Table 1 presents the comparison of the material particle distance at the point of impact on the DP (coordinates at point

A_2) and at the point of impact on the OC (coordinates at point A_3).

In the first phase of re-direction with regard to time t_2, which with an increase in the velocity v_1 decreases, the magnitude of velocity component v_{2x} ncreases and the magnitude of velocity component v_{2y} decreases. In the case $\varphi_1 = 0°$, the velocity component v_{2x} is, as a result of impact on the DP, equal to the velocity v_1. In the second phase, after it bouncing off the DP, the material particle reaches the velocity $v_{2'}$, which, compared with the impact velocity on the DP, is always smaller. This fact arises by the damping of the impact of the material particle by the DP.

By increasing the radius of a drive pulley R, the velocity $v_{2'}$ increases. The size of velocity component v_{3x} for the material particle increases in the third phase of shifting after the impact on the OC with increasing velocity v_1. With a larger radius of the drive pulley R, the size of velocity component v_{3x} for the material particle decreases, which leads to a decrease in the kinetic energy E_k of the material particle after the impact.

Figures 8 and 9 present the comparison of material particle velocity changes during three phases of shifting for both velocities.

Table 1. Evaluation of coordinates $[x_2, y_2]$ and $[x_3, y_3]$ for velocities v_1.

| Radius of carrying-up drum | x and y coordinate at point A_2 | | | | x and y coordinate at point A_3 | | | |
| | $v_0 = 2\,\mathrm{m\,s^{-1}}$ | | $v_0 = 4\,\mathrm{m\,s^{-1}}$ | | $v_0 = 2\,\mathrm{m\,s^{-1}}$ | | $v_0 = 4\,\mathrm{m\,s^{-1}}$ | |
$R[m]$	$x_2(t_2)$	$y_2(t_2)$	$x_2(t_2)$	$y_2(t_2)$	$x_3(t_3)$	$y_3(t_3)$	$x_3(t_3)$	$y_3(t_3)$
0.0500	0.550	0.371	0.550	0.093	0.137	0.5	0.548	0.5
0.0625	0.563	0.388	0.563	0.097	0.134	0.5	0.539	0.5
0.0800	0.580	0.413	0.580	0.103	0.131	0.5	0.528	0.5
0.1000	0.600	0.441	0.600	0.110	0.127	0.5	0.515	0.5
0.1250	0.625	0.479	0.625	0.120	0.123	0.5	0.500	0.5
0.1575	0.658	0.530	0.658	0.133	0.118	0.5	0.481	0.5
0.2000	0.700	0.601	0.700	0.150	0.112	0.5	0.458	0.5
0.2500	0.750	0.690	0.750	0.172	0.106	0.5	0.434	0.5
0.3150	0.815	0.815	0.815	0.204	0.099	0.5	0.405	0.5
0.4000	0.900	0.993	0.900	0.248	0.092	0.5	0.373	0.5
0.5000	0.711	1.435	1.000	0.307	0.062	0.5	0.341	0.5

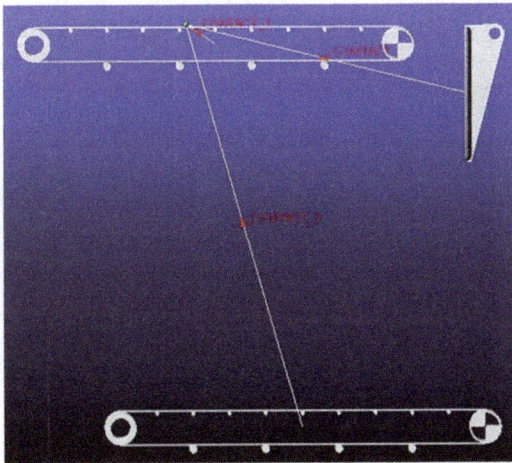

Figure 10. View of re-direction model in the ADAMS environment.

3 Simulation experiments with energy calculation model of transfer with perpendicular and inclined damping plate for velocities $v_0 = (2\,\mathrm{m\,s^{-1}}; 4\,\mathrm{m\,s^{-1}})$

The aim of experimenting with the transfer model was the verification of the trajectory of material particle movement in accordance with analytical calculation. We selected the distance of DP from IC $d_{\mathrm{okr}} = 0.5\,\mathrm{m}$ and the distance between the IC and OC as input parameters. The starting point for determination of the height of the damping plate was the total height of shifting d_3.

By reason of the absence of the type of material (sand) taken into account in comparison of the energy model, it was associate the type of material – steel for material particle from offered possibilities of the program.

As an alternative, we chose fixed-level supports (incoming belt), along which the material particle moves with an associated speed v_1.

Table 2 presents constructive parameters of components for the transfer model with associated input values.

With the use of construction parameters, a simplified transfer model was created, consisting of two fixed and rigid horizontal trays and the DP placed perpendicularly towards the trays and material particle, or a steel ball (Fig. 10).

For assurance of the material particle's impact on the DP and its impact with the bottom horizontal support, we defined these contacts in the transfer model:

1. contact of material particle with horizontal support as IC,

2. contact of material particle with the DP and the coefficient of restitution 0.5,

3. contact of material particle with the horizontal support as OC.

v_1 velocity was assigned for the material particle, and the change in velocity during the simulation is affected by gravitational acceleration $g = 9.81\,\mathrm{m\,s^{-2}}$.

For the purpose of verification of the model functionality for the designed and created transfer model, a simulation was realized in which the material particle impacted on the DP and fell out on the bottom horizontal support.

3.1 Perpendicular damping plate

Figure 11 shows how the material particle, after leaving the IC, changes its trajectory of movement with dependence on changes in input velocity v_1.

The general evaluation of the analytical calculation of results showing the values of coordinates $[x_2, y_2]$ of the material particle at the moment of its impact on the DP for standardized velocities v_1 demonstrated the consistence after the simulation experiments were carried out.

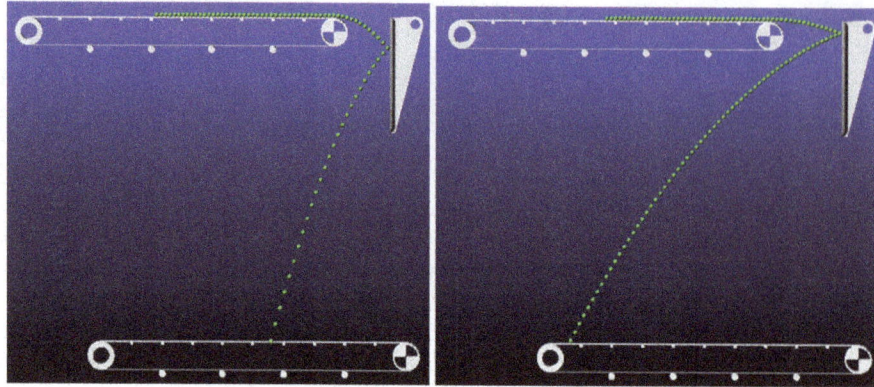

Figure 11. Trajectory of the material particle's movement at $v_1 = 2\,\mathrm{m\,s}^{-1}$ (left), $v_1 = 4\,\mathrm{m\,s}^{-1}$ (right).

Table 2. Parameters of the re-direction model in the workbench of the program ADAMS.

Title	Parameters of re-direction model							
	Length [mm]	Height [mm]	Width [mm]	Radius [mm]	Weight [kg]	Material density [kg mm^{-3}]	Modulus of elasticity [N mm^{-2}]	Poisson number [–]
Horizontal support (IC)	1000	50	400	x	x	x	x	x
Horizontal support (OC)	2000	50	400	x	x	x	x	x
Damping plate	50	2520	500	x	x	x	x	x
Material particle	x	x	x	20	0.26	$7.80.10^{-6}$	$2.07.10^{5}$	0.29

With the increase in velocity v_1, the material particle, after its separation from IC, gradually achieves a lower vertical distance y_2 at the moment of its impact on the DP. Due to the absence of a drive pulley in the simulated model, the distance x_2 was the same as in the analytical calculation for all velocities v_1, with the angle $\varphi_1 = 0°$.

It results from Fig. 11 that, with increasing velocity v_1, the angle φ_2 increases. This fact is evidenced by the results obtained by analytical calculation.

With gradually increasing velocity v_1, the kinetic energy of the moving material particle increases; Table 3 presents its values at the moment of the impact on the DP and the impact on horizontal support OC. By calculation of kinetic energy, the coefficient of restitution was reflected as $\varepsilon = 0.5$.

3.2 Inclined damping plate

With regard to the transfer model with the material particle velocity $v_1 = 2\,\mathrm{m\,s}^{-1}$, the simulation created the impact of material particle on the DP with the angle of inclination $\alpha = 20°$ and also $\alpha = 40°$. In both cases, the material particle continued in the movement oriented at the positive direction of the x axis, as shown in Fig. 12. To achieve the effect of lower energy intensity in the case of the belt conveyor with the same direction and orientation of movement, the better alternative is the situation displayed on the right in Fig. 12.

The distance from the horizontal support IC defined as DP 0.1 m at the inclination angle $\alpha \leq 20°$ for the velocity

$v_1 = 1\,\mathrm{m\,s}^{-1}$ indicated a sharp shock and bounce from the DP, after which the material particle continued in the negative direction and orientation of the x axis. At the angle of inclination $\alpha = 40°$ of the DP and the velocity of material particles $v_1 = 2\,\mathrm{m\,s}^{-1}$, the particles featured the identical direction of impact movement on horizontal support OC (Fig. 12).

The effect of the energy efficiency of the belt conveyor is increased if, with this velocity v_1, we choose a larger angle of the DP, for example $\alpha = 60°$, by which the material particle almost "matches" the position of inclination of the DP and moves in a positive direction and orientation of the x axis (Fig. 13). With a larger angle of inclination of the DP, the material particle does not impact on the DP. Therefore, if it is applied only for purpose of the direction of transported material, it is possible to exclude the DP.

From simulation experiments of energy calculation model, it results that, with higher velocity v_1 and angle of inclination of DP, we get the effect of smaller resistance of OC as a result of material particle impact in the positive direction and orientation of the x axis.

4 Design of recommendations for the practical application of transport systems at the point of rebound

By changing selected input parameters it was found that the energy intensity and instantaneous engine power of OC are

Table 3. Values of the kinetic energy for selected velocities v_1.

Velocity of the material particle	Kinetic energy at impact E_{k1} [Nmm]	Kinetic energy at impact E_{k2} [Nmm]
$v_1 = 2\,\mathrm{m\,s^{-1}}$	17 675.36	19 997.61
$v_1 = 4\,\mathrm{m\,s^{-1}}$	34 846.60	93 569.14

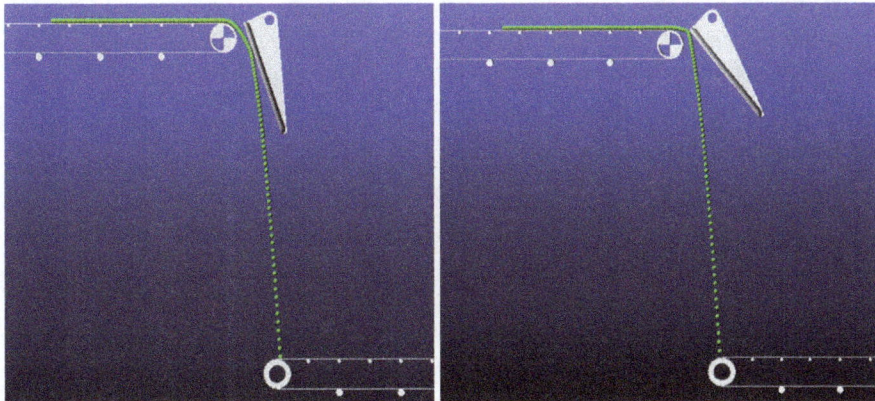

Figure 12. Trajectory of the material particle movement at $v_1 = 1\,\mathrm{m\,s^{-1}}$, $\alpha = 20°$ (left), $v_1 = 2\,\mathrm{m\,s^{-1}}$, $\alpha = 40°$ (right).

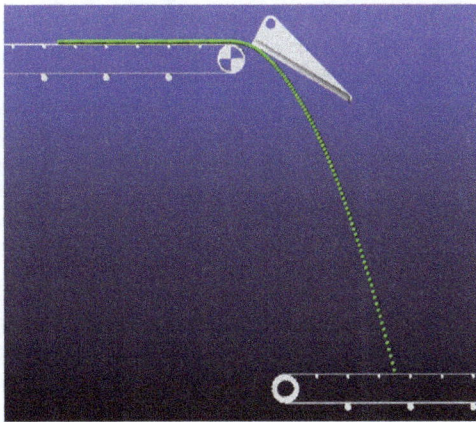

Figure 13. Trajectory of the material particle movement at $v_1\,\mathrm{m\,s^{-1}}$, $\alpha = 60°$.

affected by the horizontal distance of the DP from the IC d_1 and the coefficient of restitution ε. With $0 < \varepsilon \le 0.5$, we determined the decrease in kinetic energy of material particle E_k after the impact, and with $0.5 < \varepsilon < 1$ the increase in E_k. Because the value of the coefficient of restitution depends on the mathematical and physical properties of transported material and the character of the DP material, operators of conveying systems can not affect the coefficient of restitution by constructive change in the place of rebound. Therefore, optimization of d_1 was only considered for the value of the coefficient of restitution $\varepsilon = 0.5$.

The vertical distance from the point of particle impact on the DP to the point at which the material particle falls onto the OC d_2, the magnitude of the kinetic energy of material particle E_k after the impact is not affected. This distance affects the velocity of the material particle's impact v_3, which decreases by $d_2 < 0.5\,\mathrm{m}$, and it results in a decrease in kinetic energy of the material particle after the impact. Conversely, the velocity of material particle reflection v_3 and kinetic energy of material particle at the moment of impact increases by $d_2 > 0.5\,\mathrm{m}$. With regard to d_2 and vertical distance y_2, which the material particle has at the moment of impact on the damping plate, engineers and operators can determine the height of the DP.

The lowest instantaneous engine power of OC is achieved if the material particle, after rebound from the perpendicularly placed DP, falls onto the OC in a vertical direction, i.e. $\varphi_{2'} = 0°$. This recommendation is applicable to all standard velocities of conveyor belts.

All the same, it is necessary to consider to what extent there is an increase in the value d_{okr} from a practical point of view, i.e. that with regard to the vertical distance between conveyors d_{dop} it does not create a convergence of material particle to the DP after its separation from the drive pulley of IC. Tables 4 and 5 present change of the total height of shifting d_3 with regard to d_{okr} for both velocities.

Globally, in order to achieve a lower energy intensity of OC, a lower velocity of IC should be selected with a larger radius of the drive pulley along with the highest allowable distance of DP from the IC in the conveying system with material shifting from IC to the OC by application of DP; in addition, to achieve a lower total height of material shifting between IC and OC, it is suitable to choose a higher velocity

Table 4. Change in rebound height d_3 with regard to d_{okr} by $v_1 = 2\,\mathrm{m\,s^{-1}}$.

R [m]	Change in the total height of rebound d_3 with regard to d_{okr} [m]					
	$d_{okr}=0.5$	$d_{okr}=0.6$	$d_{okr}=0.7$	$d_{okr}=0.8$	$d_{okr}=0.9$	$d_{okr}=1$
0.0500	0.87	1.02	1.19	1.39	1.61	1.85
0.0625	0.89	1.04	1.21	1.41	1.64	1.88
0.0800	0.91	1.07	1.25	1.45	1.68	1.93
0.1000	0.94	1.10	1.28	1.49	1.73	1.98
0.1250	0.98	1.14	1.33	1.55	1.79	2.05
0.1575	1.03	1.20	1.40	1.62	1.87	2.14
0.2000	1.10	1.28	1.49	1.73	1.98	2.27
0.2500	1.19	1.39	1.61	1.85	2.12	2.42
0.3150	1.31	1.53	1.76	2.02	2.31	2.62
0.4000	1.49	1.73	1.98	2.27	2.57	2.90
0.5000	2.03	2.38	2.77	3.19	3.65	4.15

Table 5. Change in rebound height d_3 with regard to d_{okr} by $v_1 = 4\,\mathrm{m\,s^{-1}}$.

R [m]	Change in the total height of rebound d_3 with regard to d_{okr} [m]					
	$d_{okr}=0.5$	$d_{okr}=0.6$	$d_{okr}=0.7$	$d_{okr}=0.8$	$d_{okr}=0.9$	$d_{okr}=1$
0.0500	0.59	0.63	0.67	0.72	0.78	0.84
0.0625	0.60	0.63	0.68	0.73	0.78	0.85
0.0800	0.60	0.64	0.69	0.74	0.79	0.86
0.1000	0.61	0.65	0.70	0.75	0.81	0.87
0.1250	0.62	0.66	0.71	0.76	0.82	0.89
0.1575	0.63	0.68	0.73	0.78	0.84	0.91
0.2000	0.65	0.70	0.75	0.81	0.87	0.94
0.2500	0.67	0.72	0.78	0.84	0.91	0.98
0.3150	0.70	0.76	0.82	0.88	0.95	1.03
0.4000	0.75	0.81	0.87	0.94	1.02	1.10
0.5000	0.81	0.87	0.94	1.02	1.10	1.19

of IC with a smaller radius of drive pulley and the lowest allowable DP distance from IC.

5 Conclusions

Belt conveying has an unsubstitutable place in many technological processes of the mining and processing industry. There is a constant need for its continuous improvement, including questions relating to construction and, in turn, the design of new or improved types of conveyor belts. Major attention should be paid to the places of movement where the material is transported off the conveyor belt. It is at this place where serious situations can occur which result in belt degradation or complete breakdown. The most frequent situations are a conveyor belt breakdown and high wear. In order to increase the effectivity of belt conveying, much attention should be paid to construction solution of shifting places in order to minimize the occurrence of serious situations resulting in conveyor belt breakdown. This, of course, is not the only way to deal with this. The other issue to be considered is the direction and orientation of material impact.

The direction and orientation of material impact have a direct influence on the conveyor's energy intensity. For the effective operation of conveyor belts, attention must be paid to the construction of the place of impact, especially in OC.

Acknowledgements. This work is a part of projects VEGA 1/0258/14, VEGA 1/0063/16, and KEGA 014STU-4/2015.

Edited by: A. Konuralp

References

Alspaugh, M. A.: Latest Developments in Belt Conveyor Technology, in MINExpo 2004, p. 11, Las Vegas, NV, USA, available at: http://www.overlandconveyor.com/pdf/ (last access: August 2016) 2004.

Benjamin, B. C., Donecker, P., Huque, S., and Rozentals, J.: The Transfer Chute Design Manual For Conveyor Belt Systems, Aust. Bulk Handl. Rev., 1–8, 2010.

Bertrand, F., Leclaire, L. A., and Levecque, G.: DEM-based models for the mixing of granular materials, Chem. Eng. Sci., 60, 2517–2531, doi:10.1016/j.ces.2004.11.048, 2005.

Bierwisch, C., Kraft, T., Riedel, H., and Moseler, M.: Three-dimensional discrete element models for the granular statics and dynamics of powders in cavity filling, J. Mech. Phys. Solids, 57, 10–31, doi:10.1016/j.jmps.2008.10.006, 2009.

CEMA: Belt conveyors for bulk materials, sixth ed., Conveyor Equipment Manufacturers Association, Naples, Florida, USA, 1–600, 2007.

Chandramohan, R. and Powell, M. S.: Measurement of particle interaction properties for incorporation in the discrete element method simulation, Miner. Eng., 18, 1142–1151, doi:10.1016/j.mineng.2005.06.004, 2005.

David, J. and Kruses, P. E.: Conveyor Belt Transfer Chute Modeling and Other Applications using The Discrete Element Method in the Material Handling Industry, available at: http://www.ckit.co.za/secure/conveyor/papers/troughed/conveyor/conveyor.htm, last access: August 2016.

Dewicki, G.: Bulk Material Handling and Processing – Numerical Techniques and Simulation of Granular Material, Bulk Solids Handl., 23, 2–5, 2003.

Di Renzo, A. and Di Maio, F. P.: Comparison of contact-force models for the simulation of collisions in DEM-based granular flow codes, Chem. Eng. Sci., 59, 525–541, doi:10.1016/j.ces.2003.09.037, 2004.

Donohue, T. J., Roberts, A. W., Wheeler, C. A., and McBride, W.: Computer Simulations as a Tool for Investigating Dust Generation in Bulk Solids Handling Operations, Part. Part. Syst. Charact., 26, 265–274, 2010.

Grima, A. P. and Wypych, P. W.: Development and validation of calibration methods for discrete element modelling, Granul. Matter, 13, 127–132, doi:10.1007/s10035-010-0197-4, 2010.

Grima, A. P. and Wypych, P. W.: Discrete element simulations of granular pile formation: Method for calibrating discrete element models, Eng. Comput., 28, 314–339, doi:10.1108/02644401111118169, 2011.

Gröger, T. and Katterfeld, A.: Application of the Discrete Element Method in Materials Handling – Part 3: Transfer Stations, Bulk Solids Handl., 27, 158–167, 2007.

Hastie, D. B. and Wypych, P. W.: Experimental validation of particle flow through conveyor transfer hoods via continuum and discrete element methods, Mech. Mater., 42, 383–394, doi:10.1016/j.mechmat.2009.11.007, 2010.

Hou, Y. and Meng, Q.: Dynamic characteristics of conveyor belts, J. China Univ. Min. Technol., 18, 629–633, doi:10.1016/S1006-1266(08)60307-7, 2008.

Huque, A. G. and McLean, S. T.: Methods to predict material trajectories from belt conveyors and impact plates Part 1: Review, Bulk Solids Handl., 22, 348–354, 2002.

Jensen, R. P., Bosscher, P. J., Plesha, M. E., and Edil, T. B.: DEM simulation of granular media–structure interface: effects of surface roughness and particle shape, Int. J. Numer. Anal. Methods Geomech., 23, 531–547, 1999.

Kessler, F. and Prenner, M.: DEM – Simulation of Conveyor Transfer Chutes, FME Trans., 37, 185–192, 2009.

Mazurkiewicz, D.: Analysis of the ageing impact on the strength of the adhesive sealed joints of conveyor belts, J. Mater. Process. Technol., 208, 477–485, 2008.

McIlvenna, P. and Mossad, R.: Two Dimensional Transfer Chute Analysis Using a Continuum Method, in: Third International Conference on CFD in the Minerals and Process Industries CSIRO, 547–552, Melbourne, Australia, available at: available at: http://www.cfd.com.au/cfd_conf03/papers/064Mci.pdf (last access: August 2016), 2003.

Minkin, A.: Analysis of transfer stations of belt conveyors with help of discrete element method (DEM) in the mining industry, Transp. Logist., 12, 1–6, 2012.

Roberts, A. W.: Chute Performance and Design for Rapid Flow Conditions, Chem. Eng. Technol., 26, 163–170, doi:10.1002/ceat.200390024, 2003.

Roberts, A. W.: Chute Design Considerations for Feeding and Transfer, available at: http://login.totalweblite.com/Clients/doublearrow/beltcon2001/3.chutedesignconsiderationsforfeedingandtransfer.pdf (last acces: August 2016), unpublished data.

Scott, O. J.: Chute Design Considerations for Feeding and Transfer, in Conveyor Transfer Chute Design in Modern Concepts in Belt Conveying and Handling Bulk Solids, The Institute for Bulk Materials Handling Research, University of Newcastle, 809–820, 1992.

Scott, O. J. and Choules, P. R.: The use of impact plates in conveyor transfers, Tribol. Int., 26, 353–359, doi:10.1016/0301-679X(93)90072-9, 1993.

Seifried, R., Schiehlen, W., and Eberhard, P.: Numerical and experimental evaluation of the coeffcient of restitution for repeted imacts, Int. J. Impact Eng., 32, 508–524, 2005.

Spaans, C.: The calculation of the main resistence of belt conveyors, Bulk Solids Handl., 11, 1999.

Wensrich, C. M.: Evolutionary optimisation in chute design, Powder Technol., 138, 118–123, doi:10.1016/j.powtec.2003.08.062, 2003.

Wensrich, C. M. and Wheeler, C. A.: Evolutionary optimisation in loading chute design, in: 8th International Conference on bulk materials storage, handling and transportation, University of Wollongong, Australia, 2004.

Zhang, S. and Xia, X.: A New Energy Calculation Model of Belt Conveyor, in AFRICON, 2009, 1–6, Nairobi, available at: http://timetable.cput.ac.za/_other_web_files/_cue/ICUE/2009/PPT/Presentation-XiaX.pdf, 2009.

Zhang, S. and Xia, X.: Modeling and energy efficiency optimization of belt conveyors, Appl. Energy, 88, 3061–3071, 2011.

12

Comparative experimental investigation and gap flow simulation in electrical discharge drilling using new electrode geometry

Ali Tolga Bozdana[1] **and Nazar Kais Al-Karkhi**[2]

[1]Associate Professor, Mechanical Engineering Department, University of Gaziantep, Gaziantep, Turkey
[2]Lecturer, Automated Manufacturing Engineering, University of Baghdad, Baghdad, Iraq

Correspondence to: Ali Tolga Bozdana (bozdana@gantep.edu.tr)

Abstract. This study presents experimental and numerical investigation on the effectiveness of electrode geometry on flushing and debris removal in Electrical Discharge Drilling (EDD) process. A new electrode geometry, namely side-cut electrode, was designed and manufactured based on circular electrode geometry. Several drilling operations were performed on stainless steel 304 using rotary tubular electrodes with circular and side-cut geometries. Drilling performance was characterized by Material Removal Rate (MRR), Electrode Wear Rate (EWR), and Tool Wear Ratio (TWR). Dimensional features and surface quality of drilled holes were evaluated based on Overcut (OC), Hole Depth (HD), and Surface Roughness (SR). Three-dimensional three-phase CFD models were built using ANSYS FLUENT software to simulate the flow field at interelectrode gap. Results revealed that the overall performance of side-cut electrode was superior due to improved erosion rates and flushing capabilities, resulting in production of deep holes with good dimensional accuracy and surface quality.

1 Introduction

Electrical Discharge Drilling (EDD) is an electro-thermal process used for drilling small holes on electrically conductive materials. It is based on the eroding effect of electric sparks occurring between tool electrode and workpiece (Bozdana and Ulutas, 2016). Drilling operations are conducted by tubular electrodes through which the dielectric fluid is flowing for washing the debris away from the machining zone. Adequate flushing at interelectrode gap is of significant importance in drilling of holes with high aspect ratios. In production of deep holes, proper circulation of dielectric and effective debris removal are difficult to achieve, which directly affect drilling performance and hole features.

Numerous methods have been contrived in order to improve flushing in EDD applications. Some researchers tried to manipulate the relative movement between electrode and workpiece to stimulate the removal of debris away from machining zone. Orbital movement of tool electrode in radial direction was employed for manipulating side interelectrode gap (Yu et al., 2009; Bamberg and Heamawatanachai, 2009).

Bottom interelectrode gap was also manipulated by applying ultrasonic vibrations in vertical direction (Yu et al., 2009; Jahan et al., 2012; Shabgard and Alenabi, 2015). Moreover, it is reported in literature that increasing electrode rotation (Yahagi et al., 2012; Yadav and Yadava, 2015; Dwivedi and Choudhury, 2016) and dielectric pressure (Munz et al., 2013; Selvarajan et al., 2015, 2016) provided improvements in flushing and machining performance. On the other hand, only few studies were conducted for investigating the effect of electrode shape in EDD process. Nastasi and Koshy (2014) added geometric features of helical and radial slots on cylindrical copper electrodes to improve gap flushing in drilling of 6061 aluminum alloy. Plaza et al. (2014) studied the effects of helix angle and flute depth on helical electrodes to improve debris removal when increasing the hole depth in drilling of Ti-6Al-4V. Another work on using helical electrode with use of ultrasonic vibrations applied on workpiece was conducted by Hung et al. (2006) in drilling of nickel alloy Hymu 80.

In addition to experimental investigations found in the literature, there are few studies on modeling the flow characteristics at electrode-workpiece interaction. Nastasi and Koshy

Table 1. Chemical composition (in wt %) of stainless steel 304.

C: 0.08 (max)	Mn: 2 (max)	Cr: 18–20	S: 0.03 (max)
N: 0.1	Si: 0.75 (max)	Ni: 8–10.5	P: 0.045 (max)

Table 2. Drilling parameters.

Peak current	24 A
Pulse-on time	44 µs
Pulse-off time	5 µs
Voltage	50 V
Capacitance	1476 µF
Dielectric fluid	Deionized water
Dielectric pumping pressure	30 kg cm^{-2}
Electrode rotation speed	200 rpm

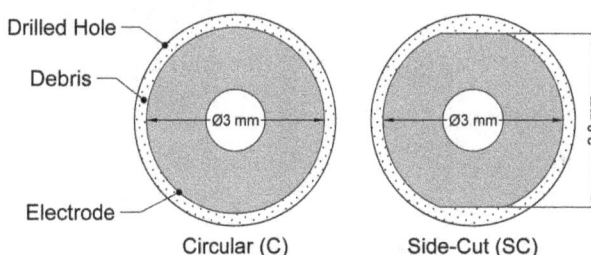

Figure 1. Circular and side-cut electrode geometries.

(2014) modeled the flow fields in frontal and lateral gaps using ANSYS CFX to optimize tool electrode comprising geometric features. Xie et al. (2015a) presented 2-D flow model in ultrasonically assisted EDM process using Computational Fluid Dynamics (CFD). Their model involved phases of kerosene dielectric and debris at bottom and side regions of interelectrode gap. Wang and Han (2014a, b) proposed 3-D model of flow field with liquid, gas, and solid phases to analyze the machining gap during electrode jump in EDM. The variations in bubble volume at interelectrode gap during ultrasonically assisted EDM were studied by Kong et al. (2015). 3-D geometrical modeling of flow field at interelectrode gap was established using FLUENT software. In the study of Xie et al. (2015b), 2-D model of flow field in ultrasonic assisted EDM was constructed with liquid and debris phases. Debris distribution and velocity variations at bottom and side gaps during ultrasonic vibration cycle were investigated through the numerical simulations. Zhang et al. (2015) developed 2-D model to simulate movement and distribution of the debris in EDM with self-adaptive electrode movement. The results indicated that the quantity of debris flushed away was limited and most of debris in the gap was aggregated at interelectrode gap.

This study presents experimental and numerical investigations on EDD of stainless steel 304 using circular and side-cut electrodes. Several drilling operations were performed in order to examine the effectiveness of electrode geometry on machining performance as well as dimensional accuracy and surface quality of drilled holes. Numerical analyses were carried out based on 3-D three-phase CFD models for simulation of flushing capabilities of electrodes.

2 Experimental work

Tubular brass electrodes with cylindrical and side-cut electrode geometries were used in this study. Circular electrode is commercially available, and side-cut electrode was manufactured by milling two sides of circular electrode (Fig. 1). Both electrodes have the identical inner and outer diameters of Ø 1 and Ø 3 mm, respectively. Therefore, it is possible to make reliable comparison between performance flushing and sparking characteristics of these electrodes.

Workpiece material was stainless steel 304 having chemical composition given in Table 1. Holes were drilled on block specimens of 50 × 35 × 4 mm, as illustrated in Fig. 2. They were produced on mating interface of specimens. After drilling, specimens were separated to perform measurements on the surfaces of drilled holes.

Table 2 presents the process parameters used in drilling operations. Five holes using both electrode geometries were drilled at five machining durations (namely 1, 2, 3, 4, and 5 min). This was done to investigate the effectiveness of electrode geometries on drilling performance and hole characteristics at different hole depths.

Drilling operations were conducted on JS AD-20 hole drilling EDM machine (Fig. 3). Vertical movement of the electrode was achieved by servo control while axis movements were displayed on coordinate display. Machining settings were selected on the control panel. The dielectric fluid was filtered and pumped through hollow electrode during drilling process.

Process performance was evaluated based on Material Removal Rate (MRR) and Electrode Wear Rate (EWR) as given in Eqs. (1) and (2), respectively. The weights of specimens and electrodes were measured using a digital scale with a precision of ±1 mg. Drilling time was recorded by a precision timer. In addition, Tool Wear Ratio (TWR) was employed in order to evaluate the dimensionless effect between electrode wear and material removal, as given in Eq. (3).

$$\text{MRR}\left(\text{g min}^{-1}\right) = \frac{\begin{array}{c}(\text{specimen weight before drilling})\\ -(\text{specimen weight after drilling})\end{array}}{\text{drilling time}} \quad (1)$$

$$\text{EWR}\left(\text{g min}^{-1}\right) = \frac{\begin{array}{c}(\text{electrode weight before drilling})\\ -(\text{electrode weight after drilling})\end{array}}{\text{drilling time}} \quad (2)$$

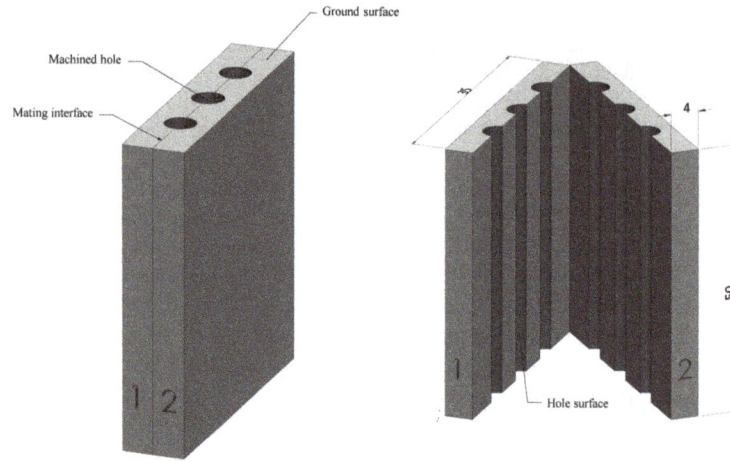

Figure 2. Sketch of specimens (all dimensions in mm).

Figure 3. Hole drilling EDM machine with experimental setup.

$$\text{TWR} = \frac{\text{EWR}}{\text{MRR}} \qquad (3)$$

Diameter and depth of drilled holes were measured on photographs taken by high-resolution scanner. Due to side-sparking between electrode and workpiece, diameter of drilled hole is always larger than electrode diameter. This is called Overcut (OC), which was calculated based on the enlargement in hole diameter with respect to electrode diameter (Eq. 4).

$$\text{OC}(\%) =$$
$$\frac{(\text{measured hole diameter}) - (\text{electrode diameter})}{\text{electrode diameter}} \times 100 \qquad (4)$$

Surface quality of holes was evaluated based on Surface Roughness (SR). On each hole surface, the roughness average (R_a) was measured using Mitutoyo SJ-401 roughness tester with a cut-off length of 0.8 mm. The measurements were repeated three times at different regions on each hole

surface, and the average of readings was taken as the roughness value.

3   Numerical analyses

Flow characteristics were analyzed based on three-phase model of flushing (i.e. water as liquid phase, bubbles as gas phase, and debris as solid phase) in ANSYS FLUENT software. The interaction between bubbles and water was simulated with Euler-Euler VOF model (Xie et al., 2015a; Wang and Han, 2014a, b) by solving the volume continuity equation as follows (Kong et al., 2015):

$$\frac{\partial}{\partial t}(\alpha_g \rho_g) + \nabla \cdot (\alpha_g \rho_g \upsilon_g) = \sum_{L=1}^{n} m_{Lg} \qquad (5)$$

where α_g is the volume fraction of gas, ρ_g is the density of gas, υ_g is the velocity of gas, m_{Lg} is the mass transfer from liquid phase to gas phase. The function DE-FIEN_MASS_TRANSFER was used to describe the transfer of water into gas (Wang and Han, 2014a). When discharge occurs, the water starts being transferred into bubbles. Processing of the water volume fraction (α_L) was computed by the following equation (Kong et al., 2015):

$$\alpha_g + \alpha_L = 1. \qquad (6)$$

For the VOF model, a single momentum conservation equation of the entire flow field was solved. The momentum equation was depending upon the volume fractions of all phases through ρ and μ (Wang and Han, 2014a):

$$\frac{\partial}{\partial t}(\rho_a \upsilon) + \nabla \cdot (\rho_a \upsilon \upsilon) = -\nabla p + \nabla \cdot \left[\mu_a \left(\nabla \upsilon + \nabla \upsilon^T \right) \right]$$
$$+ \rho_a g + F \qquad (7)$$

where ρ_a and μ_a were taken as the average of volume fraction values of all the phases, and F denotes the external body forces (i.e. the forces from interaction with the dispersed phases). The calculations of ρ_a and μ_a for two phases are as follows:

$$\rho_a = \alpha_L \rho_L + \alpha_g \rho_g \tag{8}$$

$$\mu_a = \alpha_L \mu_L + \alpha_g \mu_g \tag{9}$$

where α_L and α_g are the volume fractions of liquid and gas phases, respectively.

The debris moves with dielectric fluid in the machining gap, and hence their movement is abided by a discrete phase particle by integrating the force balance on debris, which is written in a Lagrangian reference frame. The motion equation of debris in the gap flow can be determined by Newton's second law of motion (Wang and Han, 2014a):

$$\frac{du_p}{dt} = \frac{g_z(\rho_p - \rho_L)}{\rho_p} + F_z + F_D(u_L - u_p) \tag{10}$$

where u_L and u_p are the velocities of water and debris in the direction of electrode movement, ρ_L and ρ_p are the densities of water and debris, g_z is the gravitational acceleration. The first term on the right side of Eq. (10) is the buoyancy of debris. The second term on the right side of Eq. (10) is an additional force (e.g. thermo-Phoresis force and brown force), also including forces on particles arising due to rotation of the reference frame. The third term on the right side of Eq. (10) is the drag force of debris, in which F_D was calculated as follows (Wang and Han, 2014a):

$$F_D = \frac{18\mu}{\rho_p d_p^2} \frac{C_D Re}{24} \tag{11}$$

where μ is the molecular viscosity of fluid, d_p is the diameter of debris particles, Re is Reynolds number, C_D is the coefficient of drag force.

The energy equation was also solved for the VOF model as below (Theory Guide, 2009):

$$\frac{\partial}{\partial t}(\rho E) + \nabla \cdot (\upsilon(\rho E + p)) = \nabla \cdot (k_{eff} \nabla T) + S_h. \tag{12}$$

The VOF model treats energy (E) and temperature (T) as mass-averaged variables:

$$E = \frac{\displaystyle\sum_{q=1}^{n} \alpha_q \rho_q E_q}{\displaystyle\sum_{q=1}^{n} \alpha_q \rho_q} \tag{13}$$

where E_q for each phase is based on the specific heat of that phase and the shared temperature. The properties of ρ and k_{eff} (effective thermal conductivity) are shared by phases. The source term (S_h) contains contributions from radiation as well as any other volumetric heat sources.

A three-dimensional model of the flow field during consecutive-pulse discharge was created. The deionized water (i.e. liquid phase) was introduced through the inner portion of hollow electrode. The air (i.e. the bubbles occurring due to vaporization) and the debris (i.e. the particles eroded from workpiece) were generated within the gap between electrode tip and workpiece. The model equations given in previous section were solved using a commercial CFD software package, FLUENT (v16.1). CFD simulations were done by means of GAMBIT tools in order to design the problem in geometrical configuration with appropriate mesh. Before solving fluid flow problems, FLUENT needs the domain, at which the flow takes place. Thereby, 3-D flow domains for both electrode geometries (i.e. circular and side-cut electrodes) were created in Solidworks®. After that, a geometric model of the flow field domain inside the electrode and in the machining gap was divided into a certain number of cells (intervals) using GAMBIT software and imported into FLUENT. Solid models and the corresponding meshing for both electrode geometries are shown in Figs. 4 and 5, respectively.

In order to obtain a well-posed system of equations, reasonable boundary conditions for computational domain were implemented. The gap between electrode and workpiece was initially filled with liquid (i.e. deionized water). Air bubbles and debris started to occur at bottom region of electrode, resulting from the sparks generated at the tip of electrode, which melts the metal and boils the water. The outlet boundary condition was pressure that was set as 1.013×10^5 Pa (i.e. atmospheric pressure). Pressure boundary condition was applied at top of electrode with a static reference pressure of $30 \, \text{kg} \, \text{cm}^{-2}$. The inlet velocity was calculated as:

$$\dot{m} = \rho \, u_i \, A \tag{14}$$

where $\dot{m}$ is the flow rate of water flowing through tubular electrode (measured as $0.001632 \, \text{kg} \, \text{s}^{-1}$), ρ is the density of water (i.e. $998.2 \, \text{kg} \, \text{m}^{-3}$), and A is the inner area of electrode ($0.785 \times 10^{-6} \, \text{m}^3$). Thus, the inlet velocity (u_i) was about $2 \, \text{m} \, \text{s}^{-1}$. Wall boundary conditions were defined as no-slip condition for liquid phase whereas free-slip conditions for solid and gas phases. The electrode walls (side and bottom faces) were rotated at 200 rpm. The identical boundary conditions were applied for both electrodes geometries.

Simulations were carried out using FLUENT based on the experimental setup, as illustrated in Fig. 6. The maximum machining depth was 50 mm. Side and bottom gaps were measured as suggested in literature (Wang and Han, 2014b). The difference between diameters of electrode and drilled hole was the dimension of side gap, and it was found to be about 100 μm. The dimension of bottom gap was determined based on the difference between vertical coordinates of electrode. The coordinate of electrode was recorded when spark occurred between electrode and workpiece. Without sparking, its coordinate was also recorded when electrode was in

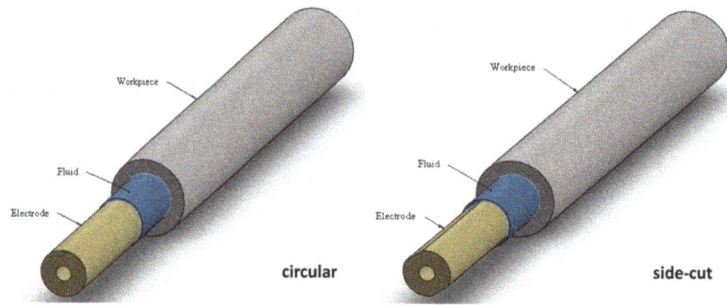

Figure 4. Solid models representing electrode-fluid-workpiece interaction.

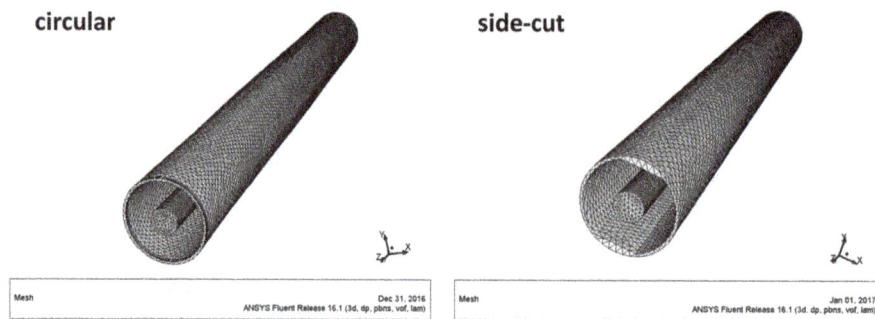

Figure 5. Meshed models representing the geometry at flushing gap.

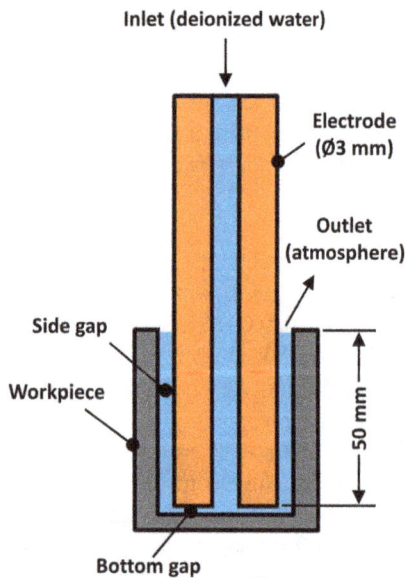

Figure 6. Sketch of setup used for numerical model.

Table 3. The properties for liquid, solid, gas phases.

Property	Liquid (water)	Solid (stainless steel 304)	Gas (air)
Density (kg m^{-3})	998.2	7000	1.225
Conductivity (W m^{-1} K^{-1})	0.6	17	0.0242
Specific Heat (J kg^{-1} K^{-1})	4182	530	1006.43
Viscosity (kg m^{-1} s^{-1})	0.001003	–	1.7894×10^{-5}

was calculated with the following formula (Fang and Manglik, 2002):

$$Re = \frac{\rho N r^2}{\mu} \tag{15}$$

where N is the angular velocity of rotating inner cylinder (rotational speed of electrode), r is the outer radius of electrode. Therefore, the maximum Reynolds number of the side gap was calculated as 450. This value is much smaller than the critical value of 2300. Thus, the flow inside the gap was said to be laminar. The properties for all phases are given in Table 3.

4 Results and discussion

Five holes were drilled using circular and side-cut electrodes at drilling times of 1, 2, 3, 4, 5 min, respectively. Figure 7

touch with workpiece. The difference between two coordinates of electrode was the dimension of bottom gap, which was determined as 480 µm.

In order to verify the flow type of flow field in the machining gap (i.e. laminar or turbulent flow), Reynolds number (Re) of inner cylinder rotation with axial flow in annuli

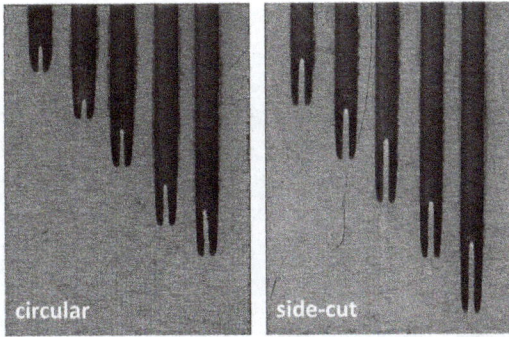

Figure 7. Photographs of drilled holes.

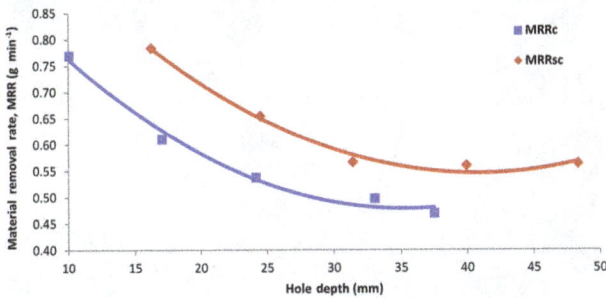

Figure 8. Results of MRR with respect to hole depth.

shows photographs of drilled holes. Table 4 presents results of MRR, EWR, TWR, OC, and SR obtained after drilling operations.

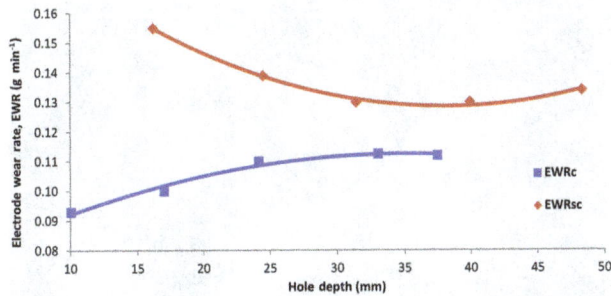

Figure 9. Results of EWR with respect to hole depth.

MRR is a significant factor in EDD process because of its vital influence on engineering economy. Higher MRR value refers to greater amount of material removal per unit time. Results of MRR for circular and side-cut electrodes are compared in Fig. 8. Side-cut electrode provides higher MRR values as compared with circular electrode, resulting in deeper holes. This is due to the reason that spark intensity was increased and flushing was improved in case of side-cut electrode. For the deepest hole, there is about 20 % increase in MRR when side-cut electrode was used. It is also observed that MRR values for side-cut electrode are not diminished after specific hole depth whereas material removal by circular

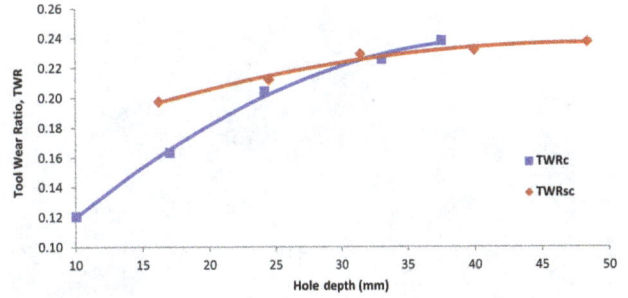

Figure 10. Results of TWR with respect to hole depth.

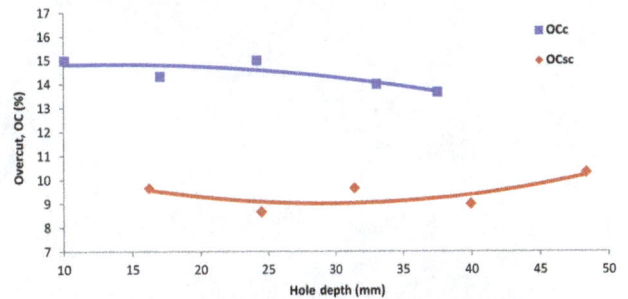

Figure 11. Results of OC with respect to hole depth.

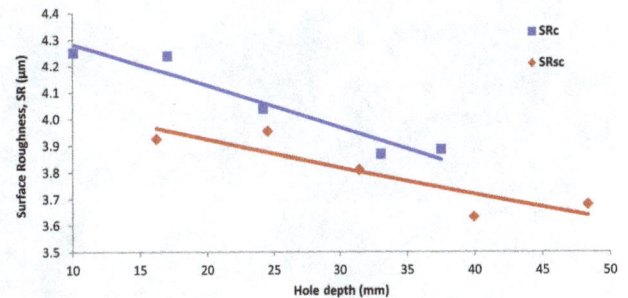

Figure 12. Results of SR with respect to hole depth.

electrode becomes lesser with increasing the depth. Moreover, MRR curves for both electrodes decrease with drilling depth since flushing the debris away from machining zone becomes more difficult.

EWR is also a significant parameter, which is related with MRR. Figure 9 shows EWR results for both electrodes. Side-cut electrode generally exhibits higher EWR, which is an expected case as the degree of erosion on both workpiece and electrode are proportional. Thus, providing high MRR, side-cut electrode causes high EWR. It is also observed that EWR for side-cut electrode decreases with an increase in hole depth while EWR for circular electrode increases due to weak flushing. This causes accumulation of debris in the machining zone, leading to increase in erosion on electrode.

TWR is usually employed as a parameter to evaluate the combined effect of MRR and EWR. It is desirable to have lower TWR value as it refers to lesser amount of electrode

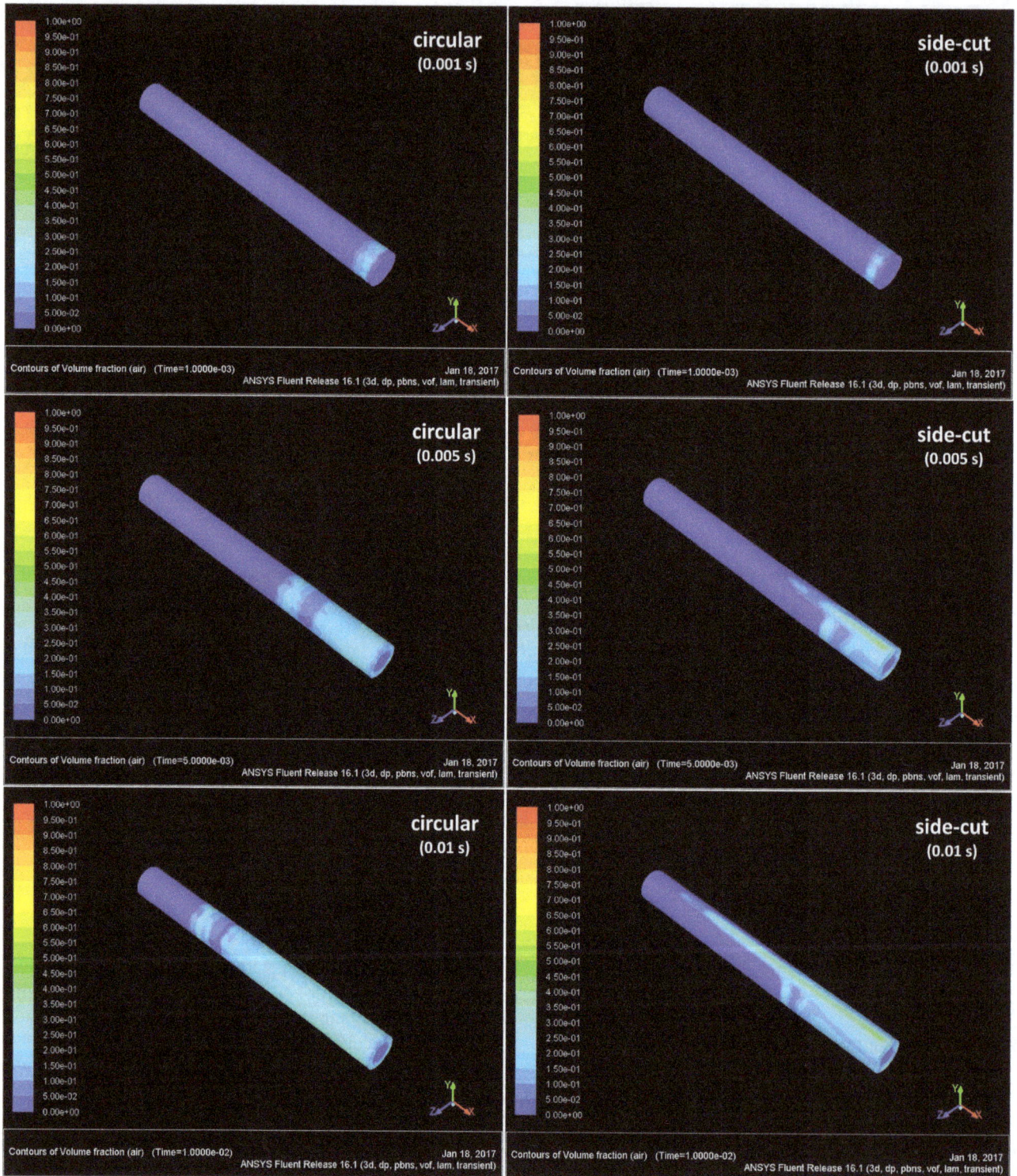

Figure 13. Volume fraction contours of bubbles at increasing time steps.

wear and greater material removal. As seen from Fig. 10, side-cut electrode causes moderate increase in TWR. However, in case of circular electrode, there is a significant increase in TWR.

Results of OC given in Fig. 11 reveal that side-cut electrode exhibits lower OC for all holes. This is due to the fact that circular electrode have greater contour along its perimeter for producing greater amount of side sparks, leading to

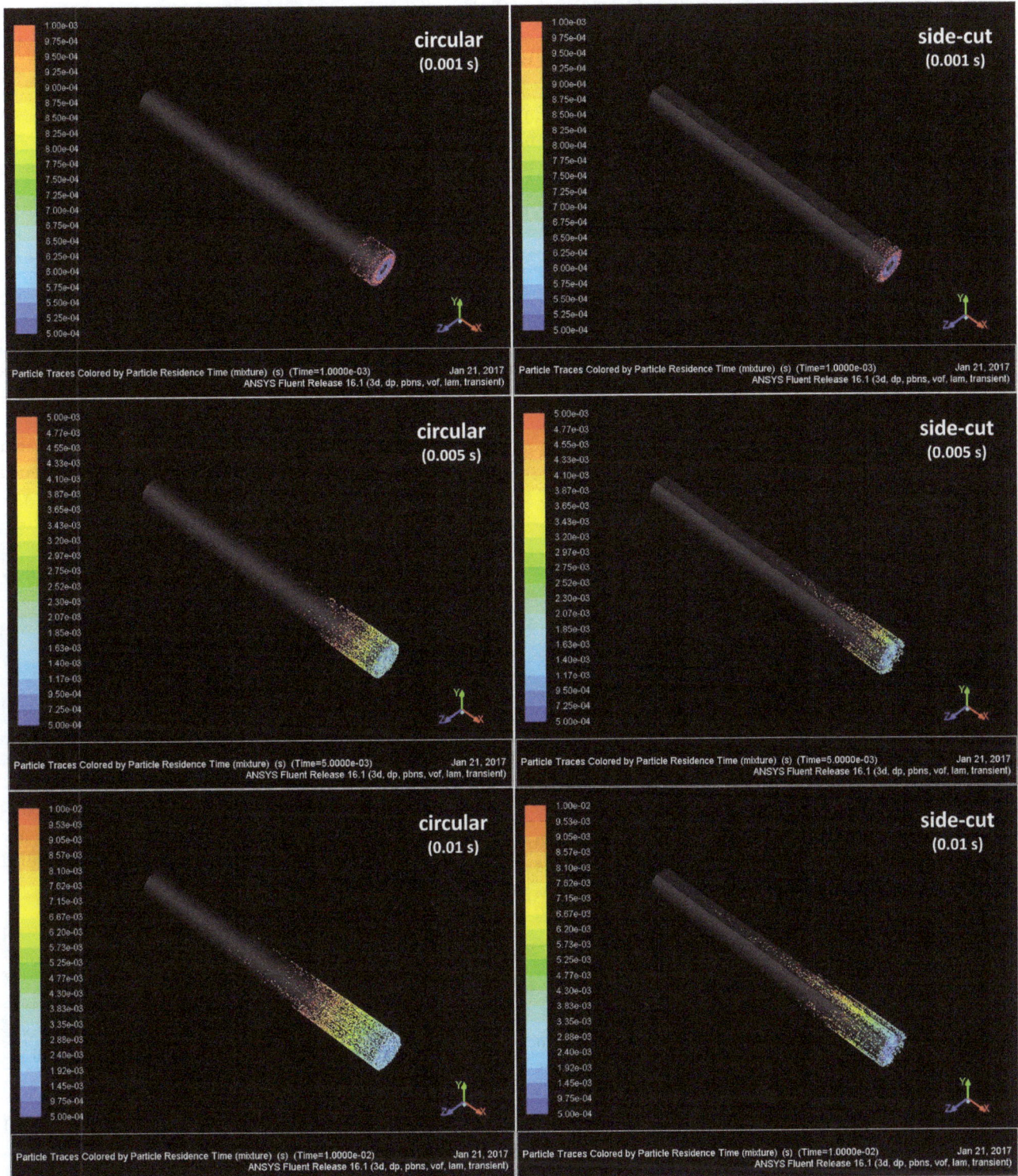

Figure 14. Debris distributions at increasing time steps.

higher degree of erosion, and hence larger enlargement in drilled hole diameter. On average, side-cut electrode provided an improvement of about 34 % in OC over circular electrode.

The comparison for SR results is given in Fig. 12. Regardless of hole depth, side-cut electrode provides smoother hole surfaces. This is because of improved flushing capabilities of side-cut electrode so that the eroded particles can be effec-

Table 4. Experimental results obtained by circular (C) and side-cut (SC) electrodes.

Hole No.	Dia. (mm)		Depth (mm)		MRR (g min^{-1})		EWR (g min^{-1})		TWR		OC (%)		SR (μm)	
	C	SC	C	SC	C	SC	C	SC	C	SC	C	SC	C	SC
1	3.45	3.29	10.06	16.20	0.770	0.784	0.093	0.155	0.121	0.198	15.00	9.67	4.25	3.93
2	3.43	3.26	17.02	24.49	0.611	0.655	0.100	0.139	0.164	0.212	14.33	8.67	4.24	3.96
3	3.45	3.29	24.17	31.42	0.538	0.567	0.110	0.130	0.204	0.229	15.00	9.67	4.04	3.81
4	3.42	3.27	33.04	39.93	0.498	0.56	0.113	0.130	0.226	0.232	14.00	9.00	3.87	3.63
5	3.41	3.31	37.49	48.34	0.470	0.564	0.112	0.134	0.238	0.238	13.67	10.33	3.89	3.68

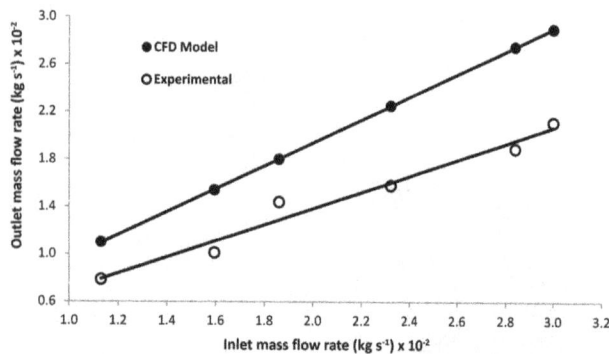

Figure 15. Validation of numerical model based on MFR.

tively washed away without sticking on workpiece surface, which results in hole surfaces free from resolidified debris.

Numerical simulations of inter-electrode gap were carried out to show the effectiveness of electrode shape on behavior of air bubbles as well as debris distribution at incremental time steps. Firstly, bubble distributions were simulated in order to define the motion of air bubbles from bottom gap to side gap. Figure 13 shows the simulation results of volume fraction at increasing time steps of 0.001, 0.005, and 0.01 s. The left column of numerals with different color fringes correspond to the volume fraction of bubbles. For instance, the regions of blue color denoted by "0" are full of water. The results in Fig. 13 reveal that circular electrode exhibits uniformly distributed of areas of bubbles around the side gap. On the other hand, the bubble distribution is nonuniform in case of side-cut electrode, owing to its shape. The regions within the sides of cut provides extra space for the bubbles to flow, leading to stimulated motion of bubbles, which results in effective flushing capabilities.

Similar effects are observed for the debris distributions, as given in Fig. 14. The left column of numerals with different color fringes correspond to the accumulated moving time of debris. Regardless of the moving direction of debris and velocity field, the dispersion degree of debris in circular electrode is higher than that in side-cut electrode. It can be seen that flushing of debris was increased when side-cut electrode was used. Simulation results reveal that side-cut electrode has a significant influence on debris exclusion from machin-

ing zone, and hence it has exhibited better flushing characteristics as compared with circular electrode.

The reliability of developed model was validated based on Mass Flow Rate (MFR). For this purpose, several tests were conducted at varying drilling depths (at a range of 1–50 mm) and inlet MFR values were measured. The corresponding outlet MFR values from models were compared with experimental values. Figure 15 shows that there is good agreement between experimental and numerical MFR values. CFD model has provided a linear correlation with coefficient of determination (R^2) of 0.969. This assures the reliability of developed model for accurate simulation of flow field through interelectrode gap.

5 Conclusions

In this research, circular and side-cut electrodes were used in hole drilling by EDD process. Several drilling operations were conducted for producing holes with different drilling times. Numerical models were developed for analysis of flushing capabilities. The results revealed that electrode shape has significant effects on machining performance as well as dimensional and surface characteristics of drilled holes. The main outcomes of this study are as follows:

- In aspect of machining performance, side-cut electrode provided higher MRR and EWR for all holes as compared with circular electrode. Similar trends of TWR were observed. The modifications on electrode geometry affected the erosion characteristics of both workpiece and electrode, leading to changes in material removal and electrode wear.

- For the assessment of dimensional features of drilled holes, deeper holes were obtained with side-cut electrode. Owing to the shape of side-cut electrode, spark intensity was increased at the bottom interelectrode gap, causing greater amount of material removal.

- Side-cut electrode also provided holes with lower OC. This is because of narrower contour along perimeter of side-cut electrode, leading to less amount of side-sparking which results in smaller enlargement on the diameter of drilled holes.

– Regarding the surface quality of drilled holes, side-cut electrode provided lower SR values. Side-cut electrode has better flushing performance due to extra space at side interelectrode gap, allowing for easier removal of debris so that smoother hole surfaces were obtained.

– Numerical results also proved that side-cut electrode has improved flushing capabilities for efficient removal of air bubbles and eroded particles.

In spite of high electrode wear, the overall performance of side-cut electrode over circular electrode was superior due to improved erosion and flushing capabilities. Experimental and numerical results revealed that side-cut electrode is preferable for producing deep holes with good dimensional accuracy and improved surface quality.

Competing interests. The authors declare that they have no conflict of interest.

Edited by: Xichun Luo

References

Bamberg, E. and Heamawatanachai, S.: Orbital electrode actuation to improve efficiency of drilling micro-holes by micro-EDM, J. Mater. Proc. Tech., 209, 1826–1834, 2009.

Bozdana, A. T. and Ulutas, T: The effectiveness of multi-channel electrodes on drilling blind holes on Inconel 718 by EDM process, Mater. Manuf. Process., 31, 504–513, 2016.

Dwivedi, A. P. and Choudhury, S. K.: Effect of tool rotation on MRR, TWR, and surface integrity of AISI-D3 steel using the rotary EDM process, Mater. Manuf. Process., 31, 1844–1852, 2016.

Fang, P. and Manglik, R. M.: The influence of inner cylinder rotation on laminar axial flows in eccentric annuli of drilling bore wells, Int. J. Transp. Phenom., 4, 257–274, 2002.

Hung, J. C., Lin, J. K., Yan, B. H., Liu, H. S., and Ho, P. H.: Using a helical micro-tool in micro-EDM combined with ultrasonic vibration for micro-hole machining, J. Micromech. Microeng., 16, 2705–2713, 2006.

Jahan, M. P., Wong, Y. S., and Rahman, M.: Evaluation of the effectiveness of low frequency workpiece vibration in deep-hole micro-EDM drilling of tungsten carbide, J. Manuf. Process., 14, 343–359, 2012.

Kong, W., Guo, C., and Zhu, X.: Simulation analysis of bubble motion under ultrasonic assisted electrical discharge machining, in: 3rd International Conference on Machinery, Materials and Information Technology Applications (ICMMITA), 28–29 November 2015, Qingdao, China, 2015.

Munz, M., Risto, M., and Haas, R.: Specifics of flushing in electrical discharge drilling, Proc. CIRP, 6, 83–88, 2013.

Nastasi, R. and Koshy, P.: Analysis and performance of slotted tools in electrical discharge drilling, Ann. CIRP: Manufact. Technol., 63, 205–208, 2014.

Plaza, S., Sanchez, J. A., Perez, E., Gil, R., Izquierdo, B., Ortega, N., and Pombo, I.: Experimental study on micro EDM-drilling of Ti6Al4V using helical electrode, Precis. Eng., 14, 821–827, 2014.

Selvarajan, L., Narayanan, C. S., and Jeyapaul, R.: Optimization of process parameters to improve form and orientation tolerances in EDM of MoSi2-SiC composites, Mater. Manuf. Process., 30, 954–960, 2015.

Selvarajan, L., Narayanan, C. S., and Jeyapaul, R.: Optimization of EDM parameters on machining Si3N4-TiN composite for improving circularity, cylindricity, and perpendicularity, Mater. Manuf. Process., 31, 405–412, 2016.

Shabgard, M. R. and Alenabi, H.: Ultrasonic assisted electrical discharge machining of Ti-6Al-4V alloy, Mater. Manuf. Process., 30, 991–1000, 2015.

Theory Guide: ANSYS FLUENT 12.0, Ansys Inc., 2009.

Wang, J. and Han, F.: Simulation model of debris and bubble movement in consecutive-pulse discharge of electrical discharge machining, Int. J. Mach. Tool. Manu., 77, 56–65, 2014a.

Wang, J. and Han, F.: Simulation model of debris and bubble movement in electrode jump of electrical discharge machining, Int. J. Adv. Manuf. Tech., 74, 591–598, 2014b.

Xie, B., Zhang, Y., Zhang, J., and Rend, S.: Numerical study of debris distribution in ultrasonic assisted EDM of hole array under different amplitude and frequency, Int. J. Hybrid Inf. Technol., 8, 151–158, 2015a.

Xie, B., Zhang, Y., Zhang, J., Dai, Y., and Liu, X.: Flow field simulation and experimental investigation of ultrasonic vibration assisted EDM holes array, Int. J. Control Autom., 8, 419–424, 2015b.

Yadav, U. S. and Yadava, V.: Experimental investigation on electrical discharge drilling of Ti-6Al-4V alloy, Mach. Sci. Technol., 19, 515–535, 2015.

Yahagi, Y., Koyano, T., Kunieda, M., and Yang, X.: Micro drilling EDM with high rotation speed of tool electrode using the electrostatic induction feeding method, Proc. CIRP, 1, 162–165, 2012.

Yu, Z. Y., Rajurkar, K. P., and Shen, H.: High aspect ratio micro-hole drilling aided with ultrasonic vibration and planetary movement of electrode by micro-ED M, Ann. CIRP: Manufact. Technol., 58, 213–216, 2009.

Zhang, W., Liu, Y., Zhang, S., Ma, F., Wang, P., and Yan, C.: Research on the gap flow simulation of debris removal process for small hole EDM machining with Ti alloy, in: 4th International Conference on Mechatronics, Materials, Chemistry and Computer Engineering (ICMMCCE), 12–13 December 2015, Xi'an, China, 2015.

13

An experimental characterization of human falling down

Libo Meng[1,2]**, Marco Ceccarelli**[3,4]**, Zhangguo Yu**[1,2,4]**, Xuechao Chen**[1,2,4]**, and Qiang Huang**[1,2,4]

[1]Intelligent Robotics Institute, School of Mechatronical Engineering, Beijing Institute of Technology, 5 Nandajie, Zhongguancun, Haidian, Beijing 100081, China

[2]Key Laboratory of Biomimetic Robots and Systems, Ministry of Education, State Key Laboratory of Intelligent Control and Decision of Complex Systems, Beijing Institute of Technology, 5 Nandajie, Zhongguancun, Haidian, Beijing 100081, China

[3]LARM: Laboratory of Robotics and Mechatronics, DICeM, University of Cassino and South Latium, Via Di Biasio 43, 03043 Cassino, Fr, Italy

[4]Beijing Advanced Innovation Center for Intelligent Robots and Systems, Beijing Institute of Technology, China

Correspondence to: Xuechao Chen (chenxuechao@bit.edu.cn)

Abstract. This paper presents results of an experimental investigation on the falling down of the human body in order to identify significant characteristics and parameters. A specific lab layout has been settled up with vision tracking system and suitable sensors to monitor information on trajectories, impact force and acceleration during the falling with elaboration procedures that make fairly easy to track and interpret the motion characteristics. We focus on the more often falling mode: forward and backward falling Tests are discussed with results from lab tests that give both behavior and values of the biomechanics of falling down of the human body. Possible protection strategies for falling based on the proposed research are talked about at the last.

1 Introduction

Tracking the motion of human is a major source of inspiration for understanding the characteristics of human motion and designing motion for robots. Gupta et al. recognize the human activities using the image sequences with action labels (Gupta et al., 2014). Roos built a walking model to predict the different risk of falling down for elder people (Roos et al., 2013). In the field of robotic, human motion data also provides a real inspiration for robot design and control. For instance, Huang et al. proposed a similarity evaluation between the human motion and robot motion and developed complex motion for the humanoid robot based on the evaluation (Huang et al., 2010). Zielinska et al. discussed the problem of using the human gaits to generate legged locomotion for biped robots. They were inspired by the biological Center of Gravity (COG) to produce leg joint trajectories (Zielinska et al., 2009). Zhao et al. design human-like motion for robotic arms using resolving the kinematic motion of human arm (Zhao et al., 2014).

In particular, falling down has attracted special interest. Such studies of human motion falling down can be found at the following researches. Robinovitch et al. have studied the human down for decades. They examined the situation of falling backward of a human. And the results showed that squatting during falling down decrease the impact severity. They also investigated the influence of two falling directions: forwards and backward, then, they used the video camera to analyze the falls in daily life (Hsiao et al., 1997; Sandler and Robinovitch, 2001; Robinovitch et al., 2004; Tan et al., 2006; Yang et al., 2013). Chen et al. proposed a biped model to capture and predict the falling cause (Chen et al., 2015). Hitcho et al. and Lee et al. studied the falling of elder person for home healthcare and hospital setting (Hitcho et al., 2004 and Lee et al., 2012). Ma et al. proposed a method of falling protected for humanoid robot inspired by the research of human falling down (Ma et al., 2014). Tomii et al. proposed a falling detection system using wireless sensors for elder people (Tomii et al., 2012).

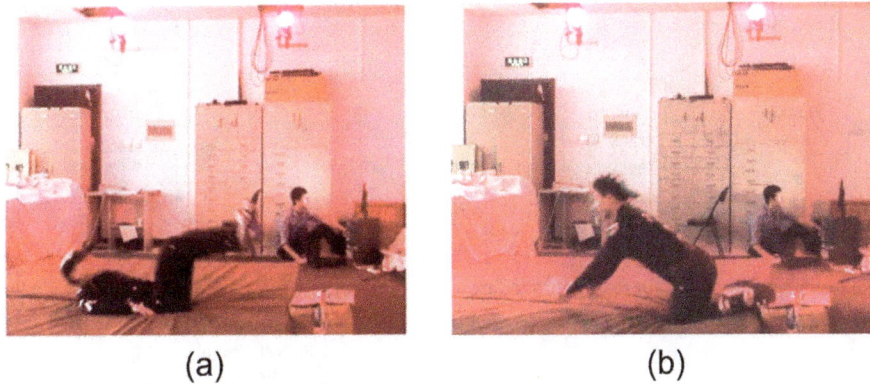

Figure 1. Human falling down in daily life: **(a)** falling backward, **(b)** falling forward.

Human motion can be acquired from different ways. Schmitz et al. used the motion capture system to measure the joints angle of human (Schmitz et al., 2014), Yang et al. introduced a method to analyze the 3-D human gait of through a markerless motion capture (Yang et al., 2014). Ayusawa et al. identified the inverse kinematics and geometric parameters using motion capture data (Ayusawa et al., 2014). Carnegie Mellon University (CMU) offered a valuable database of human motion, acquired by their vision tracking system, including walking, jumping, and sports and so on (Gross and Shi, 2001). Recently, some new studies recognize human motion including the human-object interactions and highly articulated motions using graphene strain sensors and actionlet ensemble model (Wang et al., 2014a, b). Varela et al. got excellent results from measuring the motion of knee and ankle of a human during walking using CaTraSys, a cable based parallel manipulator (Varela et al., 2015).

Above all, there have been some researches in falling down of the human, most of them focus on the aged people and the damage caused by falling. These researches are basically from the kinematic point of view which limits the application. Moreover, the studies about the impact in human fall are rarely reported. In this paper, an experimental layout consisting of motion capture system, accelerometer and force sensors is built, and a group of tests is carried out to acquire the information from human falling down. A fifteen markers model is established in the motion capture system to track the motion of the human, and two force sensor and two accelerometers are used to measure the status of the human when falling down. Thus corresponds to the two meaning of doing the experiments, namely, finding the kinematic and dynamic characteristics of motions from the necessary motion and impacting data. In fact, there are several ways to fall down for human, we examine two most common forms, falling forward and falling backwardin this paper. We use the forward and backward fall as the simplified model of human fall. The trajectories of primary points of human are evaluated, and motion strategies for reducing the impact force are summarized.

2 Characteristics of human falling down

Human falling down is common in daily life. Figure 1 shows the two most common types of a human falling down: falling backward and falling forward. In the Fig. 1a, the person fell backward, and in the Fig. 1b, the man fell forward. The fall of both of them is caused by lost balance, a possible reason for a fall in daily human life (Hsiao and Robinovitch, 1997). Lost balance leads to that human cannot recover to the standing posture, and the fall happens inevitably. Generally, a fall may consist of the following steps: Firstly, a person encounters a disturbance, such as a slip, an obstacle, or a sudden push. Secondly, the person tries to keep standing posture. If the attempt works, the person will not fall down. Otherwise, the person loses balance and falls down. Then, during the period of falling, the person uses his/her body motion, such as knee bending, waist bending and using the arm to contact the ground, to reduce the impact as far as possible. Finally, the person contacts on the ground and rolls on the ground to reduce the impact sometimes.

In this paper, we focus on the motion behavior and characteristics of a human falling in the situation when the falling cannot be avoided. In order to analyze the human falling after lost balance in detail, essential data of falling down should be acquired:

The detectable incipient motion.

- The acceleration changes during the falling.

- The trajectories of the main parts of a human.

- The motion after contacting on the ground.

- The motion strategies that can protect human from being hurt.

It is essential to recognize the detectable incipient motion, as it is the identification of whether a fall will happen or not. Through the acceleration changes, we can get the dynamic characteristics of the fall. The trunk and the head are two of the most critical components of the human body. It is reasonable to measure the variation of the acceleration of these

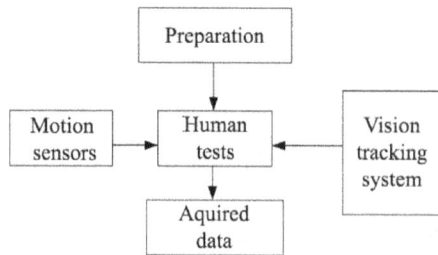

Figure 2. Structure of the experimental layout for testing procedure of human falling down.

two parts, especially at the time of impacting on the ground. The motion characteristic can be summarized by the trajectories of the main parts of the human body, such as the waist, back, knee, and ankle. Each motion, like bending the knee, can be expressed by the combination of different trajectories of body parts. The fall will not be over until the body stops to move. Thus, the motion after contacting on the ground is also essential for acquiring the characteristic of falling. We can summarize the motion and dynamic characteristics and propose some protection strategies, which can help us to understand the falling better.

3 Experimental layout

Experimental tests can be carried out to acquire the behaviors and characteristics of human motion in falling down. In this work an experimental layout with a measurement system has been settled up in the lab environment for monitoring purposes to understand configurations of human falling and to characterize numerical the motion. The measurement system consists of an optical motion capture system (from Motion Analysis Corp, Motion Analysis Corporation, 2010), two force sensors (from NITTA Corp.), and two accelerometers (from Xsens Corp.). Subjects were tested in this experimental environment by falling forward and backward. Motion trajectories were obtained by the optical motion capture system. Force sensors measured the impact force, and the acceleration of the head and chest was acquired by accelerometers. Detail experimental procedure will be introduced in the next section.

The structure of the experimental layout is designed as in the schemes in Fig. 2, four processes should be done for the tests for human, namely, preparation for the test, debugging the motion capture system and the accelerometers and the force sensors, and arranging the lab environments.

The sequence of procedure for the proposed tests is illustrated in the Fig. 3. The test includes three main sections: preparation of the test, conducting a test and the numerical analysis of the test results. The preparation consists of calibrating the sensors and the motion capture system and guiding the subjects to run the test in a standard way. Then, the tests are conducted under the direction of the operators for

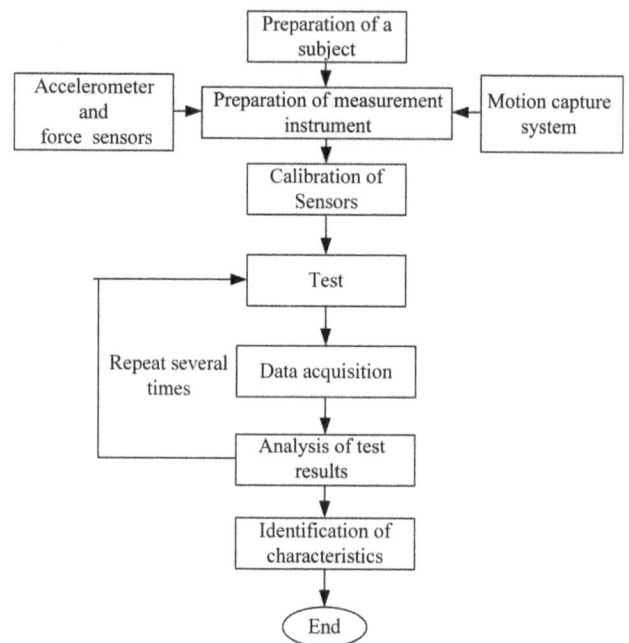

Figure 3. Flowchart for an experimental testing procedure of human falling down.

several times. During each test, the motion data are obtained from the motion tracking system. The last section is to analyze the acquired data for an evaluation of the test of falling down.

The experimental layout is explained in Fig. 4. As is shown in the Fig. 4, a gymnastic mattress with a thick of 30 cm was located on the ground as a landing surface for the fall. During all trials, the subjects stood on a stage at the front of the gymnasium mattress and then fell backward and forward. A 12-camera motion capture system was used to trace the motion trajectory of the objects from three-dimension. The frequency of Motion Capture System is 60 Hz. The resolution of this system is 0.01 mm. A regular video camera from side view was adopted to record the whole process of the falling. When human fall forward or backward, the entire chest or back impact on the mattress at different time. In order to detect the force during the whole process of the human fall, at least two force sensor should be used. During each trial, two force sensors were put on the mattress to measure the contact force between the subjects and mattress. As the frequency of the Motion Capture System is not fast enough to compute the acceleration the moment of contact, two accelerometers attached to the subjects' chest and head, which are the two of the most important parts of the human body, were used to measure the acceleration during the falling. The frequency of the acceleration sensor is 100 Hz. All the sensors were connected to one PC by CAN bus so that the acquired data can be synchronous.

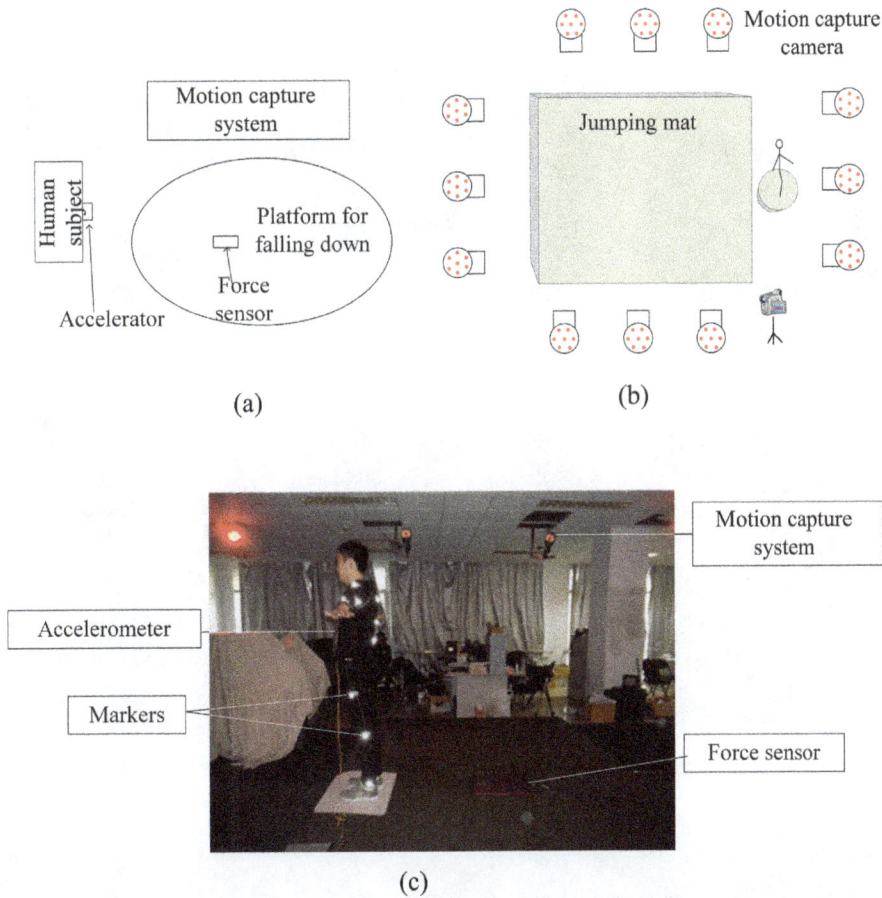

Figure 4. Experimental layout at BIT lab in Beijing: (a) a general top view scheme; (b) a scheme of lab arrangement; (c) a photo of the lab arrangement.

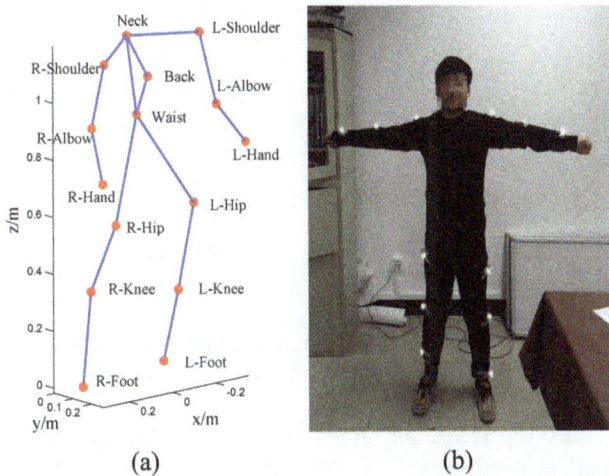

Figure 5. Mode of testing human backward falling down: (a) a scheme with sensors location; (b) a lab test.

4 Testing modes

In this paper, results are reported for the case of human falling down forward and backward. In order to obtain the data of human falling down, five healthy subjects (age 24–26 years old, height 168–173 cm, weight 65–70 kg, male and female) were recruited in the test activity for the research from Beijing Institute of Technology (see Table 1). 15 spherical markers were attached to the subject's neck, shoulders, elbows, hands, back, waist, hips, knees, and ankle. In order to track the markers without interference, a special cloth with a dark color is recommended (see Fig. 5b). This suit, including a tight coat and trousers, keeps the markers still relative to the subject. A human model based on 15 segments was built as Fig. 5 in the Motion Capture System. Basically, 15 markers can be considered enough to describe the falling motion of this work. The reason why we use the 15 markers model is that the 15 markers model has several advantages:

- Simple computation.
- Compact representation.
- Main segments of human body (leg, trunk and arm).

Figure 6. Different ways of falling forward.

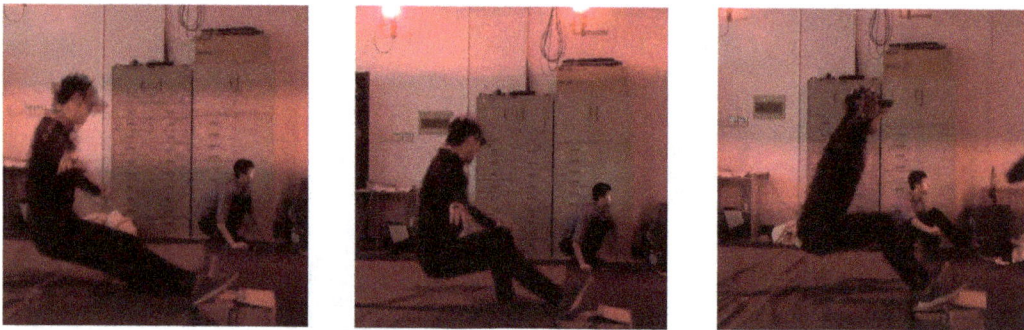

Figure 7. Different ways of falling backward.

Table 1. The parameters of the subjects.

Subjects Number	Age	Weight(kg)	Height(m)
5	24–26	65–70	1.68–1.73

– Main DOFs of human body.

In Fig. 8, a model for falling test is described. The falling test procedure is shown in the Fig. 9. Before the test, the testers interpreted the detailed procedure to the subjects as follows: All the falls were initiated from a standing position (see Fig. 9a). When a test began, the subject should be asked "are you ready?" The subject would fall down by a sudden push within a random time (1–15 s) after he or she gave the answer "Yes" (see Fig. 9b). Figures 8c and 9d illustrate the following configuration of falling down: The subject controls body motion to make a relatively safe way to contact on the mattress. After impacting on the mattress, the subject reaches the final configuration (see Fig. 9d). Since the fall is heavily affected by the pushing force, three factors are taken into consideration: (1) All the subjects shouldn't be hurt by the tests. (2) All the experiments should be end with the human falling. (3) The pushing force in different tests should be basically the same. The testers and subjects had practiced to how to push and fall for several times before the tests. Figures 6 and 7 show different ways to fall forward and backward. The third falling motion in the Figs. 6 and 7

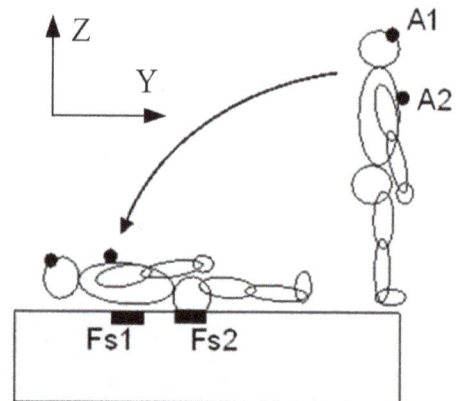

Figure 8. A model for the lab test of human falling down.

were selected as the fall motion in the final tests because that they had the smallest impact force compared the other falling styles.

Each repeated eight times for falling backward and forward, respectively.

5 Results of tests

In this section, the results of the test are reported and analyzed as based on the data acquired from the motion capture system, force sensors, and accelerometers. The markers of back, waist, hip, and ankle are selected to interpret the char-

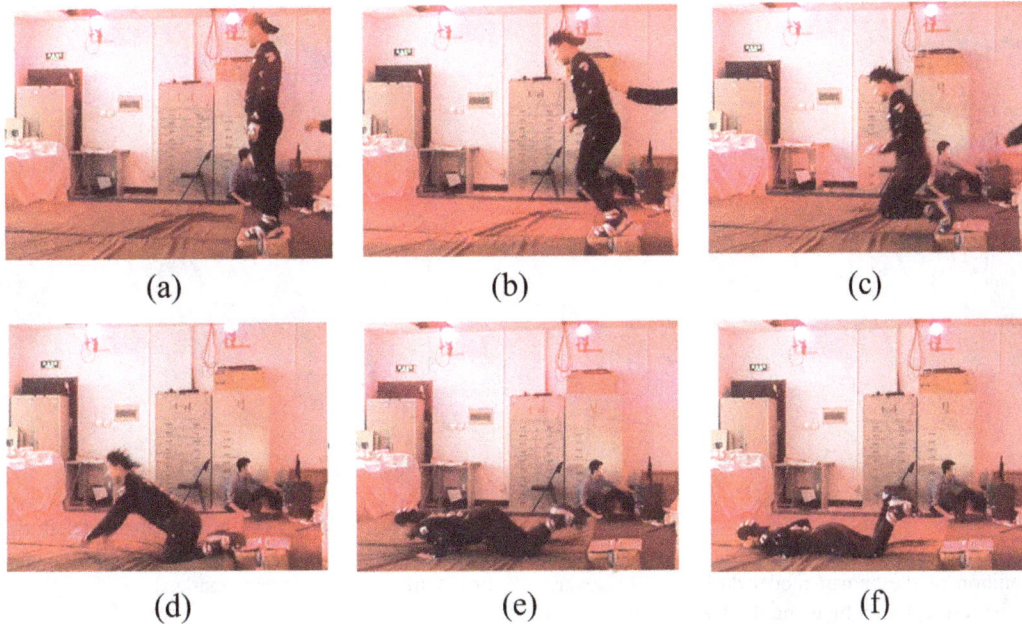

Figure 9. A snapshot of a lab test of a forward human falling down: (**a**) initial standing up; (**b**) incipient falling; (**c**) configuration during the falling; (**d**) impact on the mattress; (**e**) absorbing the impact; (**f**) final configuration.

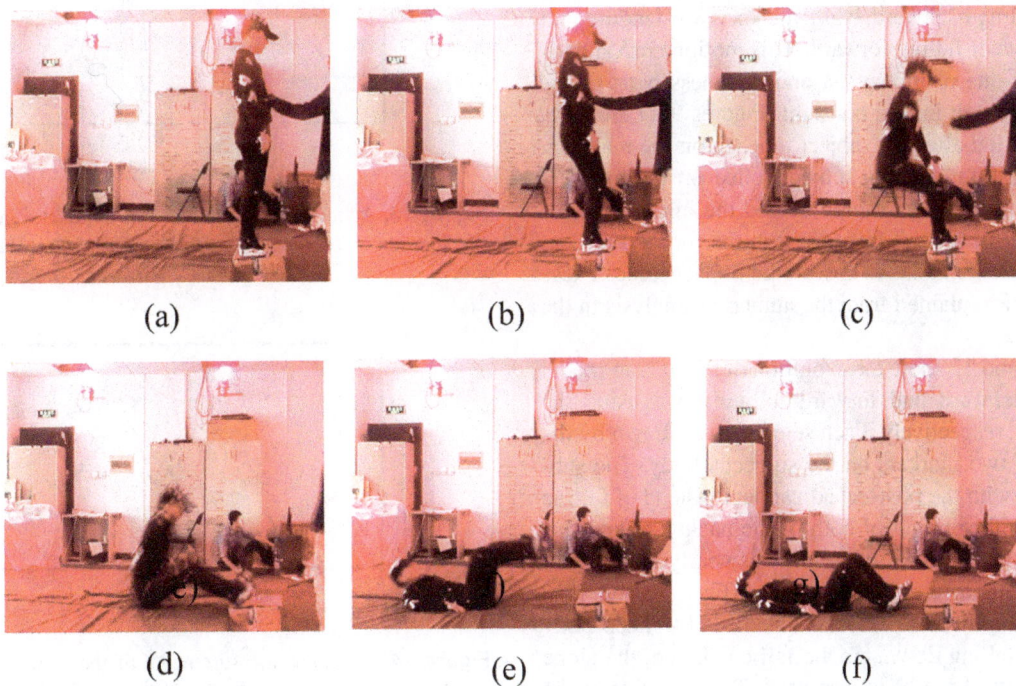

Figure 10. A snapshot of a lab test of backward human falling down: (**a**) initial standing up; (**b**) incipient falling; (**c**) configuration during the falling; (**d**) impact on the mattress; (**e**) absorbing the impact; (**f**) final configuration.

acteristic of a human falling in the sagittal plane. The body motion is to be discussed in detail in this section using numerical analysis. The results discussed in the following are from the mean of all the acquired data. We use the time the impact of the subject on the mattress as the reference time to synchronize the motion data from different subjects.

Figures 9 and 10 are snapshots of the falling backward and forward of one test. Figures 9c, d and 10c, d show the subject changed his body gait to adjust the landing parts to the

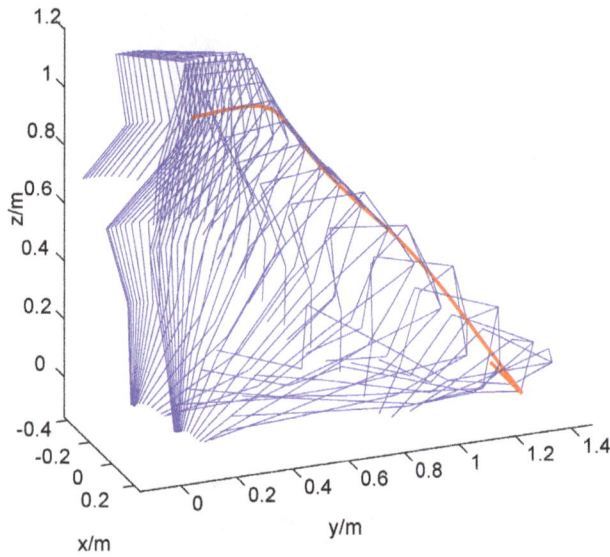

Figure 11. Evolution of the human model during the backward falling down of the test in Fig. 8 by using the body segments (red line is the trajectory of back marker).

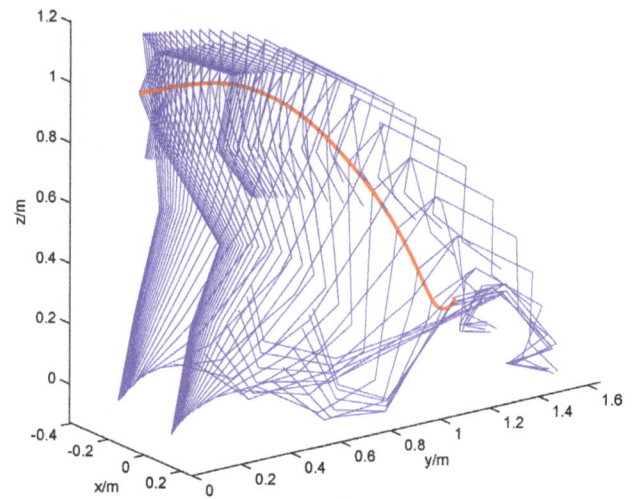

Figure 12. Evolution of the human model during the forward falling down of the test in Fig. 7 by using the body segments (red line is the trajectory of back marker).

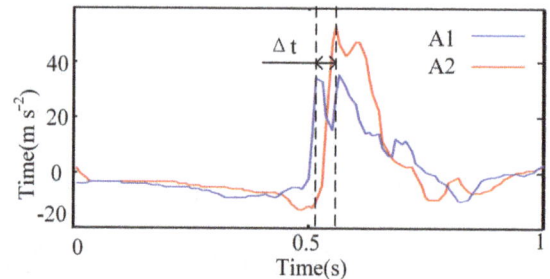

Figure 13. Acquired measurements of the backward falling down acceleration in the sensor A of the layout in Fig. 1 during the test in Fig. 8.

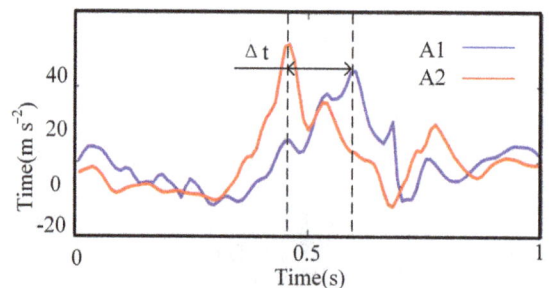

Figure 14. Acquired measurements of the forward falling down acceleration in the sensor A of the layout in Fig. 1 during the test in Fig. 7.

mattress. The hip was the most frequent landing parts among the tests of falling backward, while the knee was landing part among the tests of falling forward. This motion corresponds that the hip is one of the softest and thickness parts of the body, and has tough impact resistant. Figures 9e, f and 10e, f show the motion after the subject touched on the mattress. In the backward falling, the subject rolled on the mattress after landing. This motion could extend the impact time as well as reduce the impact force. And in the forward falling, the subject used the hands to absorb the impact force. These motions will be explained from the numerical analysis in the following part.

Figures 11 and 12 show the evolution of the human model during the backward and forward falling down using the body segment, respectively. Each segment was drawn by the trajectories of two markers on the subject's body. The subject fell down with the postural adjustment instead of falling directly. At the beginning of both of the backward and forward falling, the height of the trunk was nearly unchanged. This corresponds to the incipient phase (Figs. 9b and 10b) of the fall and indicates that the subject tried to keep body balance from falling down. As the fall continued, the slope of the trajectory of the back increased. This was due to the motion of knee bending and waist bending, which shortened the distance between the ankle and the trunk (Figs. 9c, d and 10c, d). At the end of the fall, the trajectory of the back had a sudden bounce. This corresponds the elasticity of the landing surface, and may not happen during a falling on the ground.

The acquired values of the accelerometer attached to the subject's chest and head are shown in the Figs. 11 and 13. The red and blue curve represent the acceleration the chest and head, respectively. Both of them have similar characteris-

tics. The acceleration increased after the beginning of falling down and had a massive changing when the subject impacted on the mattress. In both of the two kinds of falling, the time between the beginning of falling and impact on the mattress is about 0.5 s. Even though the landing surface is a soft mattress, the max acceleration reaches almost 6 g. It is to note

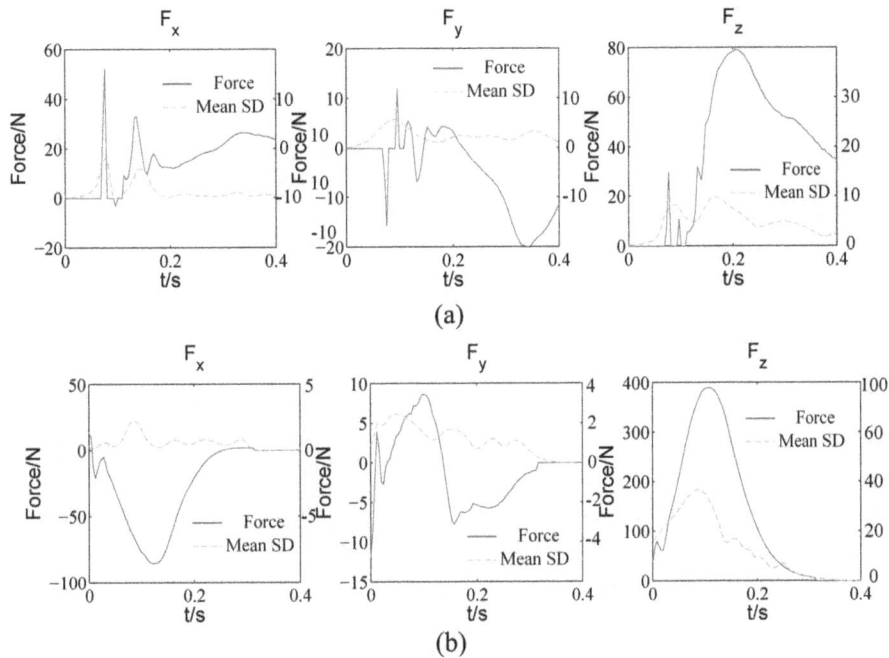

Figure 15. Acquired measurements of the backward falling down in the sensor FS1 and FS2 of the layout in Fig. 1 during the test in Fig. 8: (a) components of the force of FS1; (b) components of the force of FS2.

that the sudden acceleration change is measured over a period of about 0.2 s (Δt). This 0.2 second is nearly the same as the time of the impact between the subject and the landing mattress. Particularly, the movement range of the head was smaller than the chest. It is due that the head is the most important parts of the body, and during all the tests, the head was observed hardly to impact on the mattress.

The Figs. 15 and 16 show the force that was acquired from the two force sensors above the mattress. In order to show the repeatability of the tests, mean standard deviation was calculated (red line in each picture). The consistency is within acceptable ranges. We only focus on the moment around the impact (0.1 s before the impact, 0.3 s after the impact). In the backward falling, the subject impacted the force sensor Fs1 about 0.1 s after the force sensor Fs2, and the contact time impacting each sensor lasts about 0.2 to 0.3 s. The short period shows the rolling motion on the mattress, a protection strategy, to resist the impact injuries (Fig. 8d and e).

The force in x and y direction vibrated near the zero after impact. This indicates that slight rotation happened after the subject impacted on the mattress. In the forward falling, the force was much smaller than the force measured from the backward falling, and the magnitude of Fs1 and Fs2 were nearly the same. This is due to that the subject contacted the landing surface with the knee and then the hand, most of the impact force was absorbed. It is to note that compared to the force in the z direction, the impact force in the x and y direction are very small, this indicates that the protection should pay more attention on the collision rather than the rotation.

Nevertheless, the lateral force cannot be neglected. The time evaluation of the acquired force the impact shows that the impact is not instantaneous and it lasts with accommodation of the body with the distribution of the actions, although it is evident the first impulsive contact.

In order to emphasize the difference of the motion of main parts of the body, the trajectories of ankle, hip, knee and back were compared in the sagittal plane in Figs. 17 and 18 in backward and forward, respectively. The trajectories are in the sagittal plane. The arrows represent the motion direction. The red, blue, and green curves show the trajectories of the ankle, hip, and knee, respectively. The trajectory of the back is not a complete circle that rotates around the ankle, which indicates that the subjects used the body motion to change his/her gaits. In all the tests, the motion of the hip and the back had the nearly same trajectories. The relative significant difference exists between the trajectories knee and other makers, due to the knee bending and waist bending. In both of the forward and backward falling, the ankle lifted into the air from the ground while the body was falling down (Figs. 9e and 10e). It helped the subject to finish the rolling motion after contacted on the landing surface and reduce the impact force in the backward falling, due to that the rolling motion increases the impact time.

This motion made the waist impact before the back, thus had been improved by the time interval of two force sensors, and resisted severe damage of head.

Figures 19 and 20 show the computed angle of the knee and hip joint, respectively. The angles are computed by the

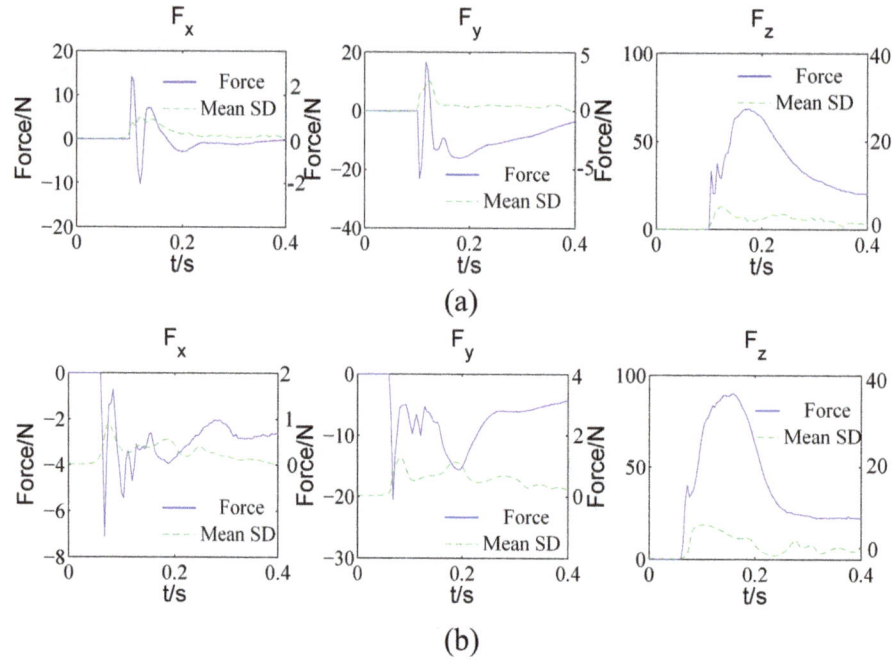

(a)

(b)

Figure 16. Acquired measurements of the forward falling down in the sensor FS1 and FS2 of the layout in Fig. 1 during the test in Fig. 8: (a) components of the force of FS1; (b) components of the force of FS2.

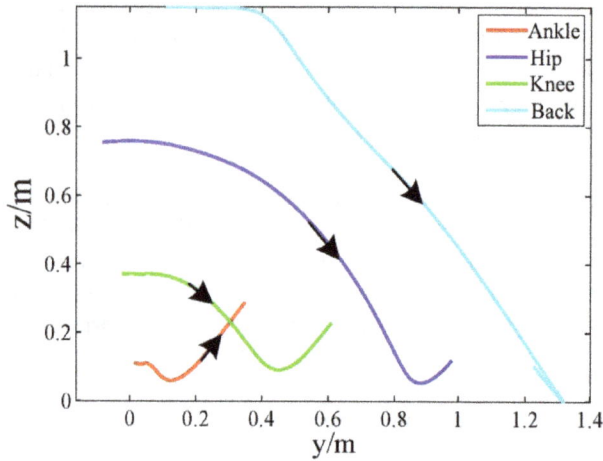

Figure 17. Trajectories of reference points of the ankle, back, hip and knee in the sagittal plane in the test of falling backward in Fig. 8.

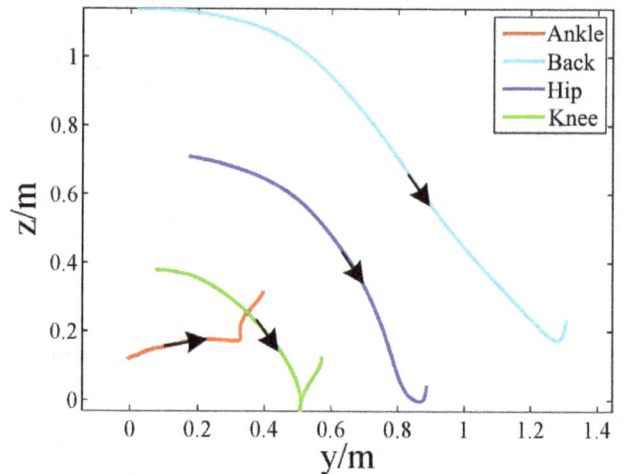

Figure 18. Trajectories of reference points of ankle, back, hip and knee in the sagittal plane in the test of falling forward in Fig. 7.

Eqs. (1) and (2):

$$\theta_{knee} = \cos^{-1} \tag{1}$$

$$\left(\frac{\begin{array}{c}(Y_{hip} - Y_{knee})^2 + (Z_{hip} - Z_{knee})^2 + (Y_{ankle} - Y_{knee})^2 \\ + (Z_{ankle} - Z_{knee})^2 - (Y_{hip} - Y_{ankle})^2 - (Z_{hip} - Z_{ankle})^2\end{array}}{2\sqrt{(Y_{hip} - Y_{knee})^2 + (Z_{hip} - Z_{knee})^2} \cdot \sqrt{(Y_{ankle} - Y_{knee})^2 + (Z_{ankle} - Z_{knee})^2}} \right)$$

$$\theta_{hip} = \cos^{-1} \tag{2}$$

$$\left(\frac{\begin{array}{c}(Y_{hip} - Y_{knee})^2 + (Z_{hip} - Z_{knee})^2 + (Y_{hip} - Y_{back})^2 \\ + (Z_{hip} - Z_{back})^2 - (Y_{back} - Y_{knee})^2 - (Z_{back} - Z_{knee})^2\end{array}}{2\sqrt{(Y_{hip} - Y_{knee})^2 + (Z_{hip} - Z_{knee})^2} \cdot \sqrt{(Y_{hip} - Y_{back})^2 + (Z_{hip} - Z_{back})^2}} \right)$$

where, the $(X_{hip}, Y_{hip}, Z_{hip})$, $(X_{knee}, Y_{knee}, Z_{knee})$, $(X_{back}, Y_{back}, Z_{back})$ stand for the coordinates of the hip knee and back, respectively.

In the backward falling, the angles variation range of the hip joint is larger than that of the knee joint. The joints reached the maximum angle at a different time. It can be concluded that the knee bending motion began firstly, and then stretched. These motions also could be seen from the Fig. 10b to d.

Table 2. Summary of acquired date for characteristics of human falling down.

Parameters	A_{ccH} (m s^{-2})	Acc$_T$(m s^{-2})	F_{1z}(N)	F_{2z}(N)	θ_{knee}(°)	θ_{waist}(°)
Backward falling	−6–38	−10–52	0–80	0–400	120–180	80–180

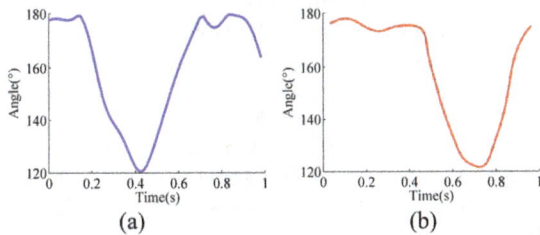

Figure 19. The computed angle of the knee joint and waist joint when falling backward in the test of Fig. 8: **(a)** the angle of knee, **(b)** the angle of waist.

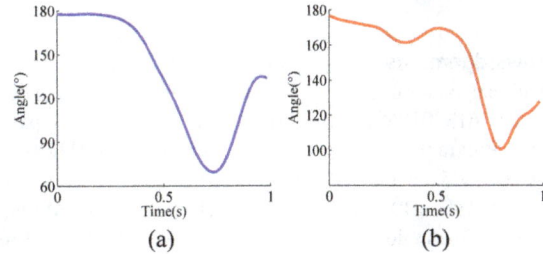

Figure 20. The computed angle of the hip joint when falling forward in the test of Fig. 7: **(a)** the angle of knee, **(b)** the angle of waist.

The case in forward falling was opposite. The angle variation of the knee joint was larger than the hip joint. The maximum angle of the knee reached about 60°, and the waist was about 80°. These two joints reached the maximum angle at nearly the same moment when the subject landed on the mattress (Fig. 9d). Then, the subject stretched the waist and knee to contact the mattress with the trunk. It resulted the force measured by the force sensors were not as large as the force in the backward falling.

The average value of the main parameters, which represent the characteristics of the human falling down, is summarized in Table 2.

The range of the acceleration of the forward and backward falling is almost the same. It is noticed that acceleration of the head (Acc$_H$) is smaller than that of the trunk (Acc$_T$) in both situations. In order to ensure the head safety, the subjects tried to protect the head from the first impact on the landing surface. This helped to reduce the sudden change of the acceleration of the head.

The main difference between the forward and the backward falling is in the knee and waist motion. In the forward falling, the knee had a larger range (θ_{knee}) of motion than the waist (θ_{waist}). In the backward falling, it is just the opposite. The data acquired from the force sensors (F_{1z} and F_{2z}) also confirm this difference. The impact force of the trunk was much smaller than that of the waist in the backward falling due to that the waist impacted on the landing surface first. While in the forward falling, both of the impact force of trunk and waist were small as a result of that the knee and arm contacted the mattress first, and absorbed most of the impact force.

Nevertheless, the combination motion of the knee and the waist generated the same results: impacting on the landing surface with the strongest and muscular parts of the body. Thus, the impact force could be resisted and absorbed, which

is the most distinct characteristic of the human falling down. We can summarize some protection strategies based on the motion and dynamic characteristics mention above as following:

– Forward falling: Using the knees as the first place to touchdown can reduce the falling momentum. The impact force of the chest can be reduced to a small value (less than 100 N) by using the hands.

– Backward falling: The hip can absorb most of the impact force (about 400 N). A rolling movement after the impact is recommended to reduce the vertical impact force. Body balance after impact can be kept by the arm motion.

– Forward and backward falling: Bending the knee and hip joint to lower the center of mass is an effective movement to reduce the impact force and acceleration changing. The arm is a secondary choice to absorb the impact force because it has relative fragile structure.

6 Conclusions

In this paper, a motion data acquired system was used for the tests of human falling. This system consists of motion capture system, accelerometer sensors, and force sensors. Two kinds of falling, backward and forward, were taken into consideration. The experimental tests and data acquisition were conducted with a detailed procedure. Results from the tests were elaborated to interpret the characteristics of the human falling down. Angles of knee and hip joints were computed for quantitative analysis the motion of human falling. Some protection strategies based on the motion and dynamic characteristics mention are proposed in order to better understand the human falling. The paper shows the characteristic motions and actions in human falling down with a numerical characterization of human parameters.

Competing interests. The authors declare that they have no conflict of interest.

Acknowledgements. This work was supported in part by the National Natural Science Foundation of China under Grant 61533004, 61320106012, 61375103, and 61321002, in part by the 863 Program of China under Grant 2015AA043202, and 2015AA042305, in part by the Key Tech. R&D Program under Grant 2015BAF13B01 and 2015BAK35B01, the Beijing Municipal Science and Technology Project under Grant D161100003016002, and in part by the "111" Project under Grant B08043.

Edited by: A. Müller

References

Ayusawa, K., Ikegami, Y., and Nakamura, Y.: Simultaneous global inverse kinematics and geometric parameter identification of human skeletal model from motion capture data, Mech. Mach. Theory, 74, 274–284, 2014.

Chen, K., Trkov, M., Yi, J., Zhang, Y., Liu, T., and Song, D.: A robotic bipedal model for human walking with slips. In Robotics and Automation (ICRA), 2015 IEEE International Conference on (pp. 6301–6306). IEEE, 2015.

Gross, R. and Shi, J.: The cmu motion of body (mobo) database, 2001.

Gupta, J. P., Dixit, P., Singh, N., and Semwal, V. B.: Analysis of Gait Pattern to Recognize the Human Activities, arXiv preprint arXiv:1407.4867, 2014.

Hitcho, E. B., Krauss, M. J., Birge, S., Claiborne Dunagan, W., Fischer, I., Johnson, S., and Fraser, V. J.: Characteristics and circumstances of falls in a hospital setting, J. Gen. Intern. Med., 19, 732–739, 2004.

Hsiao, E. T. and Robinovitch, S. N.: Common protective movements govern unexpected falls from standing height, J. Biomech., 31, 1–9, 1997.

Huang, Q., Yu, Z., Zhang, W., Xu, W., and Chen, X.: Design and similarity evaluation on humanoid motion based on human motion capture, Robotica, 28, 737–745, 2010.

Lee, Y. S. and Chung, W. Y.: Visual sensor based abnormal event detection with moving shadow removal in home healthcare applications, Sensors, 12, 573–584, 2012.

Ma, G., Huang, Q., Yu, Z., Chen, X., Hashimoto, K., Takanishi, A., and Liu, Y. H.: Bio-inspired falling motion control for a biped humanoid robot. In Humanoid Robots (Humanoids), 2014 14th IEEE-RAS International Conference on (pp. 850–855), IEEE, 2014.

Robinovitch, S. N., Brumer, R., and Maurer, J.: Effect of the "squat protective response" on impact velocity during backward falls, J. Biomech., 37, 1329–1337, 2004.

Roos, P. E. and Dingwell, J. B.: Using dynamic walking models to identify factors that contribute to increased risk of falling in older adults, Hum. Movement Sci., 32, 984–996, 2013.

Sandler, R. and Robinovitch, S.: An analysis of the effect of lower extremity strength on impact severity during a backward fall, J. Biomech. Eng., 123, 590–598, 2001.

Schmitz, A., Ye, M., Shapiro, R., Yang, R., and Noehren, B.: Accuracy and repeatability of joint angles measured using a single camera markerless motion capture system, J. Biomech., 47, 587–591, 2014.

Tan, J. S., Eng, J. J., Robinovitch, S. N., and Warnick, B.: Wrist impact velocities are smaller in forward falls than backward falls from standing, J. Biomech., 39, 1804–1811, 2006.

Tomii, S. and Ohtsuki, T.: Falling detection using multiple doppler sensors, in: e-Health Networking, Applications and Services (Healthcom), 2012 IEEE 14th International Conference on (pp. 196–201), IEEE, 2012.

Users' manual, Motion Analysis Corporation, Santa Rosa, CA, USA, 2010.

Varela, M. J., Ceccarelli, M., and Flores, P.: A kinematic characterization of human walking by using CaTraSys, Mech. Mach. Theory, 86, 125–139, 2015.

Wang, J., Liu, Z., Wu, Y., and Yuan, J.: Learning actionlet ensemble for 3D human action recognition. Pattern Analysis and Machine Intelligence, IEEE Transactions on, 36, 914–927, 2014a.

Wang, Y., Wang, L., Yang, T., Li, X., Zang, X., Zhu, M., and Zhu, H.: Wearable and highly sensitive graphene strain sensors for human motion monitoring, Adv. Funct. Mater., 24, 4666–4670, 2014b.

Yang, S. X., Christiansen, M. S., Larsen, P. K., Alkjær, T., Moeslund, T. B., Simonsen, E. B., and Lynnerup, N.: Markerless motion capture systems for tracking of persons in forensic biomechanics: an overview, Computer Methods in Biomechanics and Biomedical Engineering: Imaging & Visualization, 2, 46–65, 2014.

Yang, Y., Schonnop, R., Feldman, F., and Robinovitch, S. N.: Development and validation of a questionnaire for analyzing real-life falls in long-term care captured on video, BMC geriatrics, 13, p. 40, 2013.

Zhao, J., Xie, B., and Song, C.: Generating human-like movements for robotic arms, Mech. Mach. Theory, 81, 107–128, 2014.

Zielinska, T., Chew, C. M., Kryczka, P., and Jargilo, T.: Robot gait synthesis using the scheme of human motions skills development, Mech. Mach. Theory, 44, 541–558, 2009.

14

Shape optimization and sensitivity of compliant beams for prescribed load-displacement response

Giuseppe Radaelli[1,2] **and Just L. Herder**[1]

[1]Dept. Precision and Microsystems Engineering, Delft University of Technology,
2628 CD Delft, the Netherlands
[2]Laevo BV, 2628 CA Delft, the Netherlands

Correspondence to: Giuseppe Radaelli (g.radaelli@tudelft.nl)

Abstract. This paper presents the shape optimization of a compliant beam for prescribed load-displacements response. The analysis of the design is based on the isogeometric analysis framework for an enhanced fidelity between designed and analysed shape. The sensitivities used for an improved optimization procedure are derived analytically, including terms due to the use of nonlinear state equations and nonlinear boundary constraint equations. A design example is illustrated where a beam shape is found that statically balances a pendulum over a range of 180° with good balancing quality. The analytical sensitivities are verified by comparison with finite difference sensitivities.

1 Introduction

Mechanism synthesis can be thought of as finding a mechanism with a certain force transmission, a certain motion transmission, or both. For conventional mechanisms the analysis and synthesis of motion and forces can be performed separately. This does not hold in general for compliant mechanisms. In compliant mechanisms, which move due to deformation of slender segments (Howell, 2001), every motion is associated to a restoring force. If a certain force and motion combination is desired at a part of the mechanism which is input and output at the same time, such mechanism is sometimes called a spring (Vehar-Jutte, 2008). Not limited to the conventional coil springs, where a motion over a straight path produces a linear force characteristic, a general spring mechanism can potentially exhibit infinite types of nonlinear load-displacement responses when moved along a general trajectory, which may be non-straight. A load-displacement response, in this context, is defined as the force/moment exerted by the spring given a series of applied boundary displacements/rotations.

Applications of non-linear springs can be found in many design disciplines including prosthetics, assistive devices, MEMS and user products. Often nonlinear springs are applied as balancing mechanisms where either an external load,

e.g. a weight, or an intrinsic stiffness, e.g. in a compliant mechanism, is counteracted by such nonlinear spring (Powell and Frecker, 2005; Chen and Zhang, 2011; Hoetmer et al., 2010). Types of non-linearity that are typically interesting are constant force mechanism, bi-stable or multi-stable mechanisms and negative stiffness mechanisms (Pucheta and Cardona, 2010; Oh, 2008; Tolou, 2012).

A way of obtaining nonlinear spring behaviour is by optimization of the shape of a chosen topology of elements such as rods, beams, shells etc. Other means are to manipulate the topology of a system (Sigmund, 1997; Du and Chen, 2008) or the material properties.

In tailoring the load-response of structures and, more in general, when dealing with large deflections the non-linearity of the equilibrium equations makes optimization more challenging. The optimization procedure, which is an iterative scheme, includes at every step the solution of a nonlinear set of equations, on their turn also iterative. Clear disadvantages are the complexity of the procedure and increased computation time, but also the smoothness of the optimization function space is often compromised due to e.g. singularities and bifurcations in the solution.

Eriksson (2014) proposes a method where tracing the non-linear equilibrium at every optimisation iteration is not needed. This is done by augmenting the system of equilibrium equations such that the unknowns include the displacements and design variables, including a load parameter. By keeping the load parameter constant a sequence of responses to that loading can be obtained for different designs, without computing the whole nonlinear equilibrium every time.

Similarly, in the concept of simultaneous analysis and design (SAND) (Haftka, 1985) the analysis unknowns, e.g. the displacements, and the design variables are all treated equivalently as optimization variables. Since equilibrium is not required at every iteration step, also tracing the nonlinear equilibrium path is not needed every time. At the end of the procedure equilibrium is hopefully satisfied, and the design optimized (Ringertz, 1989).

Also in the compliant mechanisms community several approaches have been proposed to deal with geometrical nonlinearities, often present in compliant mechanisms (Bruns and Tortorelli, 2001; Joo et al., 2001; Du and Chen, 2008).

In the majority of the cases the goal is to optimize the design for the situation where the full load is applied. If, however, the nonlinear load-response itself is what must be optimized, the problem gets more involved. Examples of shape and/or topology optimization where the whole load-response or part of it is optimized can be found in Saxena and Ananthasuresh (1999), Pedersen et al. (2005), Rai et al. (2006), Jutte and Kota (2010) and Leishman and Colton (2011).

In this work we focus on shape optimization and, as such, a simple given topology is assumed. Topology and material optimization are not considered in this work. In the context of shape optimization of structures there is an increasing interest in the isogeometric analysis (IGA) paradigm (Cottrell et al., 2009). This can be considered as an alternative to finite element analysis (FEA) with some peculiar additional advantages. There is an enhanced fidelity between designed shape and analysed shape. This comes from the use of B-splines as basis functions for the computer aided geometric design (CAGD) as well as for the structural analysis. This preserves the original shapes and guarantees a high level of continuity between elements (Hughes et al., 2005).

Having the same geometrical formulation at the basis of design and analysis also gives the advantage that there is no need to spend much effort in meshing, which is done repeatedly in an optimization procedure. Instead there are some well-performing and efficient refinement algorithms that are applied in order to work with a finer discretisation of the shape for the analysis.

A third advantage is that sensitivity properties are derivable in an analytical fashion, which supposedly can improve the efficiency of an optimization procedure (Cho and Ha, 2009; Nagy et al., 2010). Also, the derivable shape sensitivities are more accurate (Koo et al., 2013), leading to preciser results.

While the work done on isogeometric shape optimization is widespread (Nagy, 2011; Wall et al., 2008), specific attention to nonlinear settings is scarce (Kiendl, 2011; Koo et al., 2013). Moreover, while typical problems where the stiffness,

the weight, the volume or stresses are optimized have been analysed extensively (Nagy et al., 2010; Hsu, 1994; Ding, 1986), there is no work known by the authors where the load-displacement of a non-linear spring is optimized within the IGA framework.

The rotation-less character of the degrees of freedom in the used isogeometric formulation gives a complication with respect to the application of rotation constraints on the beam. These constraints, imposed here by nonlinear equations on the control points by Lagrange multipliers, have a relevant impact on the derivation of the sensitivity. Together with the nonlinearity of the equilibrium equations and the unusual type of objective function, this leads to a few non-trivial problems to be dealt with in this paper.

In previous work (Radaelli and Herder, 2014) the authors have applied isogeometric shape optimization to obtain a flexible beam with a rotational load-displacement response that matches a sine. This moment-angle characteristic can be used to balance a pendulum which has a similar but opposite moment-angle characteristic.

The current work is dedicated to the derivation of the sensitivities needed for the shape optimization of a flexible beam with prescribed load-displacement. This procedure is demonstrated on the same case study of the balanced pendulum. This case study is a comprehensible but not trivial case: it requires the stiffness to go from positive, through zero, to negative. The contribution of this paper is to enhance the procedure by adding the sensitivity analysis. Special attention is paid to the formulation of an objective function for general load-displacement tracing cases and to the application of general boundary conditions as nonlinear constraint equations. Putting together these pieces in one work is a contribution that has not been found in literature, but is believed to be helpful for designers of nonlinear springs.

The rest of the paper is structured as follows: After a brief introduction to the IGA framework Sect. 2 is dedicated to the derivation of the objective function and the terms needed for the sensitivity analysis. Section 3 shows a comparison of two optimization runs on a given example problem with various optimization algorithms, with and without use of gradient information. Section 4 shows the result of a validation by comparing the analytical sensitivities with the numerically approximated sensitivities.

2   Method

The present section starts with the problem description of a compliant mechanism with tailored load-displacement response formulated as the minimization of the difference between a desired and an obtained energy-path. Following, a brief explanation is given of the basic concepts of isogeometric analysis (IGA), needed for this paper. There are many literature sources about this method. The reader is referred to Cottrell et al. (2009) for more information. In the remainder all components needed for the evaluation of the sensitivity of the objective function are treated.

2.1 Problem description and objective formulation

In the given context, a typical problem consists of obtaining an elastic system where a given force is provided along a given trajectory. Herein the terms force and trajectory can be used interchangeably with moments and rotations. Examples include constant force mechanisms (Lan et al., 2010), bi-stable and multi-stable mechanisms where the load-displacement response has multiple intersections with the zero-load line (Pucheta and Cardona, 2010; Oh, 2008), and static balancing (Chen and Zhang, 2011) where the resulting load-displacement response neutralizes as much as possible an existing load-displacement response of the system to be balanced. In these examples obtaining a certain nonlinear load-displacement response is the goal of the design process.

Provided that the considered forces are conservative, the problem can be reformulated as obtaining a given potential energy along a given trajectory. The use of energy with respect to forces often proves to be convenient due to its scalar nature. The design challenge, here reformulated as a shape optimization problem, consists thus in finding a certain elastic system which, for a series of prescribed boundary displacements and rotations, is compliant with a given potential energy-path.

A simple two-dimensional elastic beam is treated of which the shape is to be determined. Note that the method can in principle be extended to multiple beams and/or applied to other type of elements like shells or solids. In the example treated in Sect. 3 the beam is connected to the base hinge of a pendulum and the shape of the beam is optimized such that the moment characteristic balances the pendulum, see Fig. 4.

In the current framework the given energy-path is described at a discrete number of steps m resulting in a vector of potential energy values, see Fig. 1,

$$\hat{U}(\delta) \approx \hat{U} = \begin{bmatrix} \hat{\mathcal{U}}_1(\delta_1) & \hat{\mathcal{U}}_2(\delta_2) & \dots & \hat{\mathcal{U}}_m(\delta_m) \end{bmatrix} \quad (1)$$

where the hat $(\hat{\ })$ symbol refers to the target energy while the actual obtained energy-path is denoted by

$$U(\delta) \approx U = \begin{bmatrix} \mathcal{U}_1(\delta_1) & \mathcal{U}_2(\delta_2) & \dots & \mathcal{U}_m(\delta_m) \end{bmatrix}. \quad (2)$$

In the previous equations the calligraphic $\mathcal{U}$ denotes the potential energy of the system at a single configuration. δ is the imposed displacement value corresponding to the point of the trajectory.

In the case of load-displacement tailoring it is often useful to examine the shape of the energy-path without considering its amplitude. In a bending dominated problem the amplitude is scaled by sizing and material parameters through the Young's-modulus and cross-sectional properties, in the assumption that these parameters are constant over the length

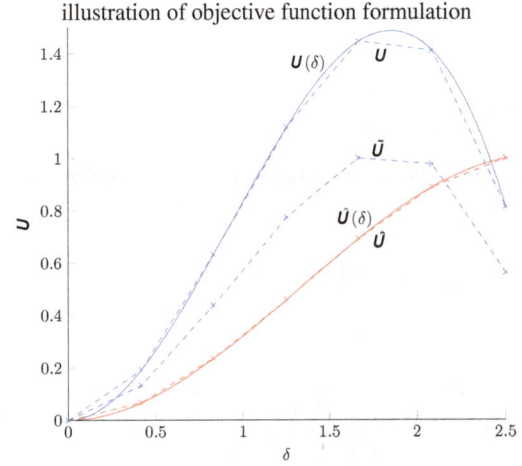

illustration of objective function formulation

Figure 1. Illustration of energy-paths for objective function. $\hat{U}(\delta) \approx \hat{U}$ is the target behaviour, $U(\delta) \approx U$ is the actual obtained behaviour and $\tilde{U}$ is the actual obtained behaviour normalized for an amplitude-independent comparison.

of the beam and that the Euler–Bernoulli assumptions hold (length $\gg$ thickness). Therefore it is useful to do the optimization on the shape of the energy-path, and once the shape of the target energy-path is achieved, one can scale the amplitude by tuning the sizing and material parameters. In particular, the width of the beam can easily be adjusted to match the desired amplitude.

For this reason the target energy-path $\hat{U}$ is required to be bounded between [0, 1]. The normalized obtained energy-path $\tilde{U}$ is defined element-wise by

$$\tilde{\mathcal{U}}_k = \frac{\mathcal{U}_k - U_{\min}}{U_{\max} - U_{\min}} \quad (3)$$

where $U_{\min}$ and $U_{\max}$ are the minimum and maximum elements of the vector U and the index $k = 1 \dots m$ represents the kth entry of the discrete energy vectors.

The proposed objective function is stated as

$$f_0 = \frac{\left(\tilde{U} - \hat{U}\right)\left(\tilde{U} - \hat{U}\right)^T}{\hat{U}\hat{U}^T} \quad (4)$$

which can be interpreted as the normalized sum of squared residuals. As a consequence of the discretisation of the energy-path, the reader should be aware of the fact that the energy difference is minimized only at those discrete points. Fluctuations between the points can theoretically not be excluded. Increasing the resolution of the discretisation helps preventing this.

The given formulation of the objective function is convenient for the sensitivity analysis. Its derivative with respect to the design vector, containing the control point positions

$x = [P_{1_x} \ P_{1_y} \ \dots \ P_{n_x} \ P_{n_y}]^T$, can be derived as

$$\frac{\mathrm{d}f_0}{\mathrm{d}x} = 2\frac{\left(\tilde{U} - \hat{U}\right)}{\hat{U}\hat{U}^T}\frac{\mathrm{d}\tilde{U}}{\mathrm{d}x} \tag{5}$$

where the derivative with respect to the normalized energy-path is

$$\frac{\mathrm{d}\tilde{U}}{\mathrm{d}x} = \frac{(U_{\max} - U_{\min})\left(\frac{\mathrm{d}U}{\mathrm{d}x} - \frac{\mathrm{d}U}{\mathrm{d}x}\Big|_{U=U_{\min}}\right)}{(U_{\max} - U_{\min})^2}$$
$$- \frac{\left(\frac{\mathrm{d}U}{\mathrm{d}x}\Big|_{U=U_{\max}} - \frac{\mathrm{d}U}{\mathrm{d}x}\Big|_{U=U_{\min}}\right)(U - U_{\min})}{(U_{\max} - U_{\min})^2} \tag{6}$$

where $\frac{\mathrm{d}U}{\mathrm{d}x}|_{U=U_{\min}}$ and $\frac{\mathrm{d}U}{\mathrm{d}x}|_{U=U_{\max}}$ are the derivatives $\frac{\mathrm{d}U}{\mathrm{d}x}$ evaluated only for the minimum and maximum entries of U. In the first term this vector, which would be one-dimensional, is replicated and tiled in order to match the dimensions of the vector $\frac{\mathrm{d}U}{\mathrm{d}x}$, from which it is subtracted. Also the $U_{\min}$ in the numerator of the second term is subtracted from all elements of the vector U.

Since the potential energy $\mathcal{U}$ at every configuration is a function of both design variables x and the state variables (displacements) $u(x)$,

$$\mathcal{U}(x, u(x)) \tag{7}$$

the total derivative of the potential energy is given by

$$\frac{\mathrm{d}\mathcal{U}(x, u)}{\mathrm{d}x} = \frac{\partial\mathcal{U}(x, \tilde{u})}{\partial x} + \frac{\partial\mathcal{U}(\tilde{x}, u)}{\partial u}\frac{\mathrm{d}u}{\mathrm{d}x}. \tag{8}$$

The partial derivative of the energy function with respect to the displacement vector u is by definition equal to the internal force vector F_i, which will be used next, but of which the derivation is omitted for the sake of conciseness. Substitution gives

$$\frac{\mathrm{d}\mathcal{U}(x, u)}{\mathrm{d}x} = \frac{\partial\mathcal{U}(x, \tilde{u})}{\partial x} + F_i\frac{\mathrm{d}u}{\mathrm{d}x}. \tag{9}$$

The partial derivative with respect to the design vector x is given in explicit form in Sect. 2.3, while the total derivatives of the displacement vector u with respect to the design vector x is elaborated in Sect. 2.4.

Note that Eq. (9) must be evaluated at every converged load step solution in order to feed $\frac{\mathrm{d}U}{\mathrm{d}x}$ as columns of the matrix $\frac{\mathrm{d}U}{\mathrm{d}x}$ in Eq. (6).

2.2 IGA introduction

Isogeometric analysis is a framework with growing popularity for a number of reasons that particularly hold for shape optimization. First, the fidelity between analysed shape and designed shape. There is no approximation involved in the

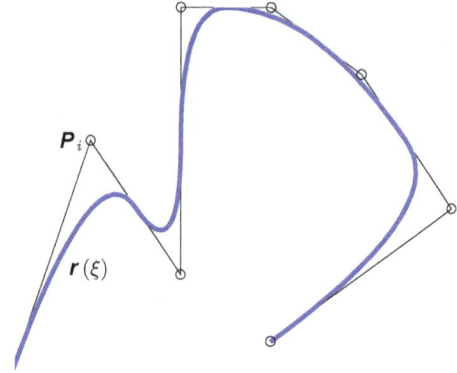

Figure 2. B-spline curve and control polygon.

discretisation of the geometry as with meshing. Instead, a refinement of the parametric description of the geometry is performed, which increases the amount of parameters without altering the geometry itself.

Secondly, and related to the first argument, not needing a conversion step between the geometric description of the design to the analysis gives speed advantages, which are of major impact in optimization where this conversion happens many times.

Third, the availability of analytic derivative information can enhance an optimization procedure significantly.

Since a lot of literature can be found about IGA, just the needed formulas are given to understand the notation. For the notation Nagy (2011) is followed. A B-spline, see Fig. 2, is defined as

$$r(\xi) = \sum_{i=1}^{n} N_{i,p}(\xi)P_i \tag{10}$$

where $P_i \in \mathbb{R}^d$ is a control point, n is the number of control points and where the pth degree basis functions are constructed recursively starting with piecewise constants

$$N_{i,0}(\xi) = \begin{cases} 1 & \text{if } \xi_i \leq \xi < \xi_{i+1}, \\ 0 & \text{otherwise} \end{cases}$$

and for $p = 1, 2, 3, \dots$, they are defined by

$$N_{i,p}(\xi) = \frac{\xi - \xi_i}{\xi_{i+p} - \xi_i}N_{i,p-1}(\xi) + \frac{\xi_{i+p+1} - \xi}{\xi_{i+p+1} - \xi_{i+1}}N_{i+1,p-1}(\xi). \tag{11}$$

In the present work a geometrically nonlinear Euler–Bernoulli beam is used. The potential energy of the beam can be written as

$$\mathcal{U} = \frac{1}{2}\int\left(EA\epsilon^2 + EI\rho^2\right)\mathrm{d}S \tag{12}$$

where the integral is evaluated numerically following a Gaussian quadrature rule using

$$\mathcal{U} \approx \mathcal{U} = \frac{1}{2}\sum_{j=1}^{n+p}\sum_{k=1}^{n_{\text{int}}}\left(EA\epsilon_k^2 + EI\rho_k^2\right)J_k\overline{J}_j\overline{w_k} \tag{13}$$

where E, A and I represent the Young's modulus, the cross-sectional area and the second moment of inertia, respectively, and are assumed constant over the length of the beam. Moreover J is the reference rod Jacobian, $\overline{J}$ and $\overline{w}$ are the Jacobian and the weight associated with the numerical integration, $n + p$ is the number of knot spans and n_{int} is the number of integration points per knot span.

The relative strain measures, i.e. the membrane strain ϵ and the bending strain ρ are given as defined in Simo (1985)

$$\epsilon = \frac{1}{2}\left(\left(\frac{ds}{dS}\right)^2 - 1\right) \tag{14}$$

$$\rho = (\kappa - K)\frac{ds}{dS}. \tag{15}$$

The differential arch length and curvature of the line of centroids of the beam are denoted by the kinematic variables dS, ds, K and κ, where the capital symbols refer to the reference state $R(S)$ and the minuscule symbols refer to the current state $r(S)$. The definitions are

$$ds(\xi) = \left|r^{(1)}\right|d\xi \tag{16}$$

$$\kappa(\xi) = \frac{[r^{(1)} \times r^{(2)}]_3}{\left|r^{(1)}\right|^3} \tag{17}$$

where the subscript 3 refers to the third component of the vector in brackets and the superscript in parenthesis denote successive differentiation w.r.t. ξ, the variable of parametrization. The definition of dS and K is similar to Eqs. (16) and (17), only replacing the curve of the current configuration $r(S)$ by the one of the reference configuration $R(S)$.

It is customary to define the geometry of a system by a relatively small set of control points which are used as optimization variables ($X = P^{\text{design}}$). A refinement step, preserving the exact shape of the curve itself, yields a larger amount of control points ($x = P^{\text{analysis}}$) that are used for the numerical analysis.

The discrete structural arrays, i.e. tangent stiffness matrix K_t and the internal force vector F_i, are derived from the energy functional of Eq. (12). The displacement of the control points of the analysis (x) are the unknowns of the equilibrium equations indicated by the vector of state variables u.

2.3 Energy derivative

Often in structural optimization the stored elastic energy, assumed equivalent to the structural compliance, is simply evaluated as $c = \frac{1}{2}F_e^T u$. While this is correct for a linear analysis, namely it represents the area under the force-displacement curve, this is not valid in general for nonlinear cases. Therefore we must go back to the definition of the strain energy given in Eq. (12) and its numerical approximation given in Eq. (13). The partial derivative of Eq. (13) with

respect to the vector x, where u is kept constant, is given by

$$\frac{\partial \mathcal{U}}{\partial x} = \frac{1}{2}\sum_{j=1}^{n+p}\sum_{k=1}^{n_{\text{int}}}\left(\left(EA\epsilon_k^2 + EI\rho_k^2\right)\frac{\partial J_k}{\partial x} \right.$$
$$\left. + \left(EA\frac{\partial\left(\epsilon_k^2\right)}{\partial x} + EI\frac{\partial\left(\rho_k^2\right)}{\partial x}\right)J_k\right)\overline{J_j}\overline{w_k}. \tag{18}$$

The partial derivatives of the squared strain measures read

$$\frac{\partial\left(\epsilon^2\right)}{\partial x} = 2\epsilon\frac{\partial\epsilon}{\partial x} = 2\epsilon\left[\frac{ds}{dS}\frac{\left(\frac{\partial ds}{\partial x}dS - \frac{\partial dS}{\partial x}ds\right)}{dS^2}\right] \tag{19}$$

and

$$\frac{\partial\left(\rho^2\right)}{\partial x} = 2\rho\frac{\partial\rho}{\partial x} = 2\rho\left[\left(\frac{\partial\kappa}{\partial x} - \frac{\partial K}{\partial x}\right)\frac{ds}{dS} + (\kappa - K)\frac{\frac{\partial ds}{\partial x}}{\frac{\partial dS}{\partial x}}\right] \tag{20}$$

where

$$\frac{\partial ds}{\partial x} = \frac{r^{(1)}R_{i,p}^{(1)}}{\left|r^{(1)}\right|}d\xi \tag{21}$$

and

$$\frac{\partial\kappa}{\partial x} = \frac{\frac{\partial}{\partial x}\left([r^{(1)} \times r^{(2)}]_3\right)\left|r^{(1)}\right|^3 - [r^{(1)} \times r^{(2)}]_3\frac{\partial}{\partial x}\left(\left|r^{(1)}\right|^3\right)}{\left|r^{(1)}\right|^6}. \tag{22}$$

Similar equations hold for $\frac{\partial dS}{\partial x}$ and $\frac{\partial K}{\partial x}$, where again the reference curve is used instead of the current curve. Furthermore

$$\frac{\partial}{\partial x}\left([r^{(1)} \times r^{(2)}]_3\right) = R_{i,p}^{(1)}\left(r_2^{(2)} - r_1^{(2)}\right)$$
$$+ \left(r_1^{(1)} - r_2^{(1)}\right)R_{i,p}^{(2)}. \tag{23}$$

The derivatives of the curves $r^{(1)}$, $r^{(2)}$ and their algorithmic implementations are readily available from e.g. Piegl and Tiller (1997).

2.4 State sensitivity

The more tedious parts of the derivation are not the partial derivatives but the total derivative of the state vector. Usually there is not an explicit relation between x and u, and thus the derivative cannot be found analytically. There are two common methods to compute them numerically. One is the direct method and the other one is the adjoint method. The direct method is used because of its simpler implementation and derivation. There is however no objection in using the adjoint method instead.

There are two aspects in the described situation that make the implementation not trivial. These two aspects have not been found combined in literature. The first aspect is that the set of equilibrium equations is nonlinear and thus requires an iterative, newton-like, procedure to solve it. The second aspect is the application of nonlinear constraint equations as Lagrange multipliers, which creates an augmented system of equations. Both aspects together lead to the following derivations. The formulation of the constraint equations in the present work will be illustrated in Sect. 2.5.

The equilibrium conditions to be solved can be formulated in terms of the tangent stiffness matrix $\mathbf{K}_t$, the internal and external force vectors F_i and F_e

$$\mathbf{K}_t(x, u)\Delta u = F_e(x, u) - F_i(x, u) \tag{24}$$

which is solved for an increment of the displacement vector Δu. While this is a good practice for the solution of the system itself, for the current derivation of the sensitivities it is more convenient to use the alternative formulation in terms of the secant stiffness matrix, where the internal force vector drops out

$$\mathbf{K}_s(x, u)u - F_e(x, u) = \mathbf{0}. \tag{25}$$

Contrary to the former formulation, here the total displacement vector u is used. This is convenient because the constraint equations will be formulated in terms of the total displacements as well. The general set of constraint equations are noted as

$$\mathcal{F}_2 = \mathbf{A}(x, u)u - b(x, u) = \mathbf{0} \tag{26}$$

and adding the constraints as Lagrange multiplier terms to the system in Eq. (25), see Felippa (2004) for a concise explanation, gives

$$\mathcal{F}_1 = \mathbf{K}_s(x, u)u + \mathbf{A}(x, u)^T \lambda - F_e(x, u) = \mathbf{0}. \tag{27}$$

The set of equations in $\mathcal{F}_1$ and $\mathcal{F}_2$ collected into the so called augmented form

$$\begin{bmatrix} \mathbf{K}_s(x, u) & \mathbf{A}(x, u)^T \\ \mathbf{A}(x, u) & \mathbf{0} \end{bmatrix} \begin{pmatrix} u \\ \lambda \end{pmatrix} = \begin{pmatrix} F_e(x, u) \\ b(x, u) \end{pmatrix} \tag{28}$$

is normally solved simultaneously for both u and the Lagrange multipliers λ which, pre-multiplied by the $\mathbf{A}$ matrix, can be interpreted as the forces needed to impose the conditions.

Taking the total derivative of both vector equations $\mathcal{F}_1$ and $\mathcal{F}_2$, applying the product rule and the chain rule and collecting $\frac{du}{dx}$ and $\frac{d\lambda}{dx}$ gives

$$\frac{d\mathcal{F}_1}{dx} = \left(\mathbf{K}_s(x, u) + \frac{\partial [\mathbf{K}_s(x, u)\tilde{u}]}{u} - \frac{\partial F_e(x, u)}{\partial u} \right) \frac{du}{dx}$$
$$+ \mathbf{A}(x, u)^T \frac{d\lambda}{x} + \frac{[d\mathbf{K}_s(x, \tilde{u})\tilde{u}]}{dx}$$
$$+ \frac{[d\mathbf{A}(x, u)^T \tilde{\lambda}]}{dx} - \frac{\partial F_e(x, \tilde{u})}{\partial x} = \mathbf{0} \tag{29}$$

and

$$\frac{d\mathcal{F}_2}{dx} = \mathbf{A}(x, u)\frac{du}{dx} + \frac{d[\mathbf{A}(x, u)\tilde{u}]}{dx} - \frac{db(x, u)}{dx} = \mathbf{0}. \tag{30}$$

Here and in the following the tilde $\tilde{(\,)}$ means that the variable is held constant during differentiation. Notice that for the terms containing $\mathbf{A}(x, u)$ and $b(x, u)$ in the present work the total derivative is directly available as will be shown in Sect. 2.5, and thus the chain rule is not being applied to those terms. The following relations between secant and tangent stiffness matrices and between the secant matrix and the internal force vector are given in Ryu et al. (1985)

$$\mathbf{K}_t(x, u) = \mathbf{K}_s(x, u) + \frac{\partial [\mathbf{K}_s(x, u)\tilde{u}]}{\partial u} \tag{31}$$

$$\frac{\partial [\mathbf{K}_s(x, \tilde{u})\tilde{u}]}{\partial x} = \frac{\partial F_i(x, \tilde{u})}{\partial x}. \tag{32}$$

Substituting Eqs. (31) and (32) into Eqs. (29) and (30) and rearranging them into matrix form gives

$$\begin{bmatrix} \mathbf{K}_t - \frac{\partial F_e(x, u)}{\partial u}(x, u) & \mathbf{A}(x, u)^T \\ \mathbf{A}(x, u) & \mathbf{0} \end{bmatrix} \begin{pmatrix} \frac{du}{dx} \\ \frac{d\lambda}{dx} \end{pmatrix} =$$
$$\begin{pmatrix} \frac{\partial F_e(x, \tilde{u})}{\partial x} - \frac{\partial F_i(x, \tilde{u})}{\partial x} - \frac{d[\mathbf{A}(x, u)^T \lambda]}{dx} \\ \frac{db(x, u)}{dx} - \frac{d[\mathbf{A}(x, u)\tilde{u}]}{dx} \end{pmatrix} \tag{33}$$

which can be solved for $\frac{du}{dx}$. Consider that in the case that the external forces are not depending on the displacements, e.g. no follower forces or pressure, the coefficient matrix on the left hand side is equal to the coefficient matrix used for the analysis steps, and is therefore already available in inverted form. This can save considerable computation time. The used constraint equations and its derivative terms are elaborated in Sect. 2.5.

2.5 Boundary constraint equations and derivatives

2.5.1 Constraint equations

In the current setting the system is loaded by applying displacements typically at the endpoints of the beam. Thereby the forces appear as reaction forces of the applied constraints instead of external forces. Once the displacement is applied, equilibrium can be found and the energy is derived. Displacements in this case can also mean rotations. In a typical example one would clamp one end of the beam, i.e. both translations and rotations zero, and apply a given motion on the other end, e.g. travelling along a curved line, or applying a rotation at a fixed point.

There is a complication that arises from the use of the isogeometric analysis method. From the rotation-less character of the control points, it follows that rotations cannot be directly applied. Instead, as described earlier in Radaelli and Herder (2014), a set of nonlinear constraints is applied that dictates the position of the second control point with respect to the first one. It is a given notion that the line connecting the first and the second control point is tangent to the beginning of the curve and, similarly, the line connecting the second-last to the last control point is tangent to the end of the curve. In the following only the beginning of the curve is considered, omitting the end of the curve. All the equations are similarly derivable replacing the subscripts 1 and 2 by $n - 1$ and n.

In general there are two types of constraints that are applied to the beam. The first is a linear set $A_1 u = b_1$ prescribing the displacement on the endpoints as

$$\begin{bmatrix} 1 & 0 & 0 & 0 & \cdots \\ 0 & 1 & 0 & 0 & \cdots \end{bmatrix} \begin{pmatrix} u_{1_x} \\ u_{1_y} \\ \vdots \end{pmatrix} = \begin{pmatrix} b_{1_x} \\ b_{1_y} \end{pmatrix} \qquad (34)$$

where b_{1_x} and b_{1_y} are the applied x and y displacements. There are cases where the applied displacement is made dependent on the design vector x. For instance in the design example given in Sect. 3, where the endpoint of the beam is pre-stressed by positioning the endpoint at the origin of the coordinate system. Here $b_{n_x} = -P_{n_x}$ and $b_{n_y} = -P_{n_y}$, and thus $b_1 = b_1(x)$.

The second type of constraints concerns the rotations and is somewhat more involved. The inclination h of the tangent line at the beginning of the curve is defined as

$$h(x, u) = \tan(\theta_0(x) + \Delta\theta(x, u)) \qquad (35)$$

where $\Delta\theta$ is the difference of the current angle θ_k from the an initial angle θ_0. The initial angle depends on the design vector x and the current angle θ_k on both the design vector x and the displacements vector u. Therefore the inclination h is is nonlinear in both x and u. The inclination h must equal the inclination of the line crossing the first two control points

$$h = \frac{(P_{2_y} + u_{2_y}) - (P_{1_y} + u_{1_y})}{(P_{2_x} + u_{2_x}) - (P_{1_x} + u_{1_x})}. \qquad (36)$$

Rewriting and separating the displacements terms gives the nonlinear set of constraints $A_{nl}(x, u) u = b_{nl}(x, u)$ as

$$\begin{bmatrix} -h(x, u) & 1 & h(x, u) & -1 & \cdots \end{bmatrix} \begin{pmatrix} u_{1_x} \\ u_{1_y} \\ u_{2_x} \\ u_{2_y} \\ \vdots \end{pmatrix}$$

$$= \begin{bmatrix} h(x, u)P_{1_x} - P_{1_y} - h(x, u)P_{2_x} + P_{2_y} \end{bmatrix}. \qquad (37)$$

The matrices A_l and A_{nl} and the vectors b_l and b_{nl} can be simply concatenated vertically to form a set of equations $Au = b$ to be added to the system as Lagrange constraints, as described in Sect. 2.4.

2.5.2 Derivatives of constraint equations

The derivatives of the constraint equations, needed in Eq. (33), will be derived next.

For the linear part of the constraints only the case where b_l is a function of x, i.e. $b_l = b_l(x)$, is mentioned. This is the case when e.g. a pre-stress proportional to a design parameter is applied to the system. Typically, if b_l is linear in x, the total derivative $\frac{db_l(x)}{dx}$ is a vector of constants.

The derivations for the nonlinear part of the constraints requires more attention. The following holds

$$\frac{db_{nl}(x, u)}{dx}$$
$$= \left(\frac{dh}{dx}(P_{1_x} - P_{2_x}) + h\left(\frac{dP_{1_x}}{dx} - \frac{dP_{2_x}}{dx}\right) - \frac{dP_{1_y}}{dx} + \frac{dP_{2_y}}{dx}\right), \qquad (38)$$

$$\frac{d[A_{nl}(x, u)\tilde{u}]}{dx} = \left(\frac{dh}{dx}(u_{2_x} - u_{1_x})\right) \qquad (39)$$

and

$$\frac{d\left[A_{nl}(x, u)^T \tilde{\lambda}\right]}{dx} = \begin{pmatrix} -\dfrac{dh}{dx}\lambda \\ \vdots \\ \dfrac{dh}{dx}\lambda \\ \vdots \end{pmatrix}. \qquad (40)$$

In these equations the derivative of h is calculated as

$$\frac{dh}{dx} = \frac{d}{dx}(\tan(\theta_0 + \Delta\theta)) = \frac{d}{dx}\left(\frac{h_0 + T_{\Delta\theta}}{1 - h_0 T_{\Delta\theta}}\right)$$
$$= \frac{\left(\frac{dh_0}{dx} + \frac{dT_{\Delta\theta}}{dx}\right)(1 - h_0 T_{\Delta\theta}) + (h_0 + T_{\Delta\theta})\left(\frac{dh_0}{dx}T_{\Delta\theta} + h_0\frac{dT_{\Delta\theta}}{dx}\right)}{(1 - h_0 T_{\Delta\theta})^2} \qquad (41)$$

where $T_{\Delta\theta}$ is a shorthand notation for $\tan(\Delta\theta)$, h_0 is the initial inclination and their derivatives are given respectively by

$$\frac{dT_{\Delta\theta}}{dx} = \sec(\Delta\theta)\frac{d\Delta\theta}{dx} \qquad (42)$$

and

$$\frac{dh_0}{dx} = \frac{d}{dx}\left(\frac{P_{2_y} - P_{1_y}}{P_{2_x} - P_{1_x}}\right). \qquad (43)$$

Now $\Delta\theta$, which is the difference between the current angle θ_k and the initial angle θ_0, is split up in the angle of the converged solution of the last iterative step θ_{k-1}, the angle that is imposed in the current iteration θ_{step}, which is known and fixed, minus the reference angle θ_0. This is a precaution measure. In fact, in the case that a rotation is applied from rest, $\Delta\theta$ is known and fixed. But in the case that the loading history is not fully known, e.g. a pre-stress is applied, than $\Delta\theta$ could contain a certain unknown rotation induced by a previous step. In order to avoid this type of error we define

$$\frac{d\Delta\theta}{dx} = \frac{d}{dx}(\theta_{k-1} + \theta_{step} - \theta_0) = \frac{d\theta_{k-1}}{dx} - \frac{d\theta_0}{dx} \qquad (44)$$

where

$$\frac{d\theta_0}{dx} = \frac{d}{dx}\left(\tan^{-1}(h_0)\right) = \frac{1}{1 + (h_0)^2}\frac{dh_0}{dx} \qquad (45)$$

is fairly simple to find, and where the following term is used which has been calculated at the end of the previous converged solution step, where the derivative of the state vector was already solved:

$$\frac{\mathrm{d}\theta_{k-1}}{\mathrm{d}x} = \frac{\partial \theta_{k-1}}{\partial x} + \frac{\partial \theta_{k-1}}{\partial u}\frac{\mathrm{d}u}{x}. \qquad (46)$$

The partial derivatives that are needed are

$$\frac{\partial \theta_{k-1}}{\partial x} = \frac{\partial}{\partial x}\left(\tan^{-1}(h_{k-1})\right) = \frac{1}{1+(h_{k-1})^2}\frac{\partial h_{k-1}}{\partial x} \qquad (47)$$

and

$$\frac{\partial \theta_{k-1}}{\partial u} = \frac{\partial}{\partial u}\left(\tan^{-1}(h_{k-1})\right) = \frac{1}{1+(h_{k-1})^2}\frac{\partial h_{k-1}}{\partial u} \qquad (48)$$

which turn out to be the same since

$$\frac{\partial h_{k-1}}{\partial x} = \frac{\partial h_{k-1}}{\partial u} = \begin{bmatrix} \frac{\left(P_{2_y}+u_{2_y}\right)-\left(P_{1_y}+u_{1_y}\right)}{\left(\left(P_{2_x}+u_{2_x}\right)-\left(P_{1_x}+u_{1_x}\right)\right)^2} \\ -\frac{1}{\left(P_{2_x}+u_{2_x}\right)-\left(P_{1_x}+u_{1_x}\right)} \\ -\frac{\left(P_{2_y}+u_{2_y}\right)-\left(P_{1_y}+u_{1_y}\right)}{\left(\left(P_{2_x}+u_{2_x}\right)-\left(P_{1_x}+u_{1_x}\right)\right)^2} \\ \frac{1}{\left(P_{2_x}+u_{2_x}\right)-\left(P_{1_x}+u_{1_x}\right)} \\ \vdots \end{bmatrix}^{T}. \qquad (49)$$

At this point Eqs. (38)–(40) are fully defined and can be used to find $\frac{\mathrm{d}u}{\mathrm{d}x}$ in Eq. (33).

2.6 Refinement term

In geometric design optimization it is a common use to define a geometry at a level with relatively few parameters which is refined to a more dense level at which the analysis is performed. Commonly this is done by meshing, while in isogeometric analysis there are so called refinement techniques, where the same spline curve is refined to a spline with more control points and/or higher order basis functions, but maintaining the exact original shape. It is not worth going much into detail here, given the amount and quality of literature on this topic, e.g. Piegl and Tiller (1997); Cottrell et al. (2009).

The Jacobian of x, the refined design vector, with respect to X, the global design vector, is needed for the sensitivity of the objective with respect to the global design vector.

$$\frac{\mathrm{d}f_0}{\mathrm{d}X} = \frac{\mathrm{d}f_0}{\mathrm{d}x}\frac{\mathrm{d}x}{\mathrm{d}X} \qquad (50)$$

The derivation of term $\frac{\mathrm{d}x}{\mathrm{d}X}$ is not within the scope of this paper, but can be found e.g. in Qian (2010).

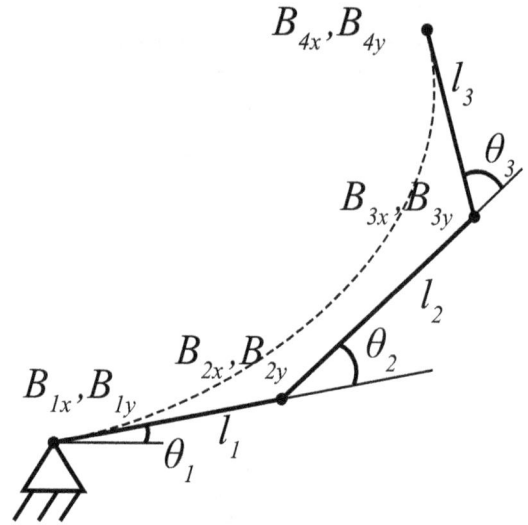

Figure 3. Transformation of coordinates of the control points in a set of generalized coordinates, described by the lengths and angles of the links of a linkage chain representing the control polygon.

2.7 Variable transform

As a last (optional) step a transformation of variables on the global design vector X to a generalized design vector q has been adopted. The latter is a vector containing lengths l and relative angles θ of the lines connecting the control points B of the global design as if it where a linkage chain, see Fig. 3. This way it becomes easy, by imposing boundaries on the search space of the optimization, to avoid loops in the curve and avoid consecutive control points lying to close to each other. The first is done by bounding the angles avoiding too sharp corners in the control polygon, and the second is realised by limiting the minimum lengths of the links. In general this avoids awkward shapes that are undesired.

The transformation is defined as

$$X = \begin{bmatrix} B_{1_x} \\ B_{1_y} \\ B_{2_x} \\ B_{2_y} \\ B_{3_x} \\ B_{3_y} \\ B_{4_x} \\ B_{4_y} \end{bmatrix} = \begin{bmatrix} q_1 \\ q_2 \\ q_1 + q_3 c(q_4) \\ q_2 + q_3 s(q_4) \\ q_1 + q_3 c(q_4) + q_5 c(q_4+q_6) \\ q_2 + q_3 s(q_4) + q_5 s(q_4+q_6) \\ q_1 + \ldots + q_7 c(q_4+q_6+q_8) \\ q_2 + \ldots + q_7 s(q_4+q_6+q_8) \end{bmatrix} \qquad (51)$$

where c and s are the shorthand notations for cos and sin, and q defined as

$$q = \begin{bmatrix} B_{1_x} & B_{1_y} & l_1 & \theta_1 & l_2 & \theta_2 & l_3 & \theta_3 \end{bmatrix}. \qquad (52)$$

Note that in a design with more control points q can be longer than shown, in that case the expression would expand in a similar fashion. For the sensitivity of the objective with re-

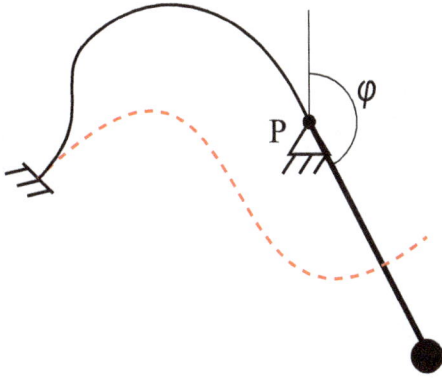

Figure 4. Topology of the system: A leaf spring (red dashed) is pre-stressed (black) by connecting one endpoint with a rigid pendulum, while the other end of the spring is clamped.

spect to the generalized variables $\frac{dX}{dq}$ is needed because

$$\frac{d f_0}{dq} = \frac{d f_0}{dX} \frac{dX}{dq}. \tag{53}$$

The full expression of $\frac{dX}{dq}$ is omitted for the sake of conciseness. Its derivation can however be considered straightforward. At this point all expressions for the sensitivity analysis are derived.

3 Design example

To asses the usefulness of the gradients in the given design problem several optimization runs are compared where one gradient-free algorithm and four gradient-based algorithms are used. The first type only needs Eq. (4) as an input, while the last four also use Eq. (53). The algorithms implemented in the *optimization toolbox* in Matlab® are used: Nelder–Mead Simplex (NMS), Trust-Region-Reflective (TRR), Interior-Point (IP), Active-Set (AS) and Sequential Quadratic Programming (SQP). See the Matlab® (2014) documentation for details on the algorithms.

Because of the high dependency with the starting point of the optimization, the runs are all performed starting at two different initial points, chosen such that the resulting behaviour would be clearly distinct.

The case study is similar to the one found in Radaelli and Herder (2014). The goal is to design a leaf spring able to statically balance a pendulum in the range from 0 to 180°, see Fig. 4. The objective is to make the reaction moment on the end of the beam follow a sinus-shaped characteristic with respect to the rotation of that point. Equivalently, knowing that the derivative of the energy with respect to the rotation equals the moment, the target energy-path $\hat{U}(\delta)$ is defined as a negative cosine function bounded in amplitude between 0 and 1, as prescribed for Eq. (3). Thus

$$\hat{U}(\varphi) = \frac{(1 - \cos(\varphi))}{2} \tag{54}$$

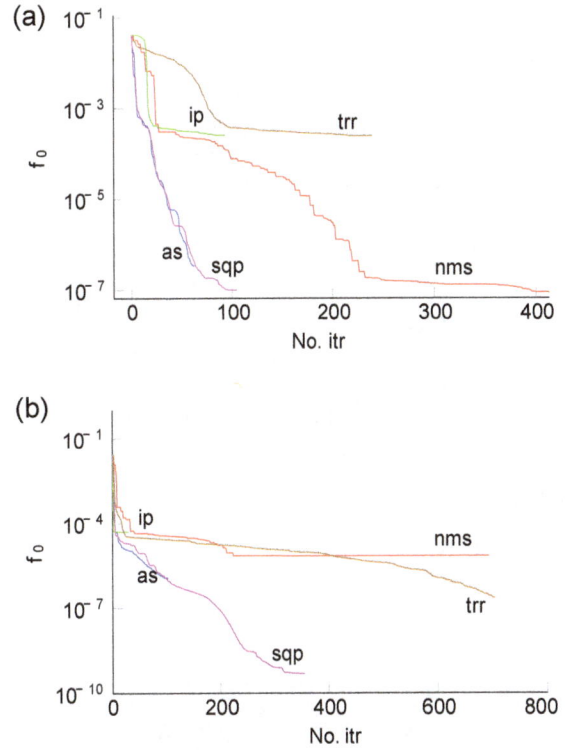

Figure 5. Objective function value vs iteration number for the algorithms Nelder–Mead Simplex (NMS), Trust-Region-Reflective (TRR), Interior-Point (IP), Active-Set (AS) and Sequential Quadratic Programming (SQP). **(a)** Example1 and **(b)** Example 2.

where φ is the rotation of the endpoint P. Before this actual working range a pre-stress step is applied where one endpoint stays clamped, and the other end is brought to the origin of the coordinate system leaving the rotation free. This guarantees that the moment before the first actual step is zero, corresponding to the needed moment when the pendulum is upright.

Some details on relevant design choices are: cross-sectional width and height are 0.01 and 0.002 m, Young's modulus is 135 GPa. The design curve is a second order B-spline with four control points and uniform knot vector. The curve is refined for analysis with 20 additional knot evenly spread over the knot vector. The used bounds on the variables of optimizations: $q_1, q_2 : [-0.3, 0.3]$, $q_3, q_5, q_7 : [.1, 0.4]$ and $q_4, q_6, q_8 : [-2, 2]$.

Figure 5a and b show the progress of the objective function plotted against the iteration steps for the two initial points

$$q_0 = [-0.2 \ -0.2 \ 0.2 \ 1.5 \ 0.2 \ -1.5 \ 0.2 \ -1.5] \tag{55}$$

and

$$q_0 = [-0.2 \ -0.2 \ 0.2 \ 1.5 \ 0.2 \ -1.5 \ 0.2 \ 1.5]. \tag{56}$$

The corresponding initial shapes and the energy trajectories are shown in Figs. 6a–9a. The shapes and energy-paths after

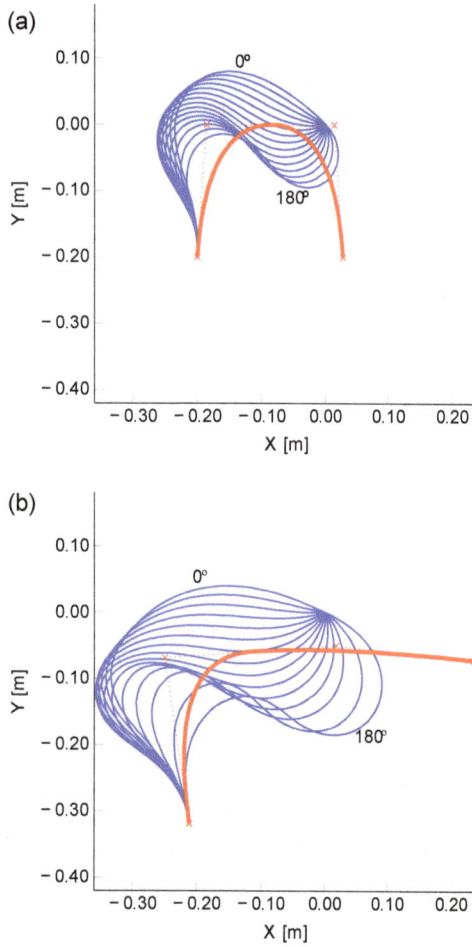

Figure 6. Example 1: shape in neutral position (red) and at the 15 increments of applied rotation of the endpoint about the origin (blue). **(a)** Initial shape and **(b)** optimum shape.

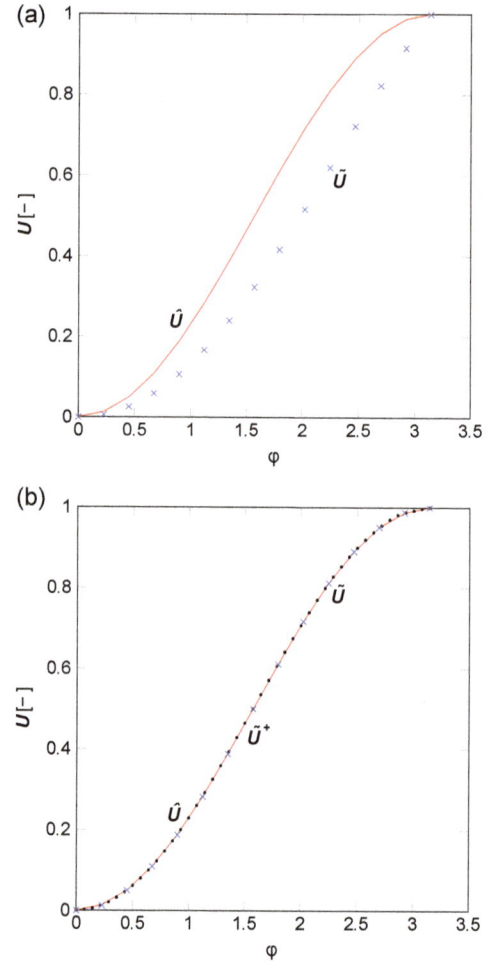

Figure 7. Example 1: target energy-path $\hat{U}$ (red line), normalized actual energy-path $\tilde{U}$ (blue crosses) and normalized actual energy path at smaller increments $\tilde{U}^{+}$ (grey dots). **(a)** Initial shape and **(b)** optimum shape.

optimization (the best result obtained) are shown in Figs. 6b–9b. The thick red lines in the shape plots is the unloaded shape of the beam, while the thin blue lines represent the deformed states at the 15 load-steps. The red line in each energy plot is the target energy-path $\hat{U}$, while the blue crosses represent the actual normalized energy-path $\tilde{U}$. The optimized results have been evaluated at smaller increments of the displacement to verify sufficient smoothness between the original increment points. The energies at the smaller increments are shown as grey dots in Fig. 7a and b.

4   Comparison with finite differences

This section is dedicated to a comparison of the result obtained in Sect. 2 with a numerical approximation of the sensitivity. The goal is twofold. One is to verify the validity of the derived equations and the other is to underline one advantage of the analytical sensitivity. Namely the independence between sensitivity and perturbation size.

In this analysis the gradient of the objective function with respect to the generalized coordinated $\frac{\mathrm{d}f_0}{\mathrm{d}q}$ is approximated by finite differences, using different perturbation sizes p ranging from 10^{-18} to 10^0. The first is near or smaller than machine precision, while the latter is in this case obviously an overly large perturbation with respect to the physical dimensions of the system.

An error norm is defined to compare the analytical sensitivities with the finite difference sensitivities. The used norm is the normalized mean error e between the values of the analytical $\frac{\mathrm{d}f_0}{\mathrm{d}q}$, Eq. (55), and its numerical approximation $\left(\frac{\mathrm{d}f_0}{\mathrm{d}q}\right)_{\mathrm{FD}}$.

$$e = \frac{\mathrm{mean}\left(\left|\left(\frac{\mathrm{d}f_0}{\mathrm{d}q}\right) - \left(\frac{\mathrm{d}f_0}{\mathrm{d}q}\right)_{\mathrm{FD}}\right|\right)}{\mathrm{mean}\left(\left|\left(\frac{\mathrm{d}f_0}{\mathrm{d}q}\right)\right|\right)}. \tag{57}$$

(a)

(b)

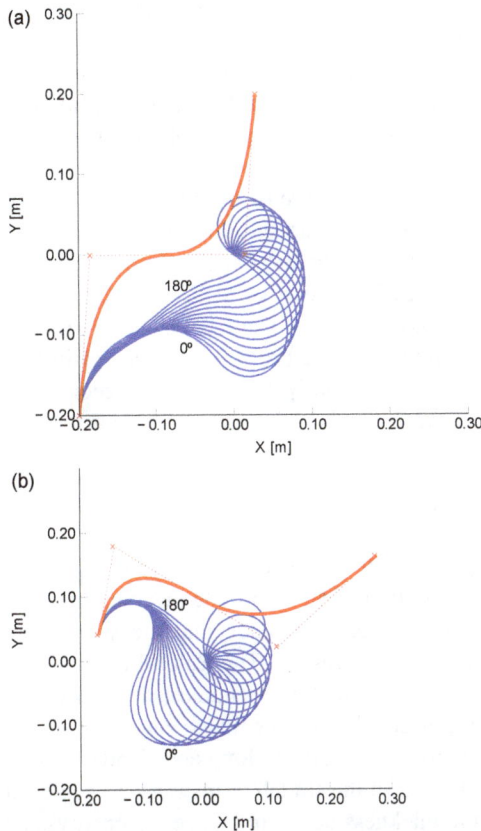

Figure 8. Example 2: shape in neutral position (red) and at the 15 increments of applied rotation of the endpoint about the origin (blue). **(a)** Initial shape and **(b)** optimum shape.

(a)

(b)

Figure 9. Example 2: target energy-path $\hat{U}$ (red line), normalized actual energy-path $\tilde{U}$ (blue crosses) and normalized actual energy path at smaller increments $\tilde{U}^+$ (grey dots). **(a)** Initial shape and **(b)** optimum shape.

The analysis is performed at two different points. One is the first starting point of the optimization shown in Sect. 3. Another is near one of the found minima. The sensitivities at an optimum point are zero. The error e includes a division by zero in that case. Therefore this error is evaluated at a point near an optimum, and not at an optimum.

Figure 10 shows the error e for a range of perturbation sizes. The blue line (crosses) represents the error at the start point q_0 and the red line (circles) represents the error at the point near the optimum.

5   Discussion

The optimization runs on the given example are considered successful in the sense that for both starting points, depending on the algorithm, a shape could be found that matches the given energy-path closely. Closely means that the significance of the remaining error is expected to be far beneath other types of errors expected in a physical realization of the concept. The optimized shapes have been analysed again with smaller increments of the rotation. Minimal differences

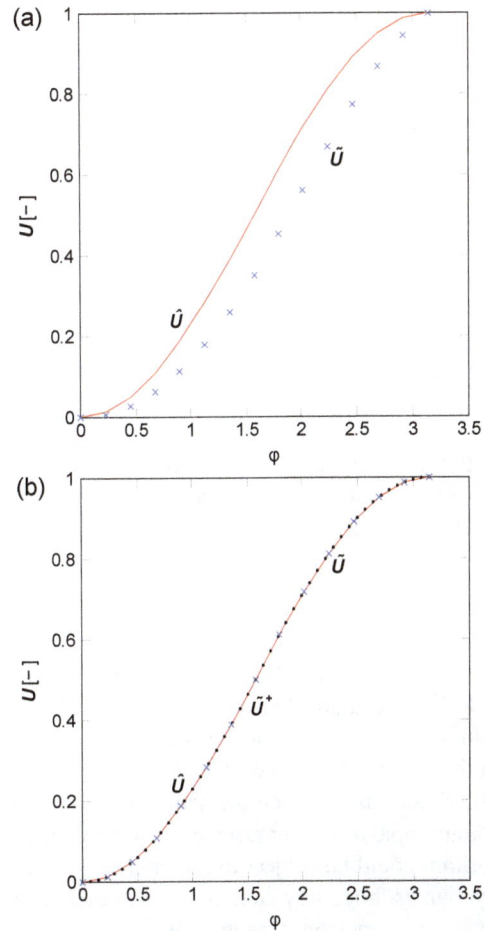

can be observed, meaning that the chosen amount of load steps was suitable for this example.

The objective function is non-convex, meaning that it cannot be guaranteed that the global minima are found in the examples. However, in the current scope it is not of practical relevance, as long as a "good-enough" minimum is obtained. In fact the nature of the objective function, which cannot become negative, tells that if the solution is close enough to zero within relevant significance the goal has been achieved.

The use of sensitivity information on the illustrated example has shown its utility in Fig. 5a and b. It can be seen that with respect to the Nelder–Mead Simplex algorithm, especially with the Active-Set and the Sequential Quadratic Programming algorithms, either a quicker (more efficient) or deeper (more effective) descent is realized, and sometimes both. It is, however, not guaranteed that this is always the case.

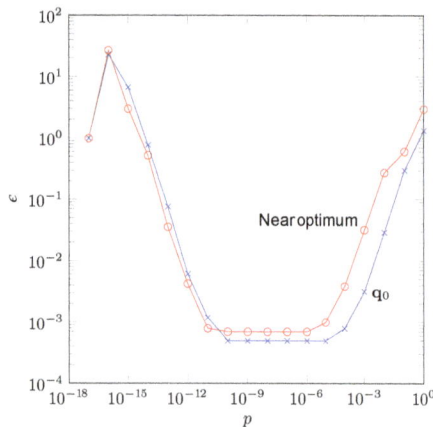

Figure 10. Error between finite-difference sensitivity and analytical sensitivities according to Eq. (57) at an initial configuration q_0 and near one of the optimum configurations, plotted on a range of perturbation sizes.

It is also noticed that the Interior-Point and the Trust-Region Reflective algorithms fail to obtain either a lower minimum or a quicker descent. This emphasizes the importance of the choice of the algorithm. Investigation on the optimal algorithm has not been the scope of this paper but is an important topic of further research. The optimal choice of the algorithm could be subject to factors like the number of design variables, boundary conditions and other constraints, and thus it may vary from case to case.

The comparison plots in Fig. 5a and b show the number of iterations on the horizontal axis. It must be noted however that for the Nelder–Mead Simplex the computation time of every iteration is shorter because it does not need to compute the sensitivity terms. The time saved at every iteration step highly depends on the performance of the code and the used hardware. To give an indication: Based on a Matlab code running on a Intel®· Core™ i7 processor the additional computation cost for the sensitivity terms is about 60 %. Nevertheless, the advantages of using sensitivity seem to hold.

The finite difference check is quite satisfying. It shows that there is a large range of perturbation sizes, from $p = 10^{-6}$ to $p = 10^{-11}$ where the difference between analytical and finite-difference sensitivities is small. On the one hand this tells that the calculations of the analytical sensitivities are correct. On the other hand it says that within this range the numerical sensitivities would give similar results, only less efficiently. However it still remains a risk that this range may vary from case to case, rising the chance to obtain imprecise or even useless numerical sensitivities.

The points where the comparison is performed are at one of the starting points of the optimization, and in the neighbourhood of the end of the optimization run. As discussed above, it would make little sense to compare the sensitivities at the end of an optimization run. Assuming this is an opti-

mum, the sensitivities would be zero. At an optimum Eq. (57) would divide by zero giving an invalid error norm. It is therefore not surprising that the error norm near the optimum solution is higher than at a point far from an optimum. Similar tendencies as in Fig. 10 have been observed starting at various initial points.

In Sect. 2.5.2 a measure is taken to include the possibility to apply pre-stress on the system, thus inducing the system into a configuration where the angles of the endpoints are not known a priori. Pre-stress is crucial in certain applications, especially those involving static balancing. However, if this is not required, then the procedure described in Sect. 2.5.2 will simplify significantly. Namely θ_{k-1} becomes fixed and thus its derivative zero. This reduces Eq. (44) to

$$\frac{\mathrm{d}\Delta\theta}{\mathrm{d}x} = -\frac{\mathrm{d}\theta_0}{\mathrm{d}x} \tag{58}$$

and the part after that, Eqs. (45)–(49), can be skipped.

In the definition of an amplitude invariant objective function the assumption was made that the sizing parameters would not affect the results. The Euler–Bernoulli conditions must be met in order for the numerical model to be valid in the first place. This means that shear terms are neglected and this is true for relatively long and slender beams. In the second place, also the stretch terms must be neglected in order for the thickness not to influence the energy-path. For a beam with the same shape but different thickness, namely, the stress condition is influenced by the stretch terms, and thus it affects the deformation and stored energy. It is verified that in the given example, multiplying the thickness by a factor of two yields a maximum amplitude deviation of 2.5 %, and multiplying by a factor of ten yields a deviation of 4 %. Care is thus needed in the choice of a suitable thickness when designing such system. The width of the spring seems in practice the most appropriate parameter to adjust after optimization without affecting the result.

6   Conclusions

The paper presents the shape sensitivity analysis for the shape optimization of beams with prescribed load-displacement response. The work is based on an isogeometric framework. It has been shown that it is possible to derive the sensitivity parameters for this type of problem analytically.

A novel and general objective function is formulated for problems with prescribed load-displacement response. The objective function is based on the potential energy of the beam determined at discrete steps of the applied displacement path.

The travelled path is imposed by application of constraints on the endpoints of the beam, involving displacements but also rotations, which is more complex due to the rotation-less character of the degrees of freedom in isogeometric analysis.

The sensitivity of the state variables is determined through the direct method, with special attention to the complications brought by nonlinear equilibrium equations and the nonlinear constraints equations, applied as Lagrange multipliers.

The effect of the load history on the sensitivity analysis, e.g. by application of pre-stress in a previous load-step, is neutralized by making smart use of the sensitivities previously calculated for the previous load steps.

The optimization is performed on an example where the optimal shape of a beam is sought that is optimal for the static balancing of a pendulum. The influence of the use of sensitivity information is shown by comparison on different optimization algorithms, with and without the use of gradients. The optimization results in very satisfying balancing springs. The use of gradients is positively influencing the efficiency of the optimization, although determining the algorithm that performs best is still an open question.

The correctness of the derived sensitivity equations is verified by comparison with finite-difference gradients. It is shown that the numerical and the analytical gradients have a good match within a certain range of perturbation sizes.

Acknowledgements. The authors would like to acknowledge STW (HTSM-2012 12814: ShellMech) for the financial support of this project.

Edited by: A. Eriksson

References

Bruns, T. E. and Tortorelli, D. A.: Topology optimization of nonlinear elastic structures and compliant mechanisms, Comput. Meth. Appl. Mech. Eng., 190, 3443–3459, doi:10.1016/S0045-7825(00)00278-4, 2001.

Chen, G. and Zhang, S.: Fully-compliant statically-balanced mechanisms without prestressing assembly: concepts and case studies, Mech. Sci., 2, 169–174, doi:10.5194/ms-2-169-2011, 2011.

Cho, S. and Ha, S.-H.: Isogeometric shape design optimization: exact geometry and enhanced sensitivity, Struct. Multidiscip. Optimiz., 38, 53–70, doi:10.1007/s00158-008-0266-z, 2009.

Cottrell, J. A., Hughes, T. J. R., and Bazilevs, Y.: Isogeometric Analysis: Toward Integration of CAD and FEA, John Wiley & Sons, Chichester, England, 2009.

Ding, Y.: Shape optimization of structures a literal survey, Comput. Struct., 24, 985–1004, 1986.

Du, Y. and Chen, L.: Topology optimization for large-displacement compliant mechanisms using element free galerkin method, Int. J. CAD/CAM, 8, 1–10, 2008.

Eriksson, A.: Constraint paths in non-linear structural optimization, Comput. Struct., 140, 39–47, doi:10.1016/j.compstruc.2014.05.003, 2014.

Felippa, C. A.: Introduction to finite element methods (ASEN 5007), University Lecture Notes, University of Colorado, http://www.colorado.edu/engineering/CAS/courses.d/IFEM.d/ (last access: 1 November 2016), 2004.

Haftka, R. T.: Simultaneous analysis and design, AIAA J., 23, 1099–1103, doi:10.2514/3.9043, 1985.

Hoetmer, K., Woo, G., Kim, C., and Herder, J.: Negative Stiffness Building Blocks for Statically Balanced Compliant Mechanisms: Design and Testing, J. Mech. Robot., 2, 041007-1-7, doi:10.1115/1.4002247, 2010.

Howell, L. L.: Compliant Mechanisms, John Wiley and Sons Inc., New York, USA, 2001.

Hsu, Y.-L.: A review of structural shape optimization, Comput. Indust., 26, 3–3, 1994.

Hughes, T., Cottrell, J., and Bazilevs, Y.: Isogeometric analysis: CAD, finite elements, NURBS, exact geometry and mesh refinement, Comput. Meth. Appl. Mech. Eng., 194, 4135–4195, doi:10.1016/j.cma.2004.10.008, 2005.

Joo, J., Kota, S., and Kikuchi, N.: Large deformation behavior of compliant mechanisms, Proceedings of DETC'01 ASME 2001 Design Engineering Technical Conference and Computers and Information in Engineering Conference, 9–12 September 2001, Pittsburgh, PA, 2001.

Jutte, C. V. and Kota, S.: Design of Single, Multiple, and Scaled Nonlinear Springs for Prescribed Nonlinear Responses, J. Mech. Design, 132, 011003, doi:10.1115/1.4000595, 2010.

Kiendl, J. M.: Isogeometric Analysis and Shape Optimal Design of Shell Structures, PhD Thesis, Munich University of Technology, Munich, Germany, 2011.

Koo, B., Ha, S.-H., Kim, H.-S., and Cho, S.: Isogeometric Shape Design Optimization of Geometrically Nonlinear Structures, Mech. Bas. Design Struct. Mach., 41, 337–358, doi:10.1080/15397734.2012.750226, 2013.

Lan, C.-C., Wang, J.-H., and Chen, Y.-H.: A compliant constant-force mechanism for adaptive robot end-effector operations, in: 2010 IEEE International Conference on Robotics and Automation, Anchorage Convention District, 3–8 May 2010, Anchorage, Alaska, USA, 2010.

Leishman, L. C. and Colton, M. B.: A Pseudo-Rigid-Body Model Approach for the Design of Compliant Mechanism Springs for Prescribed Force – Deflections, ASME, 93–102, doi:10.1115/DETC2011-47590, 2011.

Matlab®: MATLAB and Optimization Toolbox Release R2014a, The MathWorks, Inc., Natick, Massachusetts, USA, http://www.mathworks.nl/help/releases/R2014a/pdf_doc/optim/optim_tb.pdf (last access: 1 November 2016), 2014.

Nagy, A. P.: Isogeometric Design Opitmization, PhD Thesis, Delft University of Technology, Delft, the Netherlands, 2011.

Nagy, A. P., Abdalla, M. M., and Gürdal, Z.: Isogeometric sizing and shape optimisation of beam structures, Comput. Meth. Appl. Mech. Eng., 199, 1216–1230, doi:10.1016/j.cma.2009.12.010, 2010.

Oh, Y.: Synthesis of Multistable Equilibrium Compliant Mechanisms, PhD Thesis, The University of Michigan, Michigan, USA, 2008.

Pedersen, C. B., Fleck, N. A., and Ananthasuresh, G. K.: Design of a Compliant Mechanism to Modify an Actuator Characteristic to Deliver a Constant Output Force, J. Mech. Design, 128, 1101–1112, doi:10.1115/1.2218883, 2005.

Piegl, L. and Tiller, W.: The NURBS Book, 2nd Edn., Springer-Verlag, New York, NY, USA, 1997.

Powell, K. M. and Frecker, M. I.: Method for Optimization of a Nonlinear Static Balance Mechanism With Application to Ophthalmic Surgical Forceps, in: Proceedings ASME

IDETC/CIE 2005, 24–28 September 2005, Long Beach, California, USA, doi:10.1115/DETC2005-84759, 2005.

Pucheta, M. A. and Cardona, A.: Design of bistable compliant mechanisms using precision – position and rigid-body replacement methods, Mech. Mach. Theory, 45, 304–326, doi:10.1016/j.mechmachtheory.2009.09.009, 2010.

Qian, X.: Full analytical sensitivities in NURBS based isogeometric shape optimization, in: Computer Methods in Applied Mechanics and Engineering, 2010.

Radaelli, G. and Herder, J. L.: Isogeometric Shape Optimization for Compliant Mechanisms with Prescribed Load Paths, in: Proceedings ASME IDETC/CIE 2014, 17–20 August 2014, Buffalo, New York, USA, doi:10.1115/DETC2014-35373, 2014.

Rai, A., Saxena, A., Mankame, N. D., and Upadhyay, C. S.: On Optimal Design of Compliant Mechanisms for Specified Nonlinear Path Using Curved Frame Elements and Genetic Algorithm, ASME 2006 International Design Engineering Technical Conferences and Computers and Information in Engineering Conference, Vol. 2, 30th Annual Mechanisms and Robotics Conference, Parts A and B, 10–13 September 2006, Philadelphia, Pennsylvania, USA, 91–100, doi:10.1115/DETC2006-99298, 2006.

Ringertz, U. T.: Optimization of Structures with Nonlinear Response, Eng. Optimiz., 14, 179–188, doi:10.1080/03052158908941210, 1989.

Ryu, Y., Haririan, M., Wu, C., and Arora, J.: Structural design sensitivity analysis of nonlinear response, Comput. Struct., 21, 245–255, 1985.

Saxena, A. and Ananthasuresh, G. K.: Topology Synthesis of Compliant Mechanisms for Nonlinear Force – Deflection and Curved Path Specifications, J. Mech. Design, 123, 33–42, doi:10.1115/1.1333096, 1999.

Sigmund, O.: On the Design of Compliant Mechanisms Using Topology Optimization, Mech. Struct. Mach., 25, 493–524, doi:10.1080/08905459708945415, 1997.

Simo, J.: A finite strain beam formulation. The three-dimensional dynamic problem. Part I, Comput. Meth. Appl. Mech. Eng., 49, 55–70, doi:10.1016/0045-7825(85)90050-7, 1985.

Tolou, N.: Statically Balanced Compliant Mechanisms for MEMS and Precision Engineering, PhD Thesis, Delft University of Technology, Delft, the Netherlands, 2012.

Vehar-Jutte, C.: Generalized Synthesis Methodology of Nonlinear Springs for Predescribed Load-Displacement Functions, PhD Thesis, The University of Michigan, Michigan, USA, 2008.

Wall, W. A., Frenzel, M. A., and Cyron, C.: Isogeometric structural shape optimization, Comput. Meth. Appl. Mech. Eng., 197, 2976–2988, doi:10.1016/j.cma.2008.01.025, 2008.

A novel 2-DOF planar translational mechanism composed by scissor-like elements

Yi Yang, Yaping Tian, Yan Peng, and Huayan Pu
School of Mechatronic Engineering and Automation, Shanghai University,
Shanghai, 200444, China
Correspondence to: Yan Peng (pengyan@shu.edu.cn)

Abstract. The kinematic chain comprised by SLEs (scissor-like elements) has a wide range of motion, which provides a benefit for the mechanism design. A family of SLE-Pa (SLE-parallel) legs which consist of two identical SLE limbs are proposed in this paper. The mobility and kinematics are discussed for three kinds of SLE-Pa legs which are distinguished by the different positions of the middle links in legs. Through assembling these SLE-Pa legs, a novel 2-DOF planar translational mechanism is developed and its work space is studied. For the purpose of adding the recovery function, the elastic elements are installed for this mechanism. The stiffness synthesis of the mechanism is investigated for the various elastic elements and their positions. The approximation of the stiffness coefficient is also derived. Further, this kind of mechanism is applied for the design of the passive docking device. The docking procedure is simulated by Adams, and the prototype of one SLE-Pa leg is presented at the end.

1 Introduction

The simplest structure of two degrees-of-freedom (DOF) planar translational mechanism is designed by 2-PP (prismatic–prismatic) limb (Dong et al., 2004). To reduce the difficulty of manufacturing, enlarge the work space of the end-effector and improve the stiffness of the mechanism, various translational mechanisms by different kinds of kinematic limbs are constructed. Liu and Wang (2003) and Liu et al. (2004) proposed a family of 2-PPa (prismatic–parallelogram) translational mechanisms based on the parallelogram and optimally designed a PRRRP (prismatic–revolute–revolute–revolute–prismatic) 2-DOF parallel mechanism by the utilization of a performance chart (Liu et al., 2006). Wu et al. (2007) also did research on the optimal design of 2-DOF planar parallel mechanism. Kim (2007) and Pham and Kim (2013) developed two types of planar translational parallel manipulators for high-speed positioning applications. Many kinds of 2-DOF planar translational mechanisms are recommended for the design of pick-and-place robots. Huang et al. (2004, 2013) proposed a 2-D version of the delta robot with two sets of parallelograms and studied the optimal design of these 2-DOF translational parallel robots. Generally, most of the above kinematic limbs comprising the 2-DOF planar transla-

tional mechanisms were serial kinematic chains or equivalent serial kinematic chains.

The scissor-like element (SLE) is one of the most widely used units in the design of mechanisms, especially for deployable structures. The simplest planar SLE consists of two rigid segments with a revolute joint at their midpoints. The kinematic chain comprised by SLEs has a wide range of motion, which provides a benefit for the mechanism design. In the 1960s, a Spanish architect, E. P. Pienro (Escrig and Valcarcel, 1993; Kaveh and Davaran, 1996), initially employed SLEs to construct the movable theater. After that, SLEs were gradually applied from small-scale structures (Rosenfeld and Logcher, 1988; Escrig et al., 1996) to aerospace structures (Langbecker, 1999). In academic research, You and Pellegrino (1997) presented a general type of two-dimensional foldable structure consisting of different kinds of SLEs. Zhao et al. (2011) constructed foldable stairs with scissor-shape mechanisms. Bai et al. (2013) combined pantograph elements to construct scaling mechanisms for geometric figures. Kaveh et al. (1999) studied the kinematically optimal design of pantograph foldable structures. Dai and Rees (1999), Wei et al. (2010), Ding (2011), and Lu et al. (2017) analyzed the mobility of the foldable structures by screw theory from

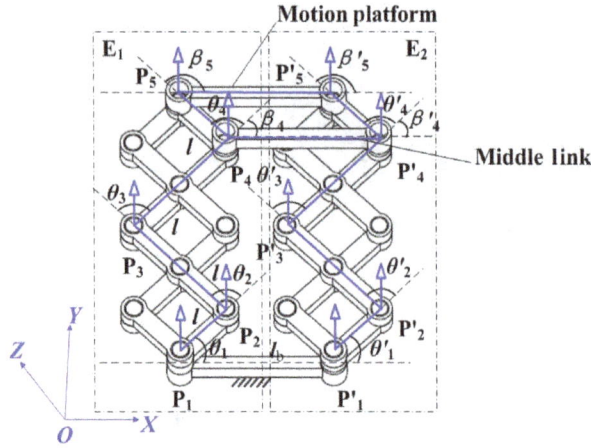

Figure 1. Structure of novel SLE-Pa leg in case 1.

the viewpoint of kinematics. All the achievements with SLEs were mainly focused on deployable mechanisms.

By the utilization of SLEs, we create a novel parallel kinematic chain, which has 2 translational degrees of freedom. Through assembling several of these kinematic chains, a translational parallel mechanism with SLEs is proposed. Based on that, a new passive docking device is invented for the purpose of joining two water pipes aboard a USV (unmanned surface vehicle). With the requirement of the recovery force in the device, the elastic components are set to this translational parallel mechanism. Hence, the stiffness of the platform in the mechanism is investigated under several conditions of the elastic components. Finally, the simulation of the docking procedure is conducted and the prototype of one SLE-Pa leg is presented.

2   Mobility of novel SLE-Pa leg

The novel kinematic chain comprises two identical SLE limbs, i.e., limbs E_1 and E_2, as shown in Fig. 1. Each SLE limb contains two full SLEs in the middle of the limb and two half-SLEs at both ends of the limb. The length of the link of the full SLE is $2l$, and one of the links of the half-SLE is l. Nodes P_1 and P_5 in limb E_1 are linked by nodes P_1' and P_5' in limb E_2, respectively. The length of $P_1 P_1'$ and $P_5 P_5'$ is both l_b. Suppose link $P_1 P_1'$ is fixed to the ground and link $P_5 P_5'$ is regarded as the motion platform. There is one middle link that connects the two corresponding nodes in the two SLE limbs by hinged joints. This kind of kinematic chain is called SLE-Pa (SLE-parallel) leg. The mobility and kinematics are discussed in the following three cases in relation to the hinged positions of the middle link.

2.1   Case 1: middle link connecting P_4 and P_4'

Screw theory is applied to analyze the mobility of motion platform $P_5 P_5'$ as shown in Fig. 1. For SLE limb E_1, the

Cartesian coordinates of joints P_1, P_2, P_3, P_4 and P_5 can be obtained as

$$\mathbf{p}_1 = \begin{bmatrix} x_a \\ y_a \\ 0 \end{bmatrix} \quad \mathbf{p}_2 = \begin{bmatrix} x_a + l_1 \cos\theta_1 \\ y_a + l_1 \sin\theta_1 \\ 0 \end{bmatrix}$$

$$\mathbf{p}_3 = \begin{bmatrix} x_a + l_1 \cos\theta_1 + 2l_1 \cos(\theta_1 + \theta_2) \\ y_a + l_1 \sin\theta_1 + 2l_1 \sin(\theta_1 + \theta_2) \\ 0 \end{bmatrix}$$

$$\mathbf{p}_4 = \begin{bmatrix} x_a + l_1 \cos\theta_1 + 2l_1 \cos(\theta_1 + \theta_2) + 2l_1 \cos(\theta_1 + \theta_2 - \theta_3) \\ y_a + l_1 \sin\theta_1 + 2l_1 \sin(\theta_1 + \theta_2) + 2l_1 \sin(\theta_1 + \theta_2 - \theta_3) \\ 0 \end{bmatrix}$$

$$\mathbf{p}_5 = \begin{bmatrix} x_a + l_1 \cos\theta_1 + 2l_1 \cos(\theta_1 + \theta_2) + 2l_1 \cos(\theta_1 + \theta_2 - \theta_3) \\ \quad + l_1 \cos(\theta_1 + \theta_2 - \theta_3 + \theta_4) \\ y_a + l_1 \sin\theta_1 + 2l_1 \sin(\theta_1 + \theta_2) + 2l_1 \sin(\theta_1 + \theta_2 - \theta_3) \\ \quad + l_1 \sin(\theta_1 + \theta_2 - \theta_3 + \theta_4) \\ 0 \end{bmatrix}.$$

Limb E_2 has the same structure as limb E_1. The positions of joints in E_2 appear to be the results of the translation of the joints from E_1. The Cartesian coordinates of joints can be calculated as

$$\mathbf{p}_i' = \mathbf{p}_i + \begin{bmatrix} l_b \\ 0 \\ 0 \end{bmatrix} \quad i = 1, 2, \ldots, 5. \tag{1}$$

The axis orientation of each joint is normal to the plane XOY, i.e., $\mathbf{s}_j = (0\ \ 0\ \ 1)$. The screws of the above joints are given as

$$\begin{cases} \$_i = \left(\ \mathbf{s}_j; \quad \mathbf{p}_i \times \mathbf{s}_j \right) & i = 1, 2, \cdots, 5 \\ \$_i' = \left(\ \mathbf{s}_j; \quad \mathbf{p}_i' \times \mathbf{s}_j \right) & i = 1, 2, \cdots, 5. \end{cases} \tag{2}$$

We choose linkages $P_1 P_2 P_3 P_4 P_4' P_3' P_2' P_1'$ and $P_1 P_2 P_3 P_4 P_5 P_5' P_4' P_3' P_2' P_1'$ to establish the screw-loop equations. For the first linkage, the screw-loop equation is

$$\$_1 \dot{\theta}_1 + \$_2 \dot{\theta}_2 - \$_3 \dot{\theta}_3 - \$_4 \dot{\beta}_4$$
$$= \$_1' \dot{\theta}_1' + \$_2' \dot{\theta}_2' - \$_3' \dot{\theta}_3' - \$_4' \dot{\beta}_4'. \tag{3}$$

According to the feature of SLE limbs, the following constraints can be obtained:

$$\theta_2 = \theta_3 = \theta_4, \quad \theta_2' = \theta_3' = \theta_4', \quad \dot{\theta}_2 = \dot{\theta}_3 = \dot{\theta}_4,$$
$$\dot{\theta}'_2 = \dot{\theta}'_3 = \dot{\theta}'_4. \tag{4}$$

Substituting Eq. (4) into Eq. (3), the screw-loop equation can be rewritten as

$$\$_1 \dot{\theta}_1 + (\$_2 - \$_3) \dot{\theta}_2 - \$_4 \dot{\beta}_4$$
$$= \$_1' \dot{\theta}_1' + (\$_2' - \$_3') \dot{\theta}_2' - \$_4' \dot{\beta}_4'. \tag{5}$$

In the same way, the screw-loop equation of linkage $P_1 P_2 P_3 P_4 P_5 P_5' P_4' P_3' P_2' P_1'$ can be derived as

$$\$_1 \dot{\theta}_1 + (\$_2 - \$_3 + \$_4)\dot{\theta}_2 - \$_5 \dot{\beta}_5$$
$$= \$_1' \dot{\theta}_1' + (\$_2' - \$_3' + \$_4')\dot{\theta}_2' - \$_5' \dot{\beta}_5'. \tag{6}$$

The above Eqs. (5) and (6) can be rearranged in matrix form as

$$
\begin{bmatrix}
\$_1 & (\$_2 - \$_3) & -\$_4 & 0 & -\$_1' \\
 & -(\$_2' - \$_3') & \$_4' & 0 & \\
\$_1 & (\$_2 - \$_3 + \$_4) & 0 & -\$_5 & -\$_1' \\
 & -(\$_2' - \$_3' + \$_4') & 0 & \$_5' &
\end{bmatrix}
\begin{bmatrix}
\dot{\theta}_1 \\ \dot{\theta}_2 \\ \dot{\beta}_4 \\ \dot{\beta}_5 \\ \dot{\theta}_1' \\ \dot{\theta}_2' \\ \dot{\beta}_4' \\ \dot{\beta}_5'
\end{bmatrix}
= 0. \tag{7}
$$

By solving Eq. (7), the velocities of joints are expressed as

$$
\begin{bmatrix}
\dot{\theta}_1 \\ \dot{\theta}_2 \\ \dot{\beta}_4 \\ \dot{\beta}_5 \\ \dot{\theta}_1' \\ \dot{\theta}_2' \\ \dot{\beta}_4' \\ \dot{\beta}_5'
\end{bmatrix}
= q_1
\begin{bmatrix}
0 \\ 1 \\ 0 \\ 1 \\ 0 \\ 1 \\ 0 \\ 1
\end{bmatrix}
+ q_2
\begin{bmatrix}
1 \\ -1 \\ 1 \\ 0 \\ 1 \\ -1 \\ 1 \\ 0
\end{bmatrix}, \tag{8}
$$

where q_1 and q_2 are arbitrary real numbers. The result indicates that the mobility of this linkage is 2 and that the velocities of joints are subject to the following relationship:

$$\dot{\theta}_1 = \dot{\theta}_1' = \dot{\beta}_4 = \dot{\beta}_4', \quad \dot{\theta}_2 = \dot{\theta}_2', \quad \dot{\beta}_5 = \dot{\beta}_5' = \dot{\theta}_1 + \dot{\theta}_2. \tag{9}$$

Equation (9) shows that the two SLE limbs are always parallel. Further, the kinematic screw of the platform $P_5 P_5'$ can be written as

$$\$_p = \$_1 \dot{\theta}_1 + (\$_2 - \$_3 + \$_4)\dot{\theta}_2 - \$_5 \dot{\beta}_5. \tag{10}$$

Substituting Eq. (8) into Eq. (10), the kinematic screw of the platform $P_5 P_5'$ can be derived as

$$
\$_p = q_1
\begin{bmatrix}
0 \\ 0 \\ 0 \\ -\sin(\theta_1 + \theta_2) \\ \cos(\theta_1 + \theta_2) \\ 0
\end{bmatrix}
+ q_2
\begin{bmatrix}
0 \\ 0 \\ 0 \\ -\sin\theta_1 \\ \cos\theta_1 \\ 0
\end{bmatrix}. \tag{11}
$$

The corresponding reciprocal screw of the platform can be derived as

$$
\$_p^r :
\begin{cases}
\$_p^{r1} = \begin{bmatrix} 0 & 0 & 1 & 0 & 0 & 0 \end{bmatrix} \\
\$_p^{r2} = \begin{bmatrix} 0 & 0 & 0 & 1 & 0 & 0 \end{bmatrix} \\
\$_p^{r3} = \begin{bmatrix} 0 & 0 & 0 & 0 & 1 & 0 \end{bmatrix} \\
\$_p^{r4} = \begin{bmatrix} 0 & 0 & 0 & 0 & 0 & 1 \end{bmatrix}
\end{cases}. \tag{12}
$$

Equations (11) and (12) indicates that the motion platform of the SLE-Pa leg has 2 pure translational degrees of freedom.

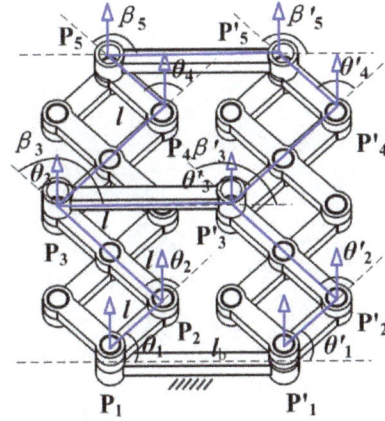

Figure 2. Structure of novel SLE-Pa leg in case 2.

2.2 Case 2: middle link connecting P_3 and P_3'

In this case, the middle link connects P_3, P_3' between two SLE limbs, as shown in Fig. 2. The screw-loop equations are established for linkages $P_1 P_2 P_3 P_3' P_2' P_1'$ and $P_1 P_2 P_3 P_4 P_5 P_5' P_4' P_3' P_2' P_1'$ with a consideration of the features of SLE limb, as follows:

$$
\begin{cases}
\$_1 \dot{\theta}_1 + \$_2 \dot{\theta}_2 - \$_3 \dot{\beta}_3 = \$_1' \dot{\theta}_1' + \$_2' \dot{\theta}_2' - \$_3' \dot{\beta}_3' \\
\$_1 \dot{\theta}_1 + (\$_2 - \$_3 + \$_4)\dot{\theta}_2 - \$_5 \dot{\beta}_5 \\
\quad = \$_1' \dot{\theta}_1' + (\$_2' - \$_3' + \$_4')\dot{\theta}_2' - \$_5' \dot{\beta}_5'
\end{cases}. \tag{13}
$$

The above Eq. (13) can be rewritten in a matrix form as

$$
\begin{bmatrix}
\$_1 & \$_2 & -\$_3 & 0 & -\$_1' \\
 & -\$_2' & \$_3' & 0 & \\
\$_1 & (\$_2 - \$_3 + \$_4) & 0 & -\$_5 & -\$_1' \\
 & -(\$_2' - \$_3' + \$_4') & 0 & \$_5' &
\end{bmatrix}
\begin{bmatrix}
\dot{\theta}_1 \\ \dot{\theta}_2 \\ \dot{\beta}_3 \\ \dot{\beta}_5 \\ \dot{\theta}_1' \\ \dot{\theta}_2' \\ \dot{\beta}_3' \\ \dot{\beta}_5'
\end{bmatrix}
= 0. \tag{14}
$$

According to Eq. (14), the velocities of joints are obtained as

$$
\begin{bmatrix}
\dot{\theta}_1 \\ \dot{\theta}_2 \\ \dot{\beta}_3 \\ \dot{\beta}_5 \\ \dot{\theta}_1' \\ \dot{\theta}_2' \\ \dot{\beta}_3' \\ \dot{\beta}_5'
\end{bmatrix}
= q_1
\begin{bmatrix}
-1 \\ 1 \\ 0 \\ 0 \\ -1 \\ 1 \\ 0 \\ 0
\end{bmatrix}
+ q_2
\begin{bmatrix}
1 \\ 0 \\ 1 \\ 1 \\ 1 \\ 0 \\ 1 \\ 1
\end{bmatrix}. \tag{15}
$$

Equation (15) shows that the mobility of this linkage is 2, and the velocities of joints are also subject to the following relationship:

$$\dot{\theta}_1 = \dot{\theta}_1', \quad \dot{\theta}_2 = \dot{\theta}_2', \quad \dot{\beta}_3 = \dot{\beta}_3' = \dot{\beta}_5 = \dot{\beta}_5' = \dot{\theta}_1 + \dot{\theta}_2. \tag{16}$$

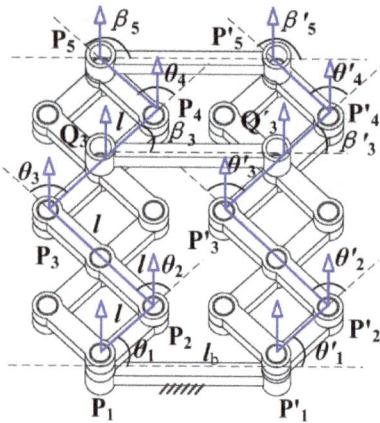

Figure 3. Structure of novel SLE-Pa leg in case 3.

The two SLEs are always parallel as well. The kinematic screw of the platform can be calculated as

$$\$_p = q_1 \begin{bmatrix} 0 \\ 0 \\ 0 \\ -\sin(\theta_1 + \theta_2) \\ \cos(\theta_1 + \theta_2) \\ 0 \end{bmatrix} + q_2 \begin{bmatrix} 0 \\ 0 \\ 0 \\ -\sin\theta_1 \\ \cos\theta_1 \\ 0 \end{bmatrix}. \tag{17}$$

The corresponding reciprocal screw of the platform can be calculated as

$$\$_p^{\mathbf{r}} : \begin{cases} \$_p^{r1} = \begin{bmatrix} 0 & 0 & 1 & 0 & 0 & 0 \end{bmatrix} \\ \$_p^{r2} = \begin{bmatrix} 0 & 0 & 0 & 1 & 0 & 0 \end{bmatrix} \\ \$_p^{r3} = \begin{bmatrix} 0 & 0 & 0 & 0 & 1 & 0 \end{bmatrix} \\ \$_p^{r4} = \begin{bmatrix} 0 & 0 & 0 & 0 & 0 & 1 \end{bmatrix} \end{cases}. \tag{18}$$

It indicates that the motion platform has 2 pure translational degrees of freedom, which is the same with the results of case 1.

2.3 Case 3: middle link connecting Q_3 and Q_3'

As shown in Fig. 3, the screw-loop equations are established for linkages $P_1 P_2 P_3 Q_3 Q_3' P_3' P_2' P_1'$ and $P_1 P_2 P_3 P_4 P_5 P_5' P_4' P_3' P_2' P_1'$. By the solution of these screw-loop equations, the velocities of joints can be obtained as

$$\begin{bmatrix} \dot{\theta}_1 \\ \dot{\theta}_2 \\ \dot{\beta}_3 \\ \dot{\beta}_5 \\ \dot{\theta}_1' \\ \dot{\theta}_2' \\ \dot{\beta}_3' \\ \dot{\beta}_5' \end{bmatrix} = q_1 \begin{bmatrix} 0 \\ 1 \\ 0 \\ 1 \\ 0 \\ 1 \\ 0 \\ 1 \end{bmatrix} + q_2 \begin{bmatrix} 1 \\ -1 \\ 1 \\ 0 \\ 1 \\ -1 \\ 1 \\ 0 \end{bmatrix}$$

$$+ q_3 \begin{bmatrix} (2l_1 \sin\theta_2 - l_b \sin(\theta_1 + \theta_2))/(-l_b \sin\theta_1) \\ (l_1 \sin\theta_2 + l_b \sin(\theta_1 + \theta_2))/(-l_b \sin\theta_1) \\ \sin(\theta_1 + \theta_2)/\sin\theta_1 \\ 0 \\ 2l_1 \sin\theta_2/(-l_b \sin\theta_1) \\ (l_1 \sin\theta_2 - l_b \sin\theta_1)/(-l_b \sin\theta_1) \\ 0 \\ 1 \end{bmatrix}. \tag{19}$$

Equation (19) shows that this SLE-Pa leg has three mobilities. Furthermore, the corresponding reciprocal screw of the platform can be calculated as

$$\$_p^{\mathbf{r}} : \begin{cases} \$_p^{r1} = \begin{bmatrix} 0 & 0 & 1 & 0 & 0 & 0 \end{bmatrix} \\ \$_p^{r2} = \begin{bmatrix} 0 & 0 & 0 & 1 & 0 & 0 \end{bmatrix} \\ \$_p^{r3} = \begin{bmatrix} 0 & 0 & 0 & 0 & 1 & 0 \end{bmatrix} \end{cases}. \tag{20}$$

Equation (20) indicates that the motion platform has an extra rotational degree of freedom along the z axis, in addition to the 2 translational degrees of freedom.

So far, the mobilities of these kinds of SLE-Pa legs in three cases have been investigated based on screw theory. The results of the other SLE-Pa legs can be deduces from the above three cases. The results are listed in Table 1.

By observing Table 1, we can conclude that if the middle link connects two corresponding corner joints of the SLE limbs, the two SLE limbs are always parallel and the motion platform has 2 pure translational degrees of freedom. If the middle link connects two corresponding middle joints of the SLE limbs, the motion platform has 3 DOFs, which includes 2 translational degrees of freedom and 1 rotational degree of freedom along the z axis.

In cases 1 and 2, suppose

$$\frac{\sin\theta_1}{\sin(\theta_1 + \theta_2)} = \frac{\cos\theta_1}{\cos(\theta_1 + \theta_2)} \tag{21}$$

and substitute it into Eqs. (11) or (17). It is found that the mobility of the SLE-Pa leg reduces to 1. At this moment, the SLE-Pa leg is at the singularity. Solving Eq. (21), the singular angle is

$$\theta_2 = 0 \text{ or } \pi, \tag{22}$$

which corresponds to the maximum or minimum length configuration of the SLE limb.

3 Kinematic analysis of novel 2-DOF translational mechanism

Through assembling the above SLE-Pa legs, a novel 2-DOF translational mechanism is invented. The 2-DOF translational mechanism is composed of three SLE-Pa legs, which are distributed around the center point by 120°, as shown in Fig. 4. As mentioned above, it is easily established that the mobility of this parallel mechanism is 2.

Table 1. Mobility of SLE-Pa legs.

No.	Cases of SLE-Pa legs	Feature of mobility
1		2 pure translational degrees of freedom.
2		2 pure translational degrees of freedom.
3		2 translational degrees of freedom + 1 rotational degree of freedom along z axis

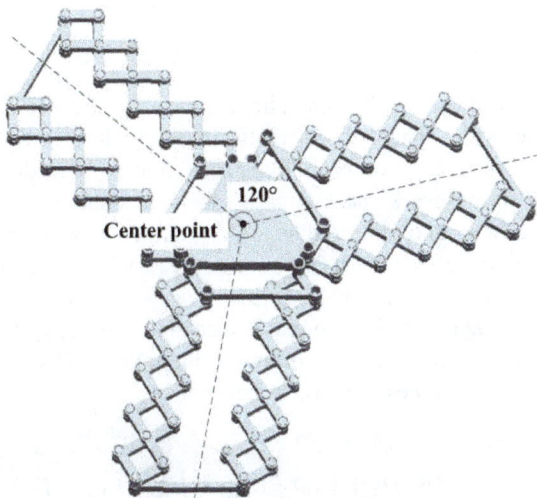

Figure 4. Two-DOF translational parallel mechanism.

In order to simplify the analysis of the kinematics, we substitute an RPR (revolute–prismatic–revolute) limb into the SLE-Pa leg and restrict the z-axis rotational degree of freedom of the motion platform. This results in the motion platform only having 2 translational degrees of freedom, which agrees with the feature of the original mechanism. The schematic diagram is found in Fig. 5.

For the kinematic chain $B_{1E}A_{1E}$, $(x_{B_{1E}}, y_{B_{1E}})$, $(x_{A_{1E}}, y_{A_{1E}})$ is its initial position. When this mechanism moves, the new coordinates of the platform are $(x_{B_{1E}} + p_x, y_{B_{1E}} + p_y)$. The displacement equations of one RPR kinematic chain can be obtained as

$$\begin{cases} (x_{B_{1E}} + p_x) - {}^1H_1\cos{}^1\theta_1 = x_{A_{1E}} \\ (y_{B_{1E}} + p_y) - {}^1H_1\sin{}^1\theta_1 = y_{A_{1E}} \end{cases} \quad (23)$$

The displacement ${}^1H_1$ and angle ${}^1\theta_1$ can be derived as

$$\begin{cases} {}^1H_1 = \sqrt{(x_{B_{1E}} + p_x - x_{A_{1E}})^2 + (y_{B_{1E}} + p_y - y_{A_{1E}})^2} \\ {}^1\theta_1 = \arccos((x_{B_{1E}} + p_x - x_{A_{1E}})/{}^1H_1) \end{cases} \quad (24)$$

Given that the length of each link of the full SLE is $2l$, the length of the link of half the SLE is l. The angle between two

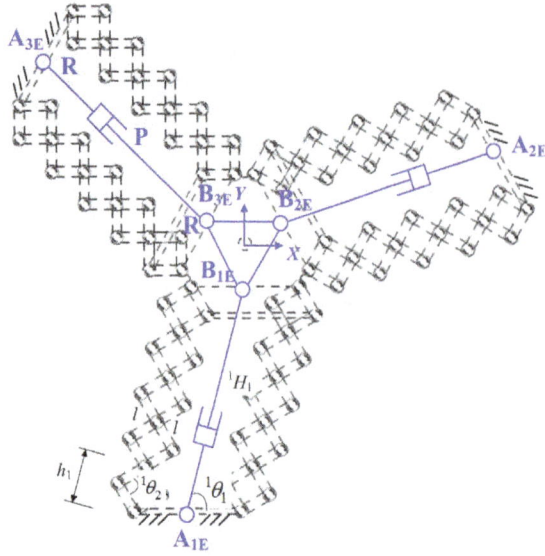

Figure 5. Simplified translational parallel mechanism.

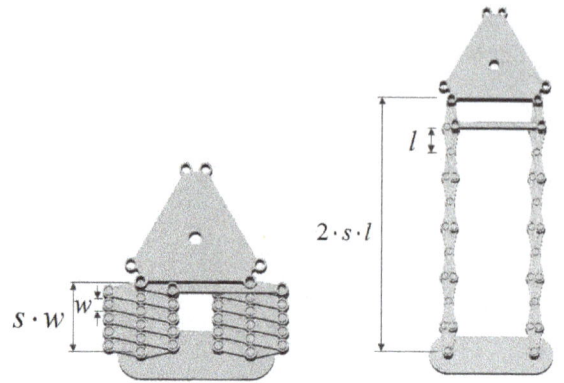

Figure 6. Minimum and maximum length of $^{i}H_1$.

links in $B_{1E}A_{1E}$ is $^{1}\theta_2$. s is the number of the parallelogram in a single SLE limb. h_1 is the length of the single parallelogram. Thus, the length of $B_{1E}A_{1E}$ is

$$^{1}H_1 = sh_1 = 2sl\sin\frac{^{1}\theta_2}{2}. \qquad (25)$$

Substitute the above equation into Eq. (24) gives

$$\begin{cases} ^{1}\theta_2 = 2\arcsin \\ \quad\left(\sqrt{(x_{B_{1E}}+p_x-x_{A_{1E}})^2+(y_{B_{1E}}+p_y-y_{A_{1E}})^2}/(2sl)\right) \\ ^{1}\theta_1 = \arccos\left((x_{B_{1E}}+p_x-x_{A_{1E}})/ \right. \\ \quad\left.\sqrt{(x_{B_{1E}}+p_x-x_{A_{1E}})^2+(y_{B_{1E}}+p_y-y_{A_{1E}})^2}\right) \end{cases} \qquad (26)$$

Through differentiating the above equations, the velocity equations can be obtained:

$$\begin{cases} ^{1}\dot\theta_2 = (\dot p_x\cos{^{1}\theta_1}+\dot p_y\sin{^{1}\theta_1})/(2sl\cos\frac{^{1}\theta_2}{2}) \\ ^{1}\dot\theta_1 = -(\dot p_x\sin{^{1}\theta_1}-\dot p_y\cos{^{1}\theta_1})/(2sl\sin\frac{^{1}\theta_2}{2}) \end{cases} \qquad (27)$$

They can be expressed in matrix form:

$$\mathbf{J}\begin{bmatrix} ^{1}\dot\theta_1 \\ ^{1}\dot\theta_2 \end{bmatrix} = \begin{bmatrix} \dot P_x \\ \dot P_y \end{bmatrix}, \qquad (28)$$

where the Jacobian matrix is

$$\mathbf{J} = \begin{bmatrix} -2sl\sin\frac{^{1}\theta_2}{2}\sin{^{1}\theta_1} & 2sl\cos\frac{^{1}\theta_2}{2}\cos{^{1}\theta_1} \\ 2sl\sin\frac{^{1}\theta_2}{2}\cos{^{1}\theta_1} & 2sl\cos\frac{^{1}\theta_2}{2}\sin{^{1}\theta_1} \end{bmatrix}. \qquad (29)$$

The determinant of the matrix is $|J| = -2s^2l^2\sin{^{1}\theta_2}$. Let $|J| = 0$; then, the singular configuration can be obtained, i.e., $^{1}\theta_2 = 0$ and π. This result agrees with Eq. (22).

For the other kinematic chains, we can calculate the angles by the transformation of coordinates. Given α_i is the rotation angle from $B_{1E}A_{1E}$ to $B_{iE}A_{iE}$, the new displacement $^{i}H_1$ and angle $^{i}\theta_1$ are

$$\begin{cases} ^{i}H_1 = \sqrt{(^{i}C_x+p_x)^2+(^{i}C_y+p_y)^2} \\ ^{i}\theta_1 = \arccos((^{i}C_x+p_x)/^{i}H_1) \\ ^{i}C_x(x_{B_{1E}}-x_{A_{1E}})\cos\alpha_i-(\gamma_{B_{1E}}-\gamma_{A_{1E}})\sin\alpha_i \\ ^{i}C_y = (x_{B_{1E}}-x_{A_{1E}})\sin\alpha_i+(\gamma_{B_{1E}}-\gamma_{A_{1E}})\cos\alpha_i \end{cases} \qquad (30)$$

Given all kinematic chains are identical, the range of each linear motion element is

$$H_{lb}\le{^{i}H_1}\le H_{ub} \quad i=1,2,\dots,n,$$

where H_{lb}, H_{ub} are the lower and upper bounds. As shown in Fig. 6, the minimum and maximum length of $^{i}H_1$ can be estimated:

$$H_{lb} \approx s\cdot w$$
$$H_{ub} \approx 2\cdot s\cdot l, \qquad (31)$$

where w is the width of each link.

The work space of the platform is the intersection of the work spaces of each kinematic chain. According to Eq. (30), the work space of each kinematic chain is a torus, whose inner and outer radius are H_{lb} and H_{ub}, respectively. The center of the torus is $(-(x_{B_{1E}}-x_{A_{1E}})\cos\alpha_i+(y_{B_{1E}}-y_{A_{1E}})\sin\alpha_i, -(x_{B_{1E}}-x_{A_{1E}})\sin\alpha_i-(y_{B_{1E}}-y_{A_{1E}})\cos\alpha_i)$, denoted by D_{iE}. The 2-DOF planar mechanism is composed of three identical chains. According to the ratio of H_{ub} to H_{lb}, there are two types of work spaces.

I. Type I: $H_{ub}/H_{lb} > \sqrt{3}$

Given $|D_{1E}D_{2E}| = |D_{2E}D_{3E}| = |D_{3E}D_{1E}| = DD$, the work space of the platform in this type can be classified as consisting of three categories. The shapes of each category can be seen in Fig. 7. The different ranges of the above three categories are displayed by the plot of DD/H_{lb} vs. H_{ub}/H_{lb} in Fig. 8.

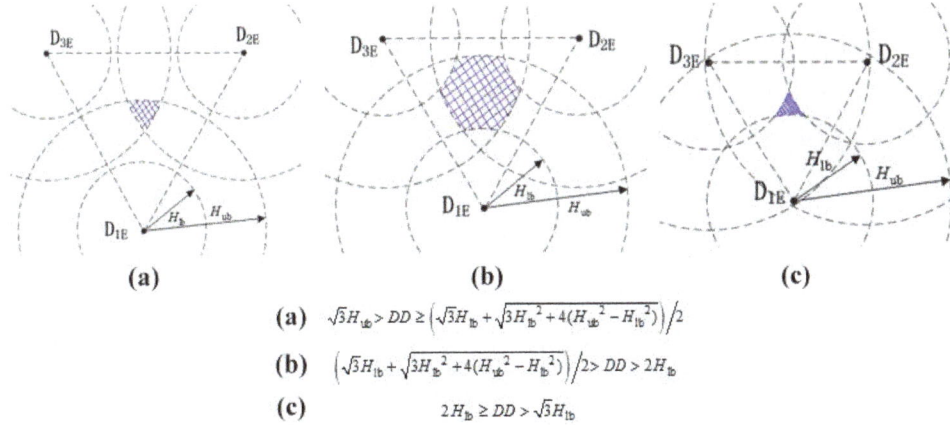

(a) $\sqrt{3}H_{ub} > DD \geq \left(\sqrt{3}H_{lb} + \sqrt{3H_{lb}^2 + 4(H_{ub}^2 - H_{lb}^2)}\right)/2$

(b) $\left(\sqrt{3}H_{lb} + \sqrt{3H_{lb}^2 + 4(H_{ub}^2 - H_{lb}^2)}\right)/2 > DD > 2H_{lb}$

(c) $2H_{lb} \geq DD > \sqrt{3}H_{lb}$

Figure 7. Three categories of type I work space.

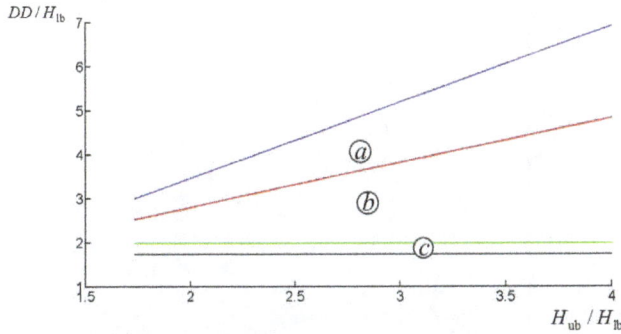

Figure 8. Ranges of three categories of type I.

II. Type II: $1 \leq H_{ub}/H_{lb} \leq \sqrt{3}$

In this type, the work space of the platform can also be classified as consisting of three categories, whose shapes are as shown in Fig. 9. The ranges of the above three categories are plotted in Fig. 10.

The numeric value of $\sqrt{3}$ is 1.73. By using the above results, the work spaces of these kinds of parallel mechanisms can be estimated. For the SLE-Pa leg, the ratio of H_{ub} to H_{lb} equals $2l/w$. In practice, the length-to-width ratio l/w of most links is generally larger than 3.0, and the ratio $2l/w$ is definitely more than 1.73. In our models for example, the length l is 27 mm, the width w is 8 mm and the ratio $2l/w$ equals 6.75. Thus, the shapes of most work spaces of these 2-DOF planar translational mechanisms belong to type I. The exact shapes depend on the specific parameters of H_{lb}, H_{ub} and the positions of SLE-Pa legs.

4 Stiffness synthesis

For the purpose of adding the recovery function, the elastic elements are installed for this mechanism. The various elastic elements and positions result in the different stiffnesses of the motion platform. In this section, we choose a torsional spring and an ordinary cylinder spring as elastic elements to discuss the variation in stiffness.

4.1 Case 1: torsional springs installed in the corner joints of SLE-Pa legs

As shown in Fig. 11, the torsional springs are all installed in the corner joints of SLE-Pa legs. Given that stiffness coefficient of each torsional spring is k_θ, the total elastic energy of the mechanism is

$$V = \sum_{i=1}^{n} \frac{1}{2}k_\theta^i \theta_2^2. \tag{32}$$

Differentiating Eq. (32) to the displacements p_x and p_y, the corresponding generalized force Q_x and Q_y of the platform can be calculated by the following equation:

$$\begin{cases} Q_x = -\dfrac{\partial V}{\partial p_x} = -\displaystyle\sum_{i=1}^{n} k_\theta \Delta^i\theta_2 \dfrac{\partial^i\theta_2}{\partial p_x} \\ Q_y = -\dfrac{\partial V}{\partial p_y} = -\displaystyle\sum_{i=1}^{n} k_\theta \Delta^i\theta_2 \dfrac{\partial^i\theta_2}{\partial p_y} \end{cases}. \tag{33}$$

Utilizing Eq. (26), the derivative of $^i\theta_2$ versus p_x and p_y can be obtained as

$$\frac{\partial^i\theta_2}{\partial p_x} = \frac{\cos{^i\theta_1}}{\cos\frac{^i\theta_2}{2}}\frac{1}{sl} \qquad \frac{\partial^i\theta_2}{\partial p_y} = \frac{\sin{^i\theta_1}}{\cos\frac{^i\theta_2}{2}}\frac{1}{sl}. \tag{34}$$

Continuing to differentiate Eq. (33) to p_x, p_y and ignoring the second-class derivative of $^i\theta_2$, we can calculate the stiffness matrix of the platform.

$$\mathbf{K}_\theta = k_\theta \begin{bmatrix} \displaystyle\sum_{i=1}^{n}\left(\frac{\partial^i\theta_2}{\partial p_x}\right)^2 & \displaystyle\sum_{i=1}^{n}\left(\frac{\partial^i\theta_2}{\partial p_x}\frac{\partial^i\theta_2}{\partial p_y}\right) \\ \displaystyle\sum_{i=1}^{n}\left(\frac{\partial^i\theta_2}{\partial p_x}\frac{\partial^i\theta_2}{\partial p_y}\right) & \displaystyle\sum_{i=1}^{n}\left(\frac{\partial^i\theta_2}{\partial p_y}\right)^2 \end{bmatrix}$$

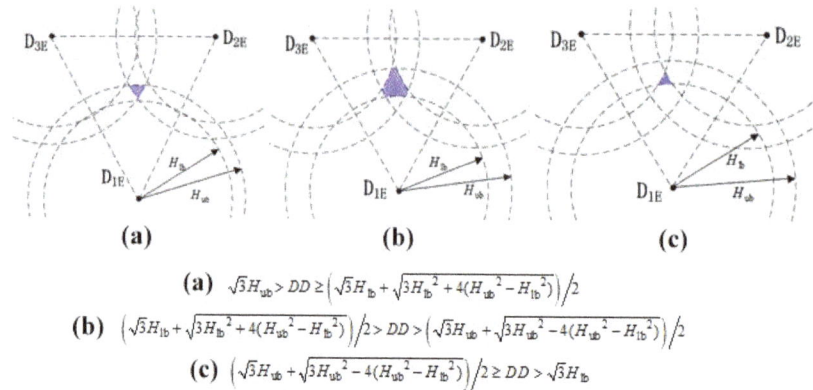

(a) $\sqrt{3}H_{ub} > DD \geq \left(\sqrt{3}H_{lb} + \sqrt{3H_{lb}^2 + 4(H_{ub}^2 - H_{lb}^2)} \right) \Big/ 2$

(b) $\left(\sqrt{3}H_{lb} + \sqrt{3H_{lb}^2 + 4(H_{ub}^2 - H_{lb}^2)} \right) \Big/ 2 > DD > \left(\sqrt{3}H_{ub} + \sqrt{3H_{ub}^2 - 4(H_{ub}^2 - H_{lb}^2)} \right) \Big/ 2$

(c) $\left(\sqrt{3}H_{ub} + \sqrt{3H_{ub}^2 - 4(H_{ub}^2 - H_{lb}^2)} \right) \Big/ 2 \geq DD > \sqrt{3}H_{lb}$

Figure 9. Three categories of type II work space.

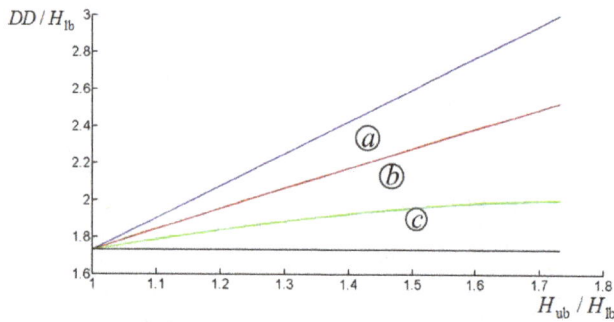

Figure 10. Ranges of three categories of type II.

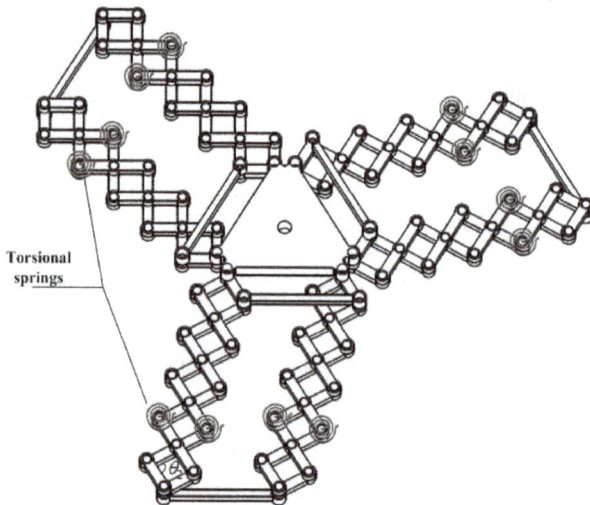

Figure 11. Torsional springs installed in the corner joints.

$$= \frac{k_\theta}{(sl)^2} \begin{bmatrix} \sum\limits_{i=1}^{n} \dfrac{\cos^{2i}\theta_1}{\cos^{2\frac{i\theta_2}{2}}} & \sum\limits_{i=1}^{n} \dfrac{\cos^i\theta_1 \sin^i\theta_1}{\cos^{2\frac{i\theta_2}{2}}} \\ \sum\limits_{i=1}^{n} \dfrac{\cos^i\theta_1 \sin^i\theta_1}{\cos^{2\frac{i\theta_2}{2}}} & \sum\limits_{i=1}^{n} \dfrac{\sin^{2i}\theta_1}{\cos^{2\frac{i\theta_2}{2}}} \end{bmatrix} \quad (35)$$

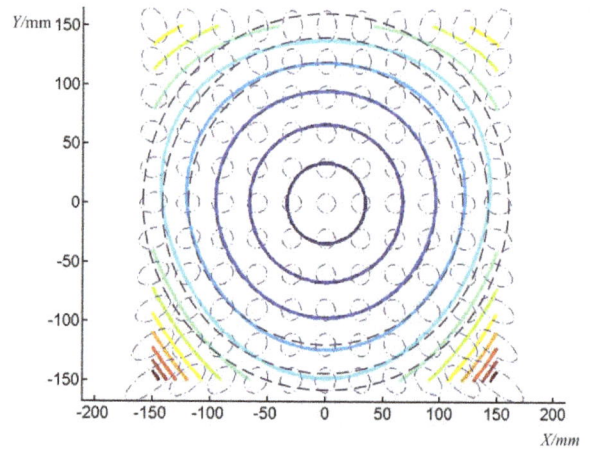

Figure 12. Stiffness contours of the platform in case 1.

Given initial θ_1 and θ_2 are 75 and 90°, the stiffness contours of the platform within the area of 150 by 150 are plotted, as shown in Fig. 12. The generalized force ellipsoids are used to evaluate the isotropy of the above stiffness matrix. The region where the contours are more intensive has higher stiffness and vice versa. If the shape of the contour approximates a circle, the stiffness of the platform in this circle can be treated as isotropy. In Fig. 12, the contours in the region of 100 by 100 are nearly circular. The generalized force ellipsoids in this region are also rounder than those outside. It appears that the stiffness here is isotropic. With the increasing area, the shapes of the contours and the generalized force ellipsoids are more different from a circle. It illustrates that the feature of stiffness becomes more anisotropic. For further discussion, we take the platform a certain distances away from the center point along ±45, ±90, ±135, 0 and 180° axes. Then we release it and let it vibrate. The responses of the displacements of the platform are plotted in Fig. 13. The responses of the displacements from left to right are under the conditions of the given distances of 50, 100 and 141 mm. The red dots represent the release position. As the distance

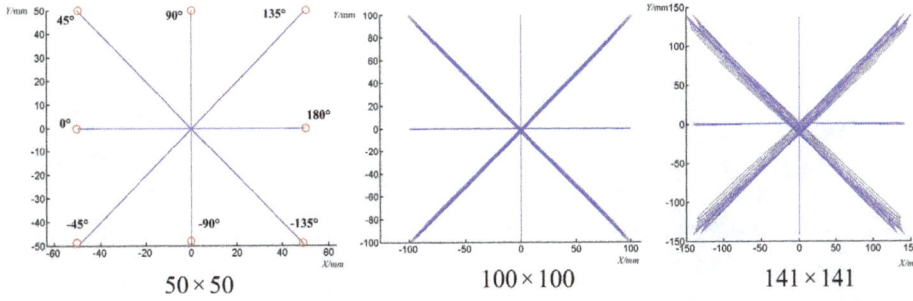

Figure 13. Responses of the displacements of the platform in case 1.

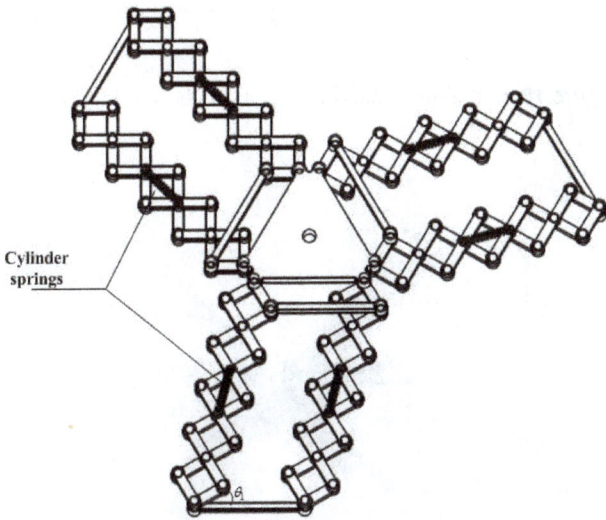

Figure 14. Cylinder springs connecting the vertical joints.

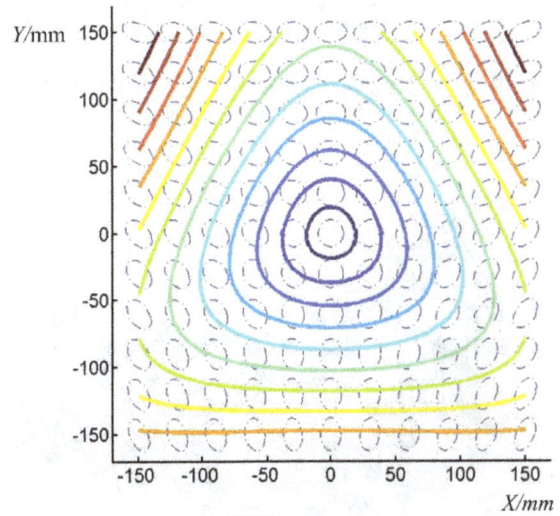

Figure 15. Stiffness contours of the platform in case 2.

increases, the response of the displacement appears whirling, which means the stiffness is becoming anisotropic. It agrees with the result of Fig. 12.

4.2 Case 2: cylinder springs connecting the vertical joints

As shown in Fig. 14, the cylinder springs connect the vertical joints in SLE-Pa legs. Given that the stiffness coefficient of each spring is k_v, the free length of the spring is h_0; the total elastic energy of the mechanism is

$$V = \sum_{i=1}^{n} \frac{1}{2} k_v ({}^i h_1 - h_0)^2 = \sum_{i=1}^{n} \frac{1}{2} k_v \left(\frac{{}^i H_1}{s} - h_0 \right)^2. \tag{36}$$

Differentiating Eq. (36) to the displacements p_x and p_y, the corresponding generalized force Q_x and Q_y of the platform can be obtained:

$$F_x = \sum_{i=1}^{n} k_v \left(\frac{{}^i H_1}{s} - h_0 \right) \frac{\cos^i \theta_1}{s}$$

$$F_y = \sum_{i=1}^{n} k_v \left(\frac{{}^i H_1}{s} - h_0 \right) \frac{\sin^i \theta_1}{s}. \tag{37}$$

The stiffness matrix of the platform can be derived as

$$\mathbf{K}_v = \frac{k_v}{s^2} \begin{bmatrix} \sum_{i=1}^{n} \cos^{2i} \theta_1 & \sum_{i=1}^{n} \cos^i \theta_1 \sin^i \theta_1 \\ \sum_{i=1}^{n} \cos^i \theta_1 \sin^i \theta_1 & \sum_{i=1}^{n} \sin^{2i} \theta_1 \end{bmatrix}, \tag{38}$$

given that initial θ_1 and θ_2 are 75 and 90°. In the area of 150 by 150, the stiffness contours and the generalized force ellipsoids are plotted, as shown in Fig. 15. It may be seen that the shapes of the contours and the generalized force ellipsoids appear very different from circles. Thus, the stiffness of the platform is anisotropic under this circumstance. To test this point further, the platform is taken 141 mm from the center point and released. The response of free vibration is plotted in Fig. 16. The whirling trajectory proves that the above analysis is correct.

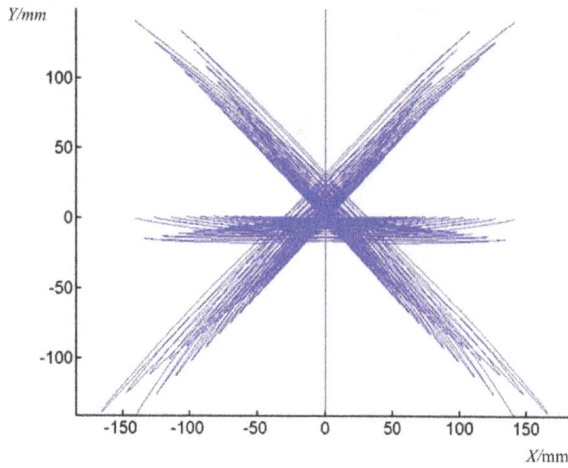

Figure 16. Responses of the displacements of the platform in case 2.

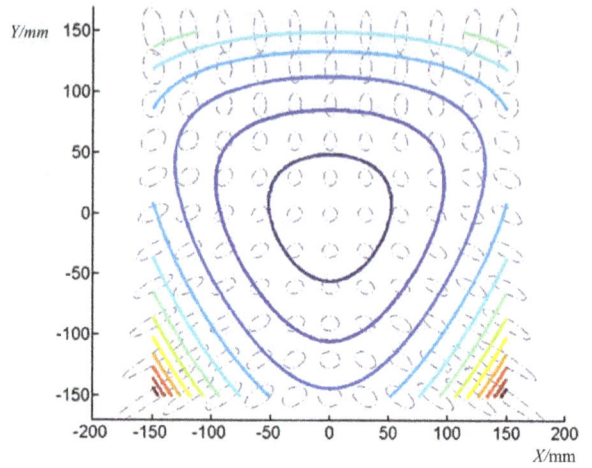

Figure 18. Stiffness contours of the platform in case 3.

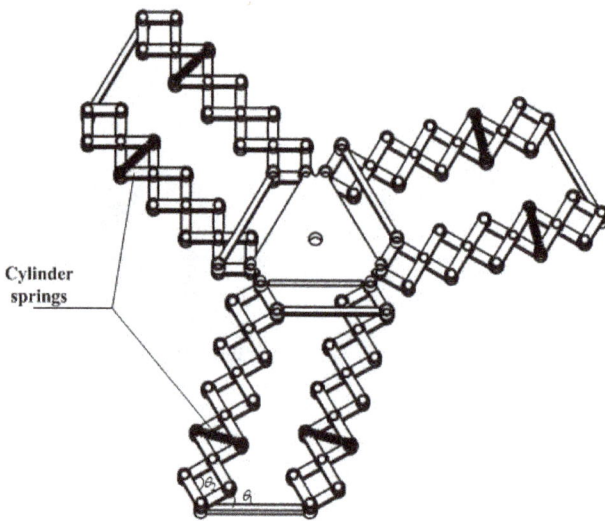

Figure 17. Cylinder springs connecting the horizontal joints.

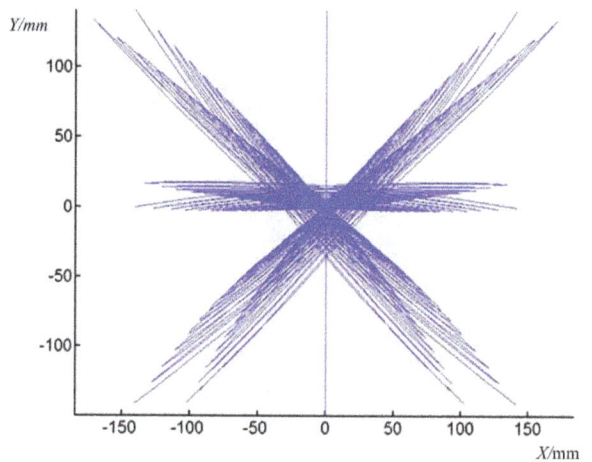

Figure 19. Responses of the displacements of the platform in case 3.

4.3 Case 3: cylinder springs connecting the horizontal joints

As shown in Fig. 17, the corresponding generalized force Q_x and Q_y of the platform can be obtained by the use of the energy method.

$$F_x = -\sum_{i=1}^{n} k_h l \sin \frac{{}^i\theta_2}{2}\left(2l\cos\frac{{}^i\theta_2}{2} - D_0\right)\frac{\partial {}^i\theta_2}{\partial p_x}$$

$$F_y = -\sum_{i=1}^{n} k_h l \sin \frac{{}^i\theta_2}{2}\left(2l\cos\frac{{}^i\theta_2}{2} - D_0\right)\frac{\partial {}^i\theta_2}{\partial p_y}, \qquad (39)$$

where k_h is the stiffness coefficient of each spring. The stiffness matrix of the platform can be derived as

$$\mathbf{K}_h = -\frac{k_h}{s^2 l}\begin{bmatrix} \sum\limits_{i=1}^{n}\left(l\cos{}^i\theta_2 - \dfrac{1}{2}D_0\cos\dfrac{{}^i\theta_2}{2}\right)\dfrac{\cos^2{}^i_2\theta_1}{\cos^2\frac{{}^i\theta_2}{2}} \\ \sum\limits_{i=1}^{n}\left(l\cos{}^i\theta_2 - \dfrac{1}{2}D_0\cos\dfrac{{}^i\theta_2}{2}\right)\dfrac{\cos{}^i\theta_1\sin{}^i\theta_1}{\cos^2\frac{{}^i\theta_2}{2}} \\ \sum\limits_{i=1}^{n}\left(l\cos{}^i\theta_2 - \dfrac{1}{2}D_0\cos\dfrac{{}^i\theta_2}{2}\right)\dfrac{\cos{}^i\theta_1\sin{}^i\theta_1}{\cos^2\frac{{}^i\theta_2}{2}} \\ \sum\limits_{i=1}^{n}\left(l\cos{}^i\theta_2 - \dfrac{1}{2}D_0\cos\dfrac{{}^i\theta_2}{2}\right)\dfrac{\sin^{2i}\theta_1}{\cos^2\frac{{}^i\theta_2}{2}} \end{bmatrix} \qquad (40)$$

The stiffness contours and the generalized force ellipsoids are drawn in Fig. 18. The feature of the stiffness is anisotropic. The response of free vibration in Fig. 19 also proves this result.

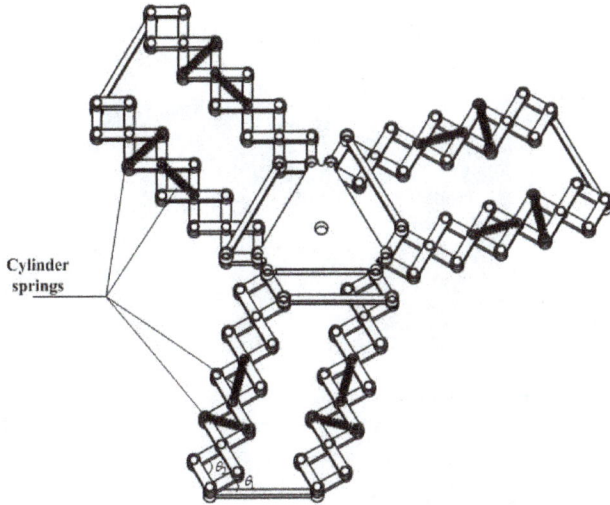

Figure 20. Cylinder springs connecting vertical joints and horizontal joints.

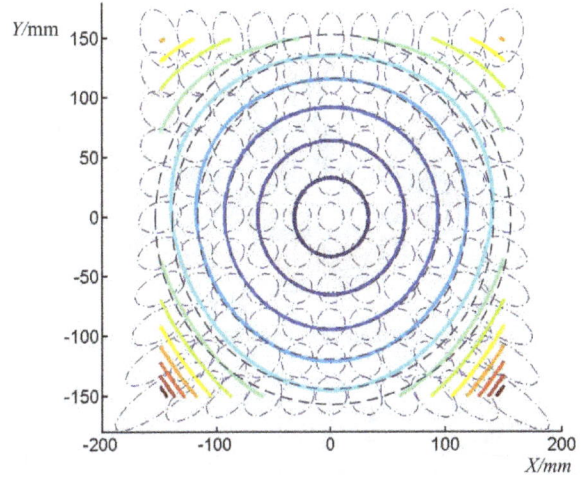

Figure 21. Stiffness contours of the platform in case 4.

4.4 Case 4: cylinder springs connecting vertical joints and horizontal joints

As shown in Fig. 20, the cylinder springs connect the vertical and horizontal joints in SLE-Pa legs. The stiffness coefficient of each spring is entirely identical, i.e., $k_v = k_h$. The stiffness of the platform can be regarded as the sum of Eqs. (38) and (40). The contours and the generalized force ellipsoids can be drawn, and the response of free vibration is displayed in Figs. 21 and 22. It is found that the stiffness of the platform appears to be isotropic in this compound connection. By the comparison of Figs. 15 and 18, the reason is found that the convexities of the contours in these two figures are the opposite of one another. This means that the recovery forces of the platform in these two cases may be complementary, which results in the stiffness being isotropic. The responses of the free vibration also illustrate this point. Through this case, it may be concluded that the isotropic stiffness can be synthesized by different anisotropic stiffness cases according to the feature of the stiffness contours.

5 Approximation of stiffness coefficient

Since the SLE-Pa legs are distributed around the center of the mechanism, $^1\theta_1, {}^2\theta_1, \ldots, {}^n\theta_1$ form an arithmetic progression if the displacement of the platform is small.

$$^i\theta_1 = {}^1\theta_1 + \frac{2\pi}{n}(i-1) \quad i = 1, 2, \ldots, n \tag{41}$$

$^i\theta_2$ in each leg seems to be equal.

$$^1\theta_2 = {}^2\theta_2 = \ldots = {}^n\theta_2 \tag{42}$$

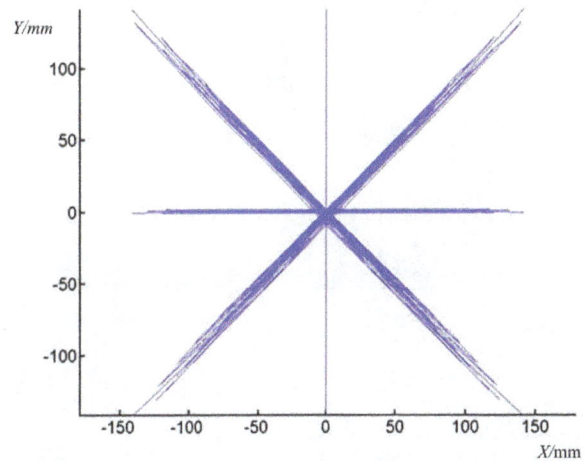

Figure 22. Responses of the displacements of the platform in case 4.

With the following condition

$$\sum_{i=1}^n \cos^2\theta_i = \frac{1}{2}n \quad \sum_{i=1}^n \sin^2\theta_i = \frac{1}{2}n$$

$$\sum_{i=1}^n \sin\theta_i \cos\theta_i = 0, \tag{43}$$

the approximation of the stiffness coefficient for case 1 in Sect. 4 can be deduced as

$$\mathbf{K}_\theta = \frac{k_\theta}{(sl)^2}\begin{bmatrix} \displaystyle\sum_{i=1}^n \frac{\cos^{2i}\theta_1}{\cos^2 \frac{i\theta_2}{2}} & \displaystyle\sum_{i=1}^n \frac{\cos^i\theta_1 \sin^i\theta_1}{\cos^2 \frac{i\theta_2}{2}} \\ \displaystyle\sum_{i=1}^n \frac{\cos^i\theta_1 \sin^i\theta_1}{\cos^2 \frac{i\theta_2}{2}} & \displaystyle\sum_{i=1}^n \frac{\sin^{2i}\theta_1}{\cos^2 \frac{i\theta_2}{2}} \end{bmatrix}$$

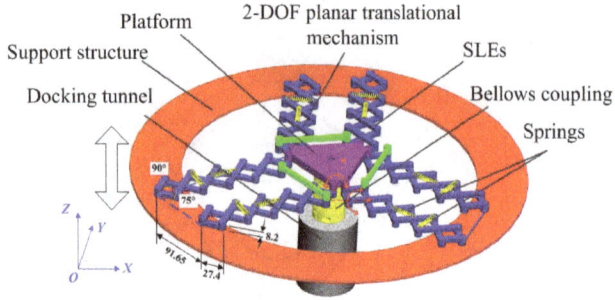

Figure 23. Passive docking device.

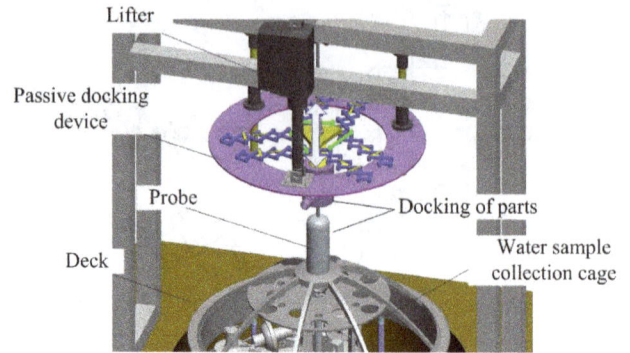

Figure 24. Docking diagram.

$$= \frac{k_\theta}{(sl)^2 \cos^2 \frac{^0\theta_2}{2}} \begin{bmatrix} \frac{1}{2}n & 0 \\ 0 & \frac{1}{2}n \end{bmatrix}. \tag{44}$$

In the same way, the approximation of the stiffness coefficient for cases 2 and 3 is

$$\mathbf{K}_v = \frac{k_v}{s^2} \begin{bmatrix} \sum_{i=1}^{n} \cos^{2i}\theta_1 & \sum_{i=1}^{n} \cos^i\theta_1 \sin^i\theta_1 \\ \sum_{i=1}^{n} \cos^i\theta_1 \sin^i\theta_1 & \sum_{i=1}^{n} \sin^{2i}\theta_1 \end{bmatrix}$$

$$= \frac{k_v}{s^2} \begin{bmatrix} \frac{1}{2}n & 0 \\ 0 & \frac{1}{2}n \end{bmatrix}, \tag{45}$$

$$\mathbf{K}_h = -\frac{k_h}{s^2 l} \frac{\left(l\cos^0\theta_2 - \frac{1}{2}D_0 \cos\frac{^0\theta_2}{2} \right)}{\cos^2 \frac{^0\theta_2}{2}} \begin{bmatrix} \frac{1}{2}n & 0 \\ 0 & \frac{1}{2}n \end{bmatrix}$$

$$= \frac{k_h}{s^2} \tan^2 \frac{^0\theta_2}{2} \begin{bmatrix} \frac{1}{2}n & 0 \\ 0 & \frac{1}{2}n \end{bmatrix}. \tag{46}$$

The approximation of the stiffness coefficient can be used for a preliminary estimate of some parameters, i.e., recovery force, vibration frequency and others, before designing the mechanism. It is found that the approximation of the stiffness coefficient in cases 1 and 3 is related to $^0\theta_2$. It means that the stiffness of the platform can be adjusted by changing $^0\theta_2$. If $k_v = k_h$ and $^0\theta_2 = 90°$, the stiffness in case 2 and 3 is equal.

6 Design of passive docking device

These novel 2-DOF planar translational mechanisms assembled by SLE-Pa legs are applied to the design of a passive docking device, which can be used for joining pipes in a water sample collection system. As shown in Fig. 23, the passive docking device comprises support structure, 2-DOF planar translational mechanism, bellows coupling and

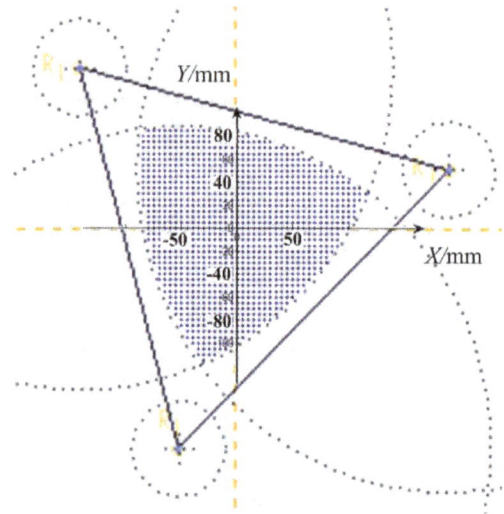

Figure 25. Work space of 2-DOF planar translational mechanism.

a docking tunnel. The bellows coupling provides x- and y-axes rotational DOFs. The 2-DOF planar translational mechanism provides x- and y-axes translational DOFs. The support structure is taken by a lifter, which moves along the z axis. Thus, the docking tunnel has a total of 5 DOFs, which has the ability to adapt to the misalignments in the docking procedure. In addition, for the purposes of a passive docking device capable of an isotropic recovery force, the cylinder springs are installed to connect vertical joints and horizontal joints. The stiffness coefficient of each cylinder spring is $0.3 \, \text{N} \, \text{mm}^{-1}$. The total mass of the platform, bellows coupling and docking tunnel is 1.247 kg.

As shown in Fig. 24, the water sample collection cage is fixed to the deck by the holding mechanism when it is pulled up from the sea. There is a docking probe at the top of the water sample collection cage. When the cage is put in place, the lifter takes the passive docking device and allows it to descend to approach the probe. In the docking procedure, the probe slowly slides into the tunnel. The 2-DOF planar translational mechanism and bellows coupling enable the tunnel to possess 4 DOFs in order to compensate for the misalign-

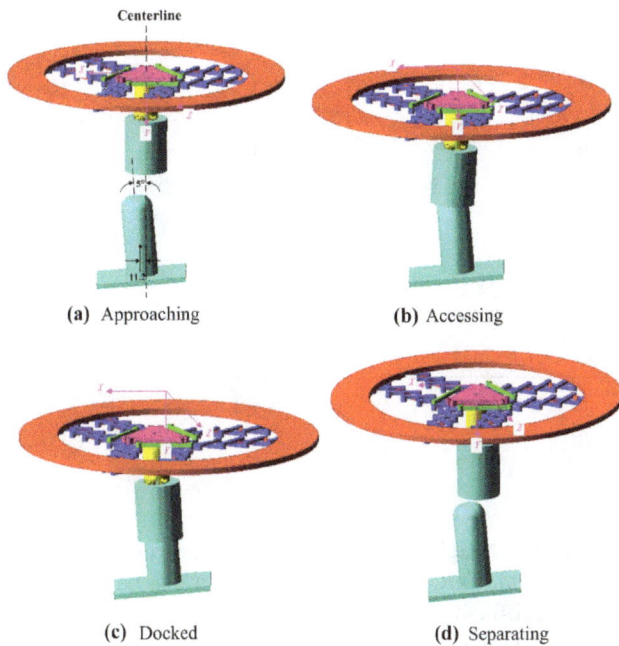

(a) Approaching **(b)** Accessing

(c) Docked **(d)** Separating

Figure 26. Docking procedure.

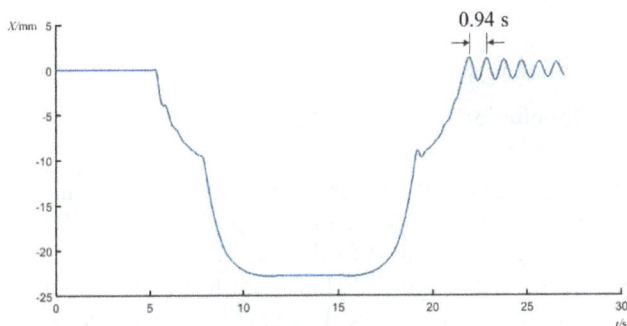

Figure 27. x-axis displacement of the platform.

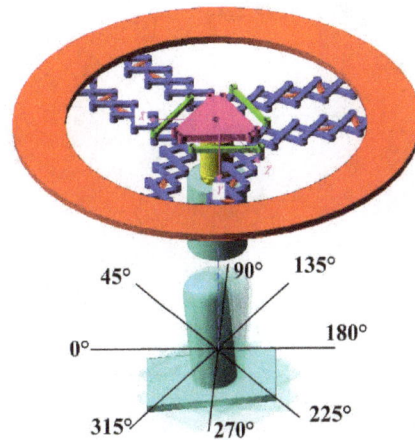

Figure 28. Eight rotation angles of the probe.

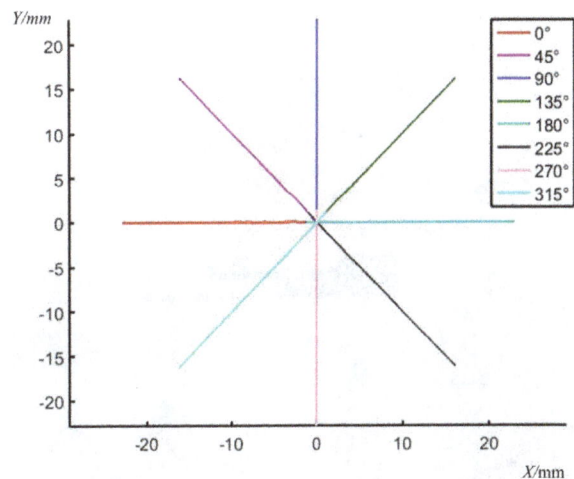

Figure 29. Eight displacement trajectories of the platform.

ments between the tunnel and probe. When the probe has entirely entered the tunnel, the sample water in the collection cage is drawn out. After this operation, the lifter takes the passive docking device and moves it up to the original place.

In the 2-DOF planar translational mechanism, the length of half an SLE is $l = 27.4$ mm, and the width is $w = 8.2$ mm. The number of SLEs in a single SLE limb is $s = 5$. The initial angles θ_1 and θ_2 are 75 and 90°, respectively. The minimum length of an SLE-Pa leg is $H_{lb} \approx s \cdot w = 41$ mm. The maximum length of an SLE-Pa leg is $H_{ub} \approx 2 \cdot s \cdot l = 274$ mm. The ratio of H_{ub} to H_{lb} is 6.68, which is larger than $\sqrt{3}$. According to Sect. 3, the shape of the work space belongs to type I(a), as shown in Fig. 25.

The docking procedure is simulated by Adams, as shown in Fig. 26. Assume that the center of the probe has an offset distance 11.2 mm from the centerline and a slip angle 5° off the horizon. At the beginning, the passive docking device

descends at a velocity of $10\,\mathrm{mm\,s^{-1}}$ to dock with the probe. When complete, the docking device stops for 3 s. Then, it ascends at a velocity of $10\,\mathrm{mm\,s^{-1}}$ until it reaches the original position. The simulation proved that the 2-DOF translational mechanism is able to produce a smooth motion in $x - y$ plane.

The x-axis displacement of the platform of the 2-DOF translational mechanism is measured in the simulation, as shown in Fig. 27. It may be seen that the platform generates vibration after the docking tunnel and the probe have been separated. This is because of the cylinder springs in the mechanism, which result in the recovery force. The period of vibration is roughly measured as 0.94 s in Fig. 27.

We can also calculate the period of vibration by the approximation of the stiffness coefficient derived in Sect. 5. By the utilization of Eqs. (45) and (46) and the consideration of two vertical and horizontal springs in each leg, the approximation of the stiffness coefficient can be obtained.

$$k_{\mathrm{E}} = 2 \times \left(\frac{n}{2} \frac{k_{\mathrm{v}}}{s^2} + \frac{n}{2} \frac{k_{\mathrm{h}}}{s^2} \right) = 0.072\,\mathrm{N\,mm^{-1}}.$$

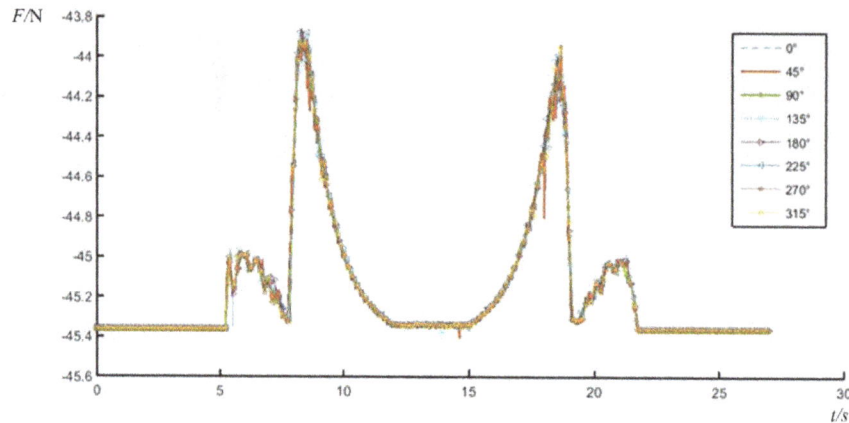

Figure 30. Actuator forces in eight cases.

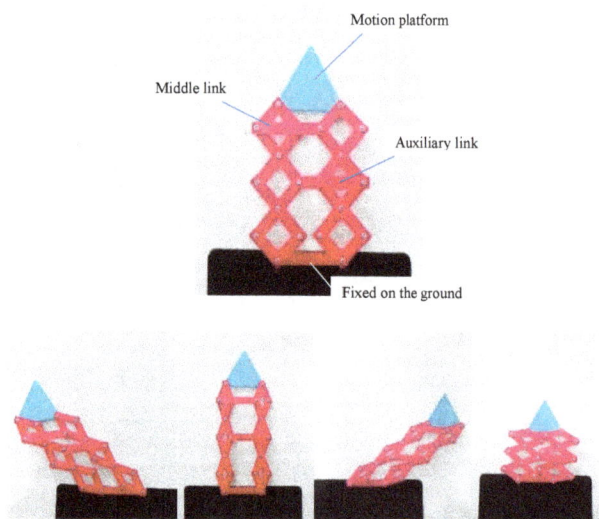

Figure 31. Prototype of one SLE-Pa leg.

The period of vibration can be then calculated as

$$T = \frac{2\pi}{\omega} = 2\pi \sqrt{\frac{m}{k_E}} = 0.827 \text{ s}.$$

This result is very close to the one of the simulation. It proves the correction of the above analysis. To illustrate the isotropy of the recovery force, the probe is rotated around the centerline to change its position and posture. The rotation angles form a tolerance of 45° arithmetic progression from 0 to 315°, as shown in Fig. 28. The docking simulation is conducted for each rotation angle, i.e., in total eight cases that need to be solved.

The eight displacement trajectories of the platform in the $x - z$ plane are calculated and plotted in Fig. 29. It shows that all the trajectories are straight lines and there is no whirling. The actuator forces in eight cases which drive the passive docking device up and down are also measured and plotted, as shown in Fig. 30. The eight force curves are almost

the same although the positions and postures of the probes are different. Through Figs. 29 and 30, it can be concluded that the stiffness of the passive docking device in the docking procedure is isotropic. The prototype of one SLE-Pa leg is also manufactured, as shown in Fig. 31. The auxiliary link in the prototype provides overconstraint which is to enhance the parallel constraint for the two SLE limbs.

7   Conclusion

A novel 2-DOF planar translational mechanism comprising SLE-Pa legs is proposed in this paper. The SLE-Pa leg consists of two identical SLE limbs. The two corresponding nodes in the two SLE limbs are hinged by one middle link. According to three different hinged positions of the middle link, the mobilities and kinematics are discussed by screw theory. The result shows that if the middle link connects two corresponding corner joints of the SLE limbs, these two SLE limbs are always parallel and the SLE-Pa leg has 2 pure translational degrees of freedom. Based on that, a novel 2-DOF translational mechanism is invented. The kinematics equations and the work spaces of the platform are given. For the purpose of this mechanism capable of recovery force, the elastic elements are installed for this mechanism. Through the analysis of the stiffness contours and the generalized force ellipsoids, the isotropy of the stiffness of the platform has been identified. By applying the above results, the passive docking device with the isotropic recovery force is invented. The docking procedure is simulated by Adams, which proves the feasibility of the 2-DOF translational mechanism and the isotropy of the device. Moreover, the prototype of one SLE-Pa leg is manufactured and presented.

Competing interests. The authors declare that they have no conflict of interest.

Acknowledgements. This project was supported by the National Natural Science Foundation of China (nos. 51675318, 51575329, 61525305).

Edited by: Chin-Hsing Kuo

References

Bai, G., Liao, Q., Li, D., and Wei, S.: Synthesis of scaling mechanisms for geometric figures with angulated-straight elements, Proc IMechE Part C: J Mechanical Engineering Science, 227, 2795–2809, 2013.

Dai, J. S. and Rees, J. J.: Mobility in metamorphic mechanisms of foldable/erectable kinds, ASME Transaction Journal of Mechanical Design, 121, 375–382, 1999.

Ding, X. L., Yang, Y., and Dai, J. S.: Topology and Kinematic Analysis of Color-Changing Ball, Mech. Mach. Theory, 46, 67–81, 2011.

Dong, J., Yuan, C., Stori, J. A., and Ferreira, P. M.: Development of a High-speed 3-axis Machine Tool using a Novel Parallel-kinematics X-Y Table, International Journal of Machine Tools and Manufacture, 44, 1355–1371, 2004.

Escrig, F. and Valcarcel, J. P.: Geometry of expandable space structures, International Journal of Space Structures, 8, 71–84, 1993.

Escrig, F., Valcarcel, J. P., and Sanchez, J.: Deployable cover on a swimming pool in Seville, Bulletin of the International Association for Shell and Spatial Structures, 37, 39–70, 1996.

Huang, T., Li, Z., Li, M., Chetwynd, D. G., and Gosselin, C. M.: Conceptual design and dimensional synthesis of a novel 2-DOF translational parallel robot for pick-and-place operations, J. Mech. Design, 126, 449–455, 2004.

Huang, T., Liu, S., Mei, J., and Chetwynd, D. G.: Optimal design of a 2-DOF pick-and-place parallel robot using dynamic performance indices and angular constraints, Mech. Mach. Theory, 70, 246–253, 2013.

Kaveh, A. and Davaran, A.: Analysis of pantograph foldable structure, Comput. Struct., 59, 131–140, 1996.

Kaveh, A., Jafarvand, A., and Barkhordari, M. A.: Optimal design of pantograph foldable structures, International Journal of Space Structures, 14, 295–302, 1999.

Kim, H. S.: Development of Two Types of Novel Planar Translational Parallel Manipulators by using Parallelogram Mechanism, Journal of the Korean Society of Precision Engineering, 24, 50–57, 2007.

Langbecker, T.: Kinematic Analysis of Deployable Scissor Structures, International Journal of Space Structures, 14, 1–15, 1999.

Liu, X. J. and Wang, J.: Some New Parallel Mechanisms Containing the Planar Four-Bar Parallelogram, Int. J. Robot. Res., 22, 717–732, 2003.

Liu, X. J., Wang, Q. M., and Wang, J.: Kinematics, Dynamics and Dimensional Synthesis of a Novel 2-DOF Translational Manipulator, J. Intell. Robot. Syst., 41, 205–224, 2004.

Liu, X. J., Wang, J., and Pritschow, G.: On the Optimal Kinematic Design of the PRRRP 2-DOF Parallel Mechanism, Mech. Mach. Theory, 41, 1111–1130, 2006.

Lu, S., Zlatanov, D., and Ding, X.: Approximation of Cylindrical Surfaces with Deployable Bennett Networks, Journal of Mechanisms and Robotics, 9, 021001, https://doi.org/10.1115/1.4035801, 2017.

Pham, V. N. and Kim, H. S.: Dynamics Analysis of a 2-DOF Planar Translational Parallel Manipulator, Journal of the Korean Society of Manufacturing Technology Engineers, 22, 185–191, 2013.

Rosenfeld, Y. and Logcher, R. D.: New concepts for deployable-collapsable structures, International Journal of Space Structures, 3, 20–32, 1988.

Wei, G., Ding, X., and Dai, J S.: Mobility and geometric analysis of the Hoberman switch-pitch ball and its variant, Journal of Mechanisms and Robotics, 2, 191–220, 2010.

Wu, J., Wang, J., Li, T., and Wang, L.: Analysis and application of a 2-DOF planar parallel mechanism, J. Mech. Design, 129, 434–437, 2007.

You, Z. and Pellegrino, S.: Foldable Bar Structures, Int. J. Solids Struct., 34, 1825–1847, 1997.

Zhao, J. S., Wang, J. Y., Chu, F., Feng, Z. J., and Dai, J. S.: Structure synthesis and statics analysis of a foldable stair, Mech. Mach. Theory, 46, 998–1015, 2011.

16

Development of a solar concentrator with tracking system

Flávia V. Barbosa[1], **João L. Afonso**[2], **Filipe B. Rodrigues**[2], **and José C. F. Teixeira**[1]

[1]Department of Mechanical Engineering, School of Engineering, University of Minho, Guimarães, 4800-058, Portugal

[2]Department of Industrial Electronics, School of Engineering, University of Minho, Guimarães, 4800-058, Portugal

Correspondence to: Flávia V. Barbosa (flavia.barbosa.ep@gmail.com)

Abstract. Solar Energy has been, since the beginning of human civilization, a source of energy that raised considerable interest, and the technology used for their exploitation has developed constantly. Due to the energetic problems which society has been facing, the development of technologies to increase the efficiency of solar systems is of paramount importance. The solar concentration is a technology that has been used for many years by the scientist, because this system enables the concentration of solar energy in a focus, which allows a significant increase in energy intensity. The receiver, placed at the focus of the concentrator, can use the stored energy to produce electrical energy through Stirling engine, for example, or to produce thermal energy by heating a fluid that can be used in a thermal cycle. The efficiency of solar concentrators can be improved with the addition of a dual axis solar tracker system which allows a significant increase in the amount of stored energy. In response to the aforementioned, this paper presents the design and construction of a solar dish concentrator with tracking system at low cost, the optical and thermal modelling of this system and a performance analysis through experimental tests. The experimental validation allows to conclude that the application of a tracking system to the concentrator is very important since a minimum delay of the solar radiation leads to important losses of system efficiency. On the other hand, it is found that the external factors can affect the final results which include the optical and geometrical properties of the collector, the absorptivity and the position of the receiver as well as the weather conditions (essentially the wind speed and clouds). Thus, the paper aims to present the benefits of this technology in a world whose the consumption of energy by fossil fuels is a real problem that society needs to face.

1 Introduction

During the last decades the huge development of the society led to an increase in energy consumption. For many years oil, coal and natural gas provided the backbone of the energy resources. The intense release of pollutant gases by these resources, including CO_2, contributed, on one hand, to the increase of the greenhouse effect, raising the melting of the polar ice caps, and on the other hand to the deterioration of the air properties, producing, in extreme cases and with other harmful gases, like NO_x, the smog (Allaby, 2003). These effects, along with the volatility of the oil price, alarmed the developed countries in the search for "clean" energetic solu-

tions and led to an intense research to find sustainable renewable alternatives. Among the various forms of energy (solar, biomass, geothermal, wave and tidal, wind and water), the solar energy is certainly the most important source of energy, because all the others depend on it. The sun continuously emits energy at a temperature of about 5800 K, being called solar radiation (Nogueira, 2010). Solar radiation can be used through solar photovoltaic and solar thermal systems. Solar photovoltaic systems directly generate electricity through photovoltaic cells, while solar thermal systems need to heat a fluid to produce thermal energy, which can be used as a heat source or in a thermal cycle, but also to produce electricity by a Stirling engine (Fernandes et al., 2009; Kalogirou, 2014).

Through the concentration of solar beams, it is possible to increase the energy density (Kalogirou, 2014) and, in turn, the heat source temperature. Since this is a technology that can be exploited for thermal and electrical conversion and that uses a completely clean and inexhaustible source of energy, it is undoubtedly a solution which can contribute to minimize the consumption of fossil fuels.

Several technologies of Concentrating Solar Power (CSP) are implemented at industrial level. The great advantage of these technologies is the combined production of heat and power (CHP) to fulfil energy demands while providing various environmental, functional and economic benefits (Ferreira et al., 2016). Four CSP technologies are available for such purpose. Solar power towers, used to produce electrical energy, present potential advantages over others CSP technologies in terms of efficiency, heat storage, performance, capacity factors and costs. Fresnel linear reflectors and parabolic through types are applied in steam supply applications for industrial processes (Fernandes et al., 2009; Simbolotti, 2013). Parabolic through solar collectors are widely implemented across the world by industry, mostly in space heating and cooling, water desalination, food processing and fish, metal industry and products, pharmaceutical processing and dairy products (Kurup and Turchi, 2015). Despite of their high efficiency (up to 30 %), solar dishes have not yet been implemented on large commercial scale. However, they have been applied for electricity generation through Stirling engines or micro-turbines (Simbolotti, 2015). Although this technology is not widely applied at industrial level, it is important to note that the use of solar dish concentrators are a promising solution for residential applications mainly because of their high net efficiency, favourable ratio of thermal to electrical power, similar to a typical heat-to-electricity demand ratio of a domestic load, and low noise and emissions (Ferreira et al., 2015).

The demand for economically viable solutions in the development branch of solar dish concentrators has been increasing. Ripasso Company developed a new hybrid Stirling which allows a solar-to-grid electricity conversion of approximately 32 %, a fuel conversion ratio of 35 % and without water needed for electricity production (Ripasso, 2016). Solartron Company developed a dual axis hybrid solar dish concentrator that provides 45 kW of thermal energy through a direct heat and free piston Stirling engine, with thermal fluid heating (i.e. in solar power plants that utilize high temperature heat such as ORC – Organic Rankine Cycle). Its main applicability is water desalination, Solar Enhanced Oil Recovery, water heating and the electricity production (Solartron, 2010).

To raise the performance of solar concentrators and make viable their implementation, it is necessary to couple a tracking system. Solar trackers position the system in order to maximize the energy capture. There are different types of solar trackers that can be organized in one or two axis motion (Mousazadeh et al., 2009). Depending on the solar concentrator type, one or another system will be favoured. However, there is no doubt that the two axis system is the most efficient, since it follows the sun as its azimuthal movement and altitude, but its complexity increases its cost (Kalogirou, 2014). Two axis tracking system is widely implemented in solar thermal tower power plant and solar dish systems, since these technologies require the concentration of solar rays in a focus. To ensure the maximum performance of the system it is extremely important to ensure a correct sun tracking with minimal errors. For that purpose it is crucial to implement a good control system, which involves the application of a controller, sensors and motors, preferably at low cost.

In Portugal, the investment in solar trackers for photovoltaic panels has increased exponentially in the last decade, despite of the performance of photovoltaic modules, with the largest market share, to be approximately 15 % (Fraunhofer ISE, 2015). Unfortunately, at a national level, the application of solar concentrators with tracking system for thermal application has not yet developed. This aspect has often been neglected in Portugal, since it does not exist any solar thermal plant or solar dish concentrators all over the country (CSP World, 2015). Since Portugal is one of the European countries with higher levels of irradiation (1300–1800 kWh m^{-2}) (SolarGis, 2010) the question that arises is: "why the solar concentrator system are not used?", since they have a much higher energy use compared to photovoltaic systems.

Taking into consideration the solar potential of Portugal and the high efficiency of solar dishes, this work pretends to demonstrate the added-value of the application of this CSP technology. The development of a solar dish concentrator with a low cost tracking system allows to increase its application and disseminate the system into a wider market. It is hoped that this work will enhance the awareness of the private and industrial users to the technology and help to disseminate it.

2 Solar concentrator

2.1 Solar dish collector

The solar dish collector, as shown in Fig. 1, is a focus collector with dual axis solar tracker, allowing solar radiation to focus on the receiver located at the dish's focal point. In order to collect maximum radiation and convert into heat, the parabolic structure completes the entire tracking of the sun's path, optimizing the energy harvesting for thermal applications. The receiver absorbs the solar energy and converts it into thermal energy through a working fluid. This thermal energy is then converted into electricity by the use of a Stirling engine or through an OCR plant. This component must be designed to ensure the maximum absorption of solar radiation and the minimum heat losses to the environment, and its structure varies according to the working fluid and depends on the system application. This technology can yield temperatures above 1500 °C (Stine and Geyer, 2001).

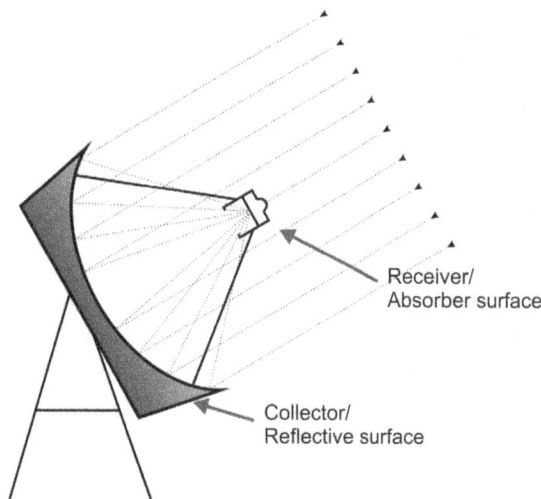

Figure 1. Solar dish collector representation.

Solar dish concentrators are used to achieve high temperatures and this concentration of solar radiation is accomplished by reflecting or refracting the incident flux on the aperture area (reflective surface) onto a smaller absorber area (receiver). To ensure maximum system efficiency, the understanding of all phenomena involved in this process is crucial. To understand the solar radiation conversion process through the collector and the thermal performance of the receiver, all the essential physical principles are discussed in this paper.

2.2 Optical modelling

Some optical parameters related to the collector need to be considered during the study of the solar dish collector.

2.2.1 Concentration ratio

The term concentration ratio is used to describe the amount of solar radiation concentrated by a particular concentrator. Two different definitions of concentration ratio are often used, briefly defined as follows (Rabl, 1976):

– Optical concentration ratio (CR_o): relation between the incident beam radiation on the reflective surface area (I_{sur}) and the radiation that reaches the receiver (I_{rec}), Eq. (1):

$$CR_o = \frac{I_{rec}}{I_{sur}}. \tag{1}$$

– Geometric concentration ratio (CR): relation between the capture area of the collector (A) and the receiver area (A_{rec}), Eq. (2):

$$CR = \frac{A}{A_{rec}}. \tag{2}$$

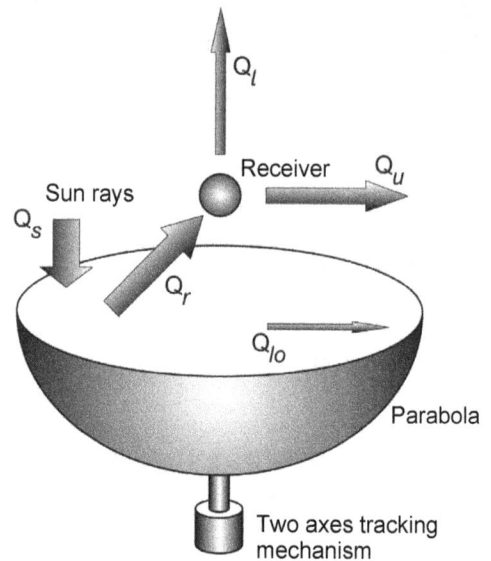

Figure 2. Energy balance of the solar dish concentrator (adapted from Thakkar, 2015).

Since the thermal losses are an important factor in the design of solar concentrators, the term of concentration ratio used along this work will be the geometric concentration ratio, CR.

2.2.2 Optical efficiency (η_o)

This parameter defines the performance of the collector through the quantification of the radiation captured by the reflective surface and the radiation that is reflected onto the receiver (Thakkar, 2015). For that, several parameters need to be considered, as defined in Eq. (3):

$$\eta_o = \rho_{s\_m} \cdot \tau_g \cdot \alpha_r \cdot S, \tag{3}$$

where ρ_{s_m} is the collector specular reflectance, τ_g is the transmittance of glass envelope covering the receiver (if present), S is the receiver shading factor (fraction of collector aperture not shielded by the receiver), and α_r is the absorptivity of the receiver (Mohamed et al., 2012).

2.3 Thermal modelling

The solar radiation incident on the collector is reflected into a single point on the receiver. Because the reflecting surface is not perfect, part of the radiation will be dissipated and the fraction absorbed by the receiver is known as optical heat. Due essentially to the heat losses by radiation and convection, part of the incident energy on the receiver will be lost to the surroundings. The energy balance of the system is illustrated by Fig. 2.

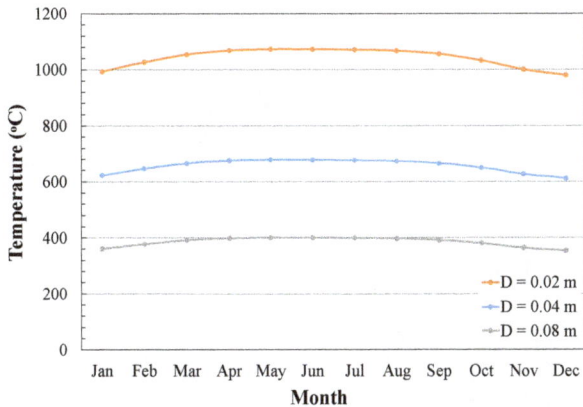

Figure 3. Maximum theoretical temperature reached by the receiver as a function of its absorber diameter.

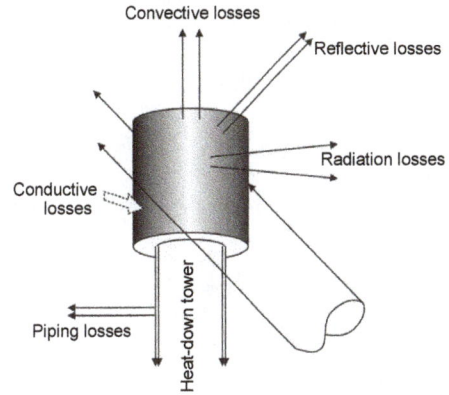

Figure 4. Receiver energy balance (adapted from Gorjan et al., 2013).

2.3.1 Useful heat

Under steady state conditions, the useful heat (Q_u) delivered by a solar collector system is equal to the energy absorbed by the heat transfer fluid, which is determined by the radiant solar energy falling on the receiver (optical heat, Q_r) minus the direct or indirect heat losses from the receiver to the surroundings (Q_l), as presented in Eq. (4) (Thakkar, 2015):

$$Q_u = Q_r - Q_l. \tag{4}$$

The optical energy absorbed by the receiver is obtained by Eq. (5) (Thakkar, 2015):

$$Q_r = A_{ac} \cdot \eta_o \cdot I_b, \tag{5}$$

where A_{ac} is the dish collector aperture area, η_o the collector's optical efficiency and I_b the direct normal insolation per unit of collector area; furthermore the heat losses from the receiver to the surroundings (Q_l) result of three heat transfer processes, conduction, convection and radiation, given by Eq. (6) (Gorjan et al., 2013).

$$Q_l = Q_{conduction} + Q_{convection} + Q_{radiation}. \tag{6}$$

2.3.2 Heat losses

Due to the concentration of solar rays in a small area, the receiver, the absorber area of this component can reach extremely high temperatures. To know the maximum theoretical temperature that the receiver can reach and in which proportion the diameter of the absorber area influences this temperature, a study was developed and presented in Fig. 3, in which three dimension were considered: the theoretical dimension of the receiver (0.02 m), the practical value adopted (0.04 m) and twice the practical value (0.08 m). These results are obtained through the application of the collector's dimensions used for the practical experiments and Eq. (7) (Steinfeld and Schubnell, 1993). From this analysis it is possible to

conclude that, as expected, an increase of the receiver's dimensions in relation to the ideal size (CR decreases with the increasing of the receiver diameter, since the collector diameter is constant) leads to a lower concentration of temperatures, i.e. a higher dissipation of heat concentration through the material.

$$T_{rec} = \left(\frac{CR \cdot Q_r}{\varepsilon \cdot \sigma} \right)^{1/4}, \tag{7}$$

where CR is the geometric concentration ratio, Q_r the optical energy absorbed by the receiver, ε the emittance of the absorber surface and σ the Stefan–Boltzmann constant ($5.670367 \times 10^{-8}\,\mathrm{W\,m^{-2}\,K^{-4}}$).

Because of its temperature, the receiver is the component responsible for the largest heat losses to the environment. There are many types of heat losses by the receiver presented in the energy balance illustrated in Fig. 4, which necessarily have to be minimized to ensure a high performance of the system (Thakkar, 2015). The main processes that affect the receiver's thermal losses are conduction, convection and radiation. However, since convection and radiation losses are much higher than conduction losses this last process was not considered in this study.

Convective heat losses

The convection losses can be obtained by Eq. (8) (Gorjan et al., 2013), in which h_{cv} is the convective heat transfer coefficient between the receiver and the environment, Eq. (9) (Cengel, 2015):

$$Q_{cv} = h_{cv} \cdot A_{rec} \cdot (T_{rec} - T_{amb}), \tag{8}$$

$$h_{cv} = \frac{Nu_D \cdot k_{ar}}{D}, \tag{9}$$

where Nu_D is the Nusselt number, k_{ar} is the thermal conductivity of the air and D the diameter of the absorber surface.

Two regimes of convection can be considered, forced and natural. For natural convection (when air is stagnant), the

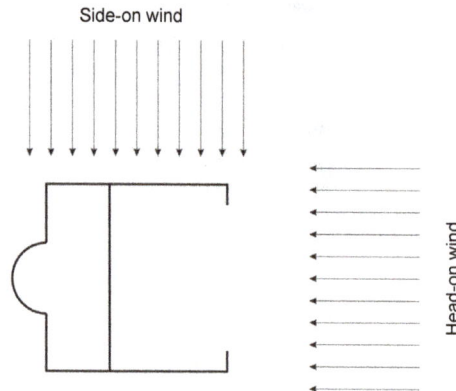

Figure 5. Effect of forced convection losses on receiver (adapted from Robbert, 1993).

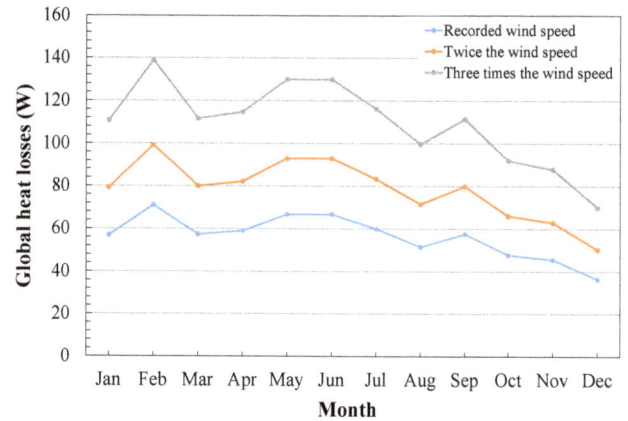

Figure 6. Theoretical analysis of the influence of the wind speed in the system heat losses.

heat transfer coefficient, $h_{cv, nat}$, can be calculated by a correlation defined by Eq. (10) for laminar flow ($10^4 < GrPr < 10^9$) and Eq. (11) for turbulent flow ($GrPr > 10^9$), where Gr and Pr are respectively the Grashof and Prandtl numbers (Holman, 1976):

$$h_{cv, nat} = 1.42 \cdot \left(\frac{\Delta T}{L}\right)^{1/4}, \tag{10}$$

$$h_{cv, nat} = 0.95 \cdot (\Delta T)^{1/3}, \tag{11}$$

where ΔT is the temperature difference between the wall and the air (°C) and L the vertical dimension (m).

Heat losses by forced convection occur due to the wind flux around the receiver. The wind flow has two direction, side-on wind and head-on wind, as it can be seen in Fig. 5.

To calculate the heat losses by convection due to side-on wind, the bottom side of the receiver has been assumed as a flat plate which is exposed to the parallel flow. The value of average Nusselt number for laminar flow ($Re < 5 \times 10^5$) on a flat plate is equal to Eq. (12), and that for turbulent flow ($5 \times 10^5 < Re < 10^7$) is calculated by Eq. (13) (Gorjan et al., 2013).

$$Nu_D = 0.0664 \cdot Re_D^{0.5} \cdot Pr^{1/3} \tag{12}$$

$$Nu_D = 0.037 \cdot Re_D^{0.8} \cdot Pr^{1/3} \tag{13}$$

In the case of head-on flow, Hess (1973) proposed the Eq. (14) which is function of the Peclet number (Pe). In this equation, it is assumed that the fluid flow is laminar and irrotational and the fluid is incompressible and inviscid.

$$Nu_D = 1.2 \cdot Pe_D^{0.5} \tag{14}$$

Radiation heat losses

These losses are related to the fourth of the temperature power of the receiver surface, Eq. (15). However, they can be minimized by increasing the receiver absorptivity and by

minimizing the absorber area. To determine the heat losses by radiation, it is first necessary to calculate the radiation coefficient, h_r, through Duffie equation given by Eq. (16), where ε is the emittance of the absorber surface, σ is the Stefan–Bolzmann constant and T_{amb} is the ambient temperature (Duffie and Beckman, 1991).

$$Q_{radiation} = h_r \cdot A_{rec} \cdot (T_{rec} - T_{amb}) \tag{15}$$

$$h_r = 4 \cdot \sigma \cdot \varepsilon \cdot T_{amb}^3 \tag{16}$$

As it is known, the wind speed influences the heat losses by convection, since the higher the wind speed, the higher the convection coefficient. To analyse the influence of this parameter in the global heat losses, a parametric study was performed (Fig. 6), in which the average wind speed recorded in Braga (city located in northern Portugal) in each month of 2015 was considered, velocities two and three times higher were also considered. The results obtained prove that the increase of the wind speed greatly influences the heat losses. For the case of twice the wind speed, the heat losses increases in average about 20 and 50 W when the speed is tripled.

.3 Thermal efficiency

The thermal efficiency in solar dish concentrators can be defined in two different ways (Gorjan et al., 2013):

- Thermal collector efficiency: ratio between the useful energy delivered to the fluid and the total energy captured by the reflective surface, as expressed by Eq. (17).

$$\eta_{concentrator} = \frac{Q_u}{A_{ac} \cdot I_b} \tag{17}$$

- Thermal receiver efficiency: ratio between the useful energy delivered from the receiver to the fluid and the energy falling on the receiver, Eq. (18).

$$\eta_{receiver} = \frac{Q_u}{A_{ac} \cdot I_b \cdot \eta_o} \tag{18}$$

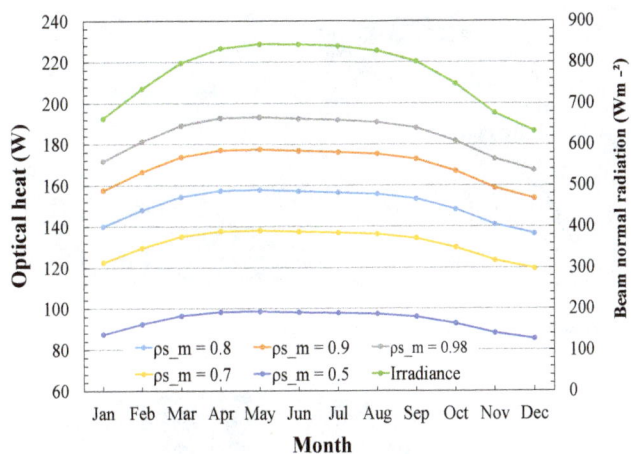

Figure 7. Optical heat variation as a function of the collector reflectivity.

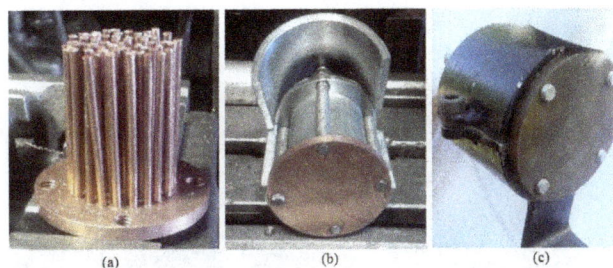

Figure 8. Solar receiver: **(a)** cooper fins, **(b)** internal structure, **(c)** external insulation.

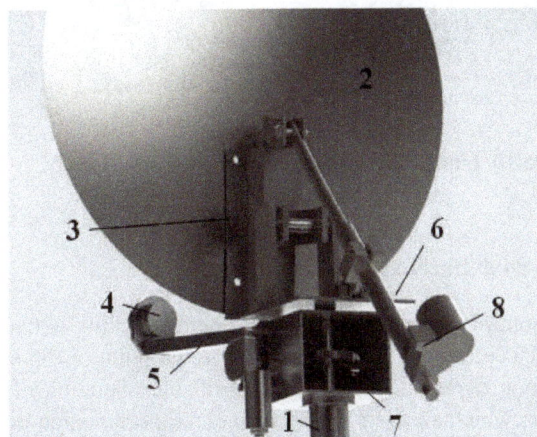

Figure 9. Solar tracker structure.

3 Solar concentrator components

3.1 Collector

The concentrator prototype is a small diameter parabola with an offset structure to minimize the efficiency losses due to the shadowing effect. As presented in Eq. (5) the optical energy of the solar concentrator is given as a function of the collector's reflectivity. To analyse into which extent this parameter affects the concentrator optical heat, a theoretical analysis was developed considering five reflectivity values (0.5, 0.7, 0.8, 0.9 and 0.98). Through the results presented in Fig. 7, it is observed that the optical energy decrease approximately 10 % per decrease of 0.1 of the reflectivity value, compared to the best scenario ($\rho_{\text{s\_m}} = 0.98$), which can lead to a loss of optical heat of 50 % for a reflectivity value of 0.5.

Due to the paramount importance of this parameter in the system performance, an experimental study will be presented below, in which two reflective surfaces were considered. One was covered with aluminium foil ($\rho_{\text{s\_m}} = 0.8$), and another was coated with galvanized silver ($\rho_{\text{s\_m}} = 0.96$).

3.2 Receiver

The type of solar receiver usually used in solar concentrators is of cavity type. However, in this project, a new receiver type, which allows the absorption of solar radiation through a surface composed by a high heat conductive material, was designed and constructed (Fig. 8). This surface transfers the heat absorbed to a working fluid (water) through a finned surface. To minimize losses to the outside, the metallic receiver is insulated with rock wool, because of its low thermal conductivity.

In solar receivers, losses by convection have a considerable importance, but particularlly in this type since the absorber area, which consists of a copper cylinder, is in direct contact with the environment. So any change in the environ-

ment will have important repercussions on the heat absorbed by the receiver. If the wind speed increases or if a cloud interfers with the solar radiation, the useful heat will be reduced, as well as the system performance.

3.3 Solar tracker structure

For the design of the solar tracker structure two forces need to be considered: the system weight and the aerodynamic force. The total mass supported by the mast, Fig. 9 (1), is obtained by adding the mass of the dish (2), the support dish (3), the receiver (4), the receiver support (5), the base support (6), the linear actuator (7) and the reduction gear (8). Thus the total mass of the system is 30.5 kg so the system weight is 301 N. Regarding the aerodynamic force, this is less predictable as it acts on the dish in different directions. However, the critical situation is when the wind flows perpendicularly to the collector when it is positioned vertically. To calculate this force, Eq. (19) was applied, considering the air density (ρ) of $1.2 \, \text{kg m}^{-3}$, an air speed (v) of $100 \, \text{km h}^{-1}$, an aerodynamic coefficient of parabolic dish (C_{D}) of about 0.95 and its area (A) as $0.342 \, \text{m}^2$.

$$F_{\text{ae}} = \frac{1}{2} \cdot \rho \cdot v^2 \cdot A \cdot C_{\text{D}} \tag{19}$$

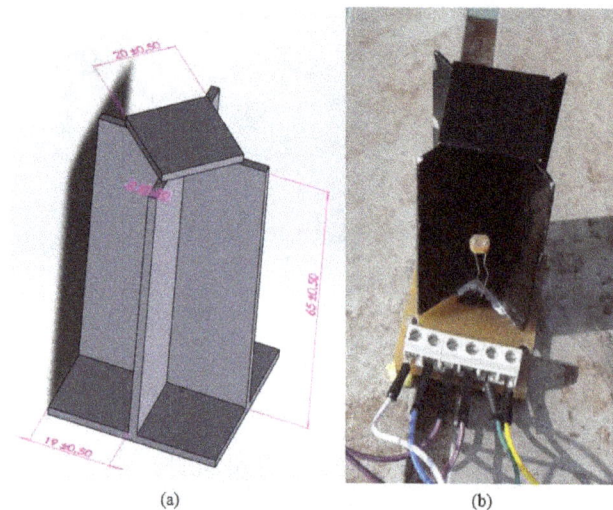

Figure 10. Luminosity sensor: (**a**) 3D model, (**b**) prototype.

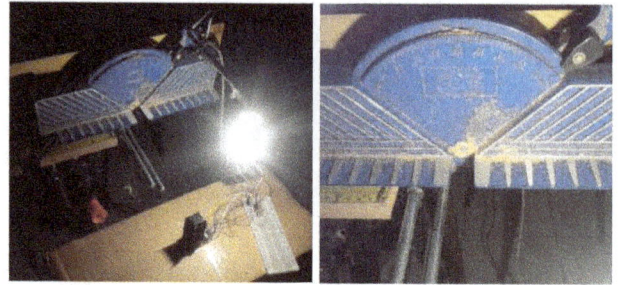

Figure 11. First experimental set up (light radiation).

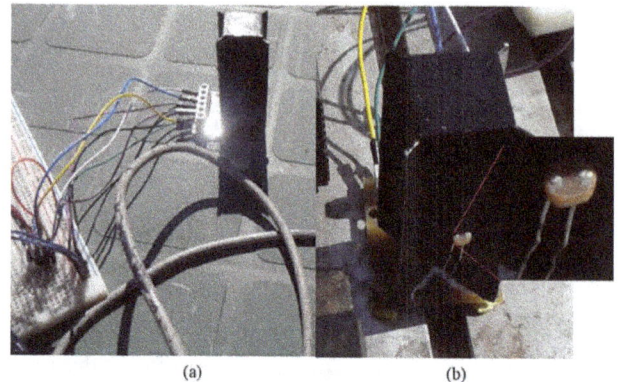

Figure 12. Second experimental set up (solar radiation).

3.4 Solar tracking controller system

The control system ensures the automatic tracking of the solar concentrator, according to the optimal point of the solar radiation captured, through a controllable operation of the motors which allows the azimuthal and tilt orientation of the collector. To ensure the tracking of the sun, it was applied a system control which uses a luminosity sensor, due to its low cost and good accuracy. The light sensor receives the information about the sun position through the incident radiation on a photosensor, the LDR. This electronic component consists of a semiconductor material whose resistance varies with the intensity of the light source (in this specific case of solar radiation, Cortez, 2013). This information is transmitted by the two pairs of LDRs (one pair for the control of the azimuth angle and another for the collector tilt) to the microcontroller which operates the motors to position the collector towards the sun. The sensor was developed by taking the example of a widely used model in academic solar concentration systems, as shown in Fig. 10.

To determine the system sensitivity an analysis of voltage variation of each LDRs as a function of the angle of incidence of radiation was performed. For this analysis, the pair of LDRs that control the azimuth movement was chosen, because this is the critical motion of the solar tracker, since the concentrator must move, for the longest day of the year (21 June), approximately 246°. The tests which allow the system sensitivity analysis can be divided in two groups: the first test performed is based on the variation of the angle of incidence of a light radiation, with a precision of one degree (Fig. 11); while the second test is based on the variation of the sun azimuth angle (Fig. 12).

Through the connection between the breadboard and the light sensor (shown in Fig. 12), the photosensors voltage were sent to the current standard Arduino development plat-

form (Fisher and Gould, 2012), which was subsequently transmitted to a computer for reading and analysis. It is important to note that, in the absence of brightness, the voltage is 0 V and the maximum value is 5 V, these values are given by the Arduino. The results show the LDRs voltage as a function of the variation of the angle of incidence of radiation. In order to facilitate the results analysis, the charts in Figs. 13 and 14 were plotted. These voltages are obtained by averaging ten values collected by the program.

The experiment starts when the voltage of both LDRs is the same (in this case approximately 2.20 V), since the incident radiation is equal in both cases (angle = 0°). When this condition is verified, one LDR starts to be shaded while the other is progressively illuminated with the increase of the angle of incidence of radiation. As it can be observed in Fig. 13, the voltage difference between LDR1 and LDR2 increases with the increase of the angle of incidence until the voltage value of both LDR starts to be constant. The voltage of the LDR illuminated reaches the maximum value (close to 5 V) while the LDR shadowed reaches the minimum value (close to 0 V). It is also verified that the voltage of the two LDR varies almost equally from East to West and from West to East. Considering these results it was concluded that the LDRs are sensitive to radiation changes, and the variation of one degree of the angle of incidence causes a readily measurable voltage difference.

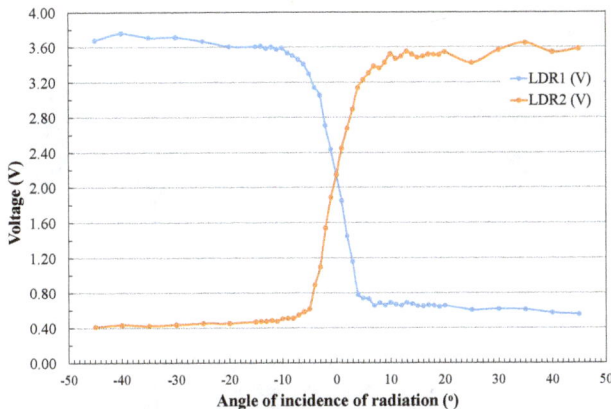

Figure 13. Variation of LDRs voltage as a function of the angle of incidence of light radiation.

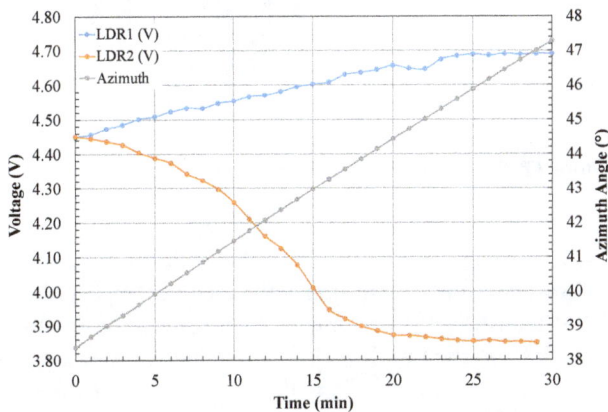

Figure 14. Variation of LDRs voltage as a function of the azimuth angle.

Maintaining the same initial conditions of the procedure explained above, the second test was performed. Through the analysis of Fig. 14 it is observed that the voltage difference between LDR1 and LDR2 is 0.21 V during the first 8 min, 0.66 V after 16 min, and 0.83 V after 24 min. It is noted a more sudden voltage change in the LDRs during the first 16 min, after which the voltage tends to stabilize. It is also observed that from a variation of approximately 6° of the azimuth angle, the voltage difference between LDR1 and LDR2 begins to level off (between 0.02 and 0.03 V). These results allow to conclude that the movement of one degree of the sun azimuth angle implies a difference between LDR1 and LDR2 of about 0.1 V. This may be a good starting point for the controller programming, since it should be considered the critical voltage difference between LDR1 and LDR2, from which the order to move the motors will be given. If this value is too low, the controller will continuously make adjustments to the system, resulting in an increased wear of motors. However, if it is too high, there will be a loss of performance in the system.

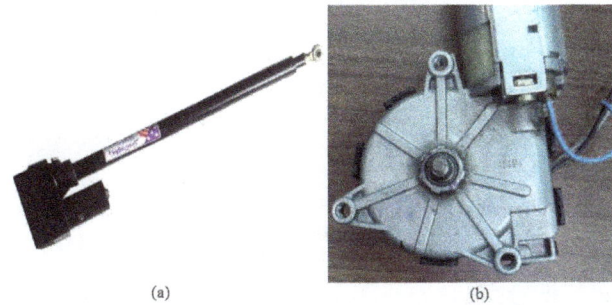

Figure 15. Tracker motors: **(a)** linear actuator, **(b)** electric motor.

3.5 Tracking motors

Motors applied in the solar tracker should allow the slope of the parabolic dish and its movement according to the azimuth angle of the sun. To ensure the collector's slope, a linear actuator was chosen, as shown in Fig. 15a, since is a simple system and it is sized according to the size of the parabolic dish. In order to perform the azimuthal trajectory, describing the movement according to the variation of the azimuth angle of the sun, it was necessary to develop a mechanical system capable to rotate a total of 246° in a day (critical situation). For this application, the gearmotor system demonstrates to be the most practicable solution for the design, since it is a worm gear system integrated into a box, within which the output shaft drives the entire system. Gearmotors are applied in some solar tracker systems, however its cost is very high. To ensure the low cost of the solar concentrator, it was decided to manufacture a gear box using a standard motor usually used in the windshield wiper and climbing systems and lowering of the car windows, as shown in Fig. 15b. To design the gearmotor system, the transmission ratio is the most important parameters. However the motor sensitivity influences this parameter and so, it needs to be controlled. To determine the motor sensitivity a test similar to the sensor sensitivity was carried out, through which it is observed that the ideal transmission ratio is 16 : 1.

3.6 A low cost system

One important aim of this project is the development of a low cost solar tracker, so during the project, the search for cost effective solutions was mandatory. The cost analysis can be divided in three groups of components: controller system, structure and motors. As mentioned above, to minimize costs and considering its good accuracy, a control system which uses a luminosity sensor was constructed and its estimated costs was about EUR 52. Tracking motors are always a huge problem in terms of tracking systems costs, mainly for the azimuthal motor. To avoid heavy costs, a simple actuator was bought and a gearmotor was constructed using a standard motor, for a grand total of EUR 370. In reference to the solar tracker structure which involve all the fixing parts,

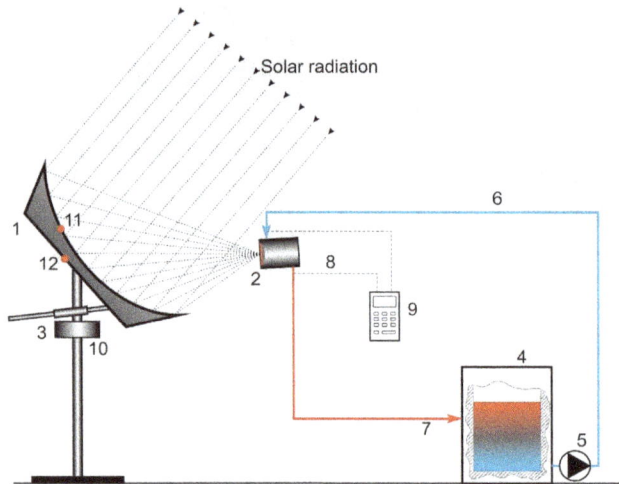

Figure 16. Experimental set up of the solar dish concentrator.

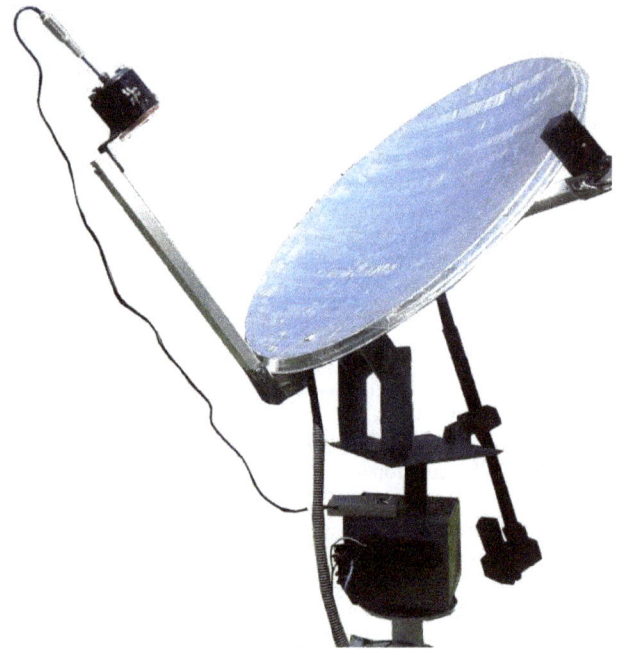

Figure 17. Solar dish concentrator prototype.

the mast, shaft, bearings, connection elements, machining, welding and painting processes, the cost amounts to only EUR 135. Proceeding to the sum of the values presented and adding the cost of the coating collector and the receiver, the total cost of the system is about EUR 750, which is really an interesting value when compared to similar structures available in the market. Therefore it is concluded that the developed system fulfil the aim, ensuring a good reliability at low cost.

4 Testing facility

To analyse the potential of the solar dish concentrator, a test facility was developed, as schematically presented in Fig. 16. Through this testing facility, experimental tests were carried out in order to evaluate the response and the efficiency of the system. The experiments were conducted in clear sky days and the velocity of the wind was registered from the meteorological information. The experimental set up consists of a collector (1) which collects the solar radiation and reflects it at the receiver (2); a dual axis solar tracker (3), controlled by the control system (10), ensures the correct tracking of the sun. A 4 L insulated water tank (4) is heated by the receiver; a pump (5), is placed into the tank to ensure the circulation of the water, at a flow rate of about $4.73 \times 10^{-5} \mathrm{m}^3 \mathrm{s}^{-1}$, and two transparent and flexible PVC tubes ensure the cold water inlet (6) at the receiver and the hot water outlet (7). Two thermocouples (8) placed in the tubes next to the receiver allow the measurement of the water temperature which can be read by the digital thermometer (9). Two other thermocouples placed on the front (11) and back (12) of the collector, enable the temperature measurement of these two surfaces. The final prototype can be observed in Fig. 17.

Considering the experimental set up shown in Fig. 16, four tests were performed:

– Experiment 1: the first test analyses the response of the system without activation of the solar tracker, in order to demonstrate the performance losses resulting from the non-activation of the solar tracker, highlighting the importance of this system in this type of applications.

– Experiment 2: this test evaluates the influence of the reflective surface. For the analysis two collectors were used, one with a lower reflectivity value (0.8), represented in the final results as "Collector 1", and another with a higher reflectivity value (0.96), namely "Collector 2". The parameters measured were the internal and external receiver temperature (without water circulation) and the efficiency of the system during the water heating throughout the day.

– Experiment 3: this test allows to evaluate the thermal efficiency as function of the receiver position.

– Experiment 4: the fourth test evaluates the thermal efficiency as function of the solar irradiance. For this purpose the results obtained on 10 August and 28 September are compared.

5 Results and discussion

5.1 Response without tracking system activation

By analysing the date in Fig. 18, it is clear that the water warms up during the first 10 min, at approximately $0.2 \,^{\circ}\mathrm{C}\,\mathrm{min}^{-1}$. From this moment the temperature in the inlet

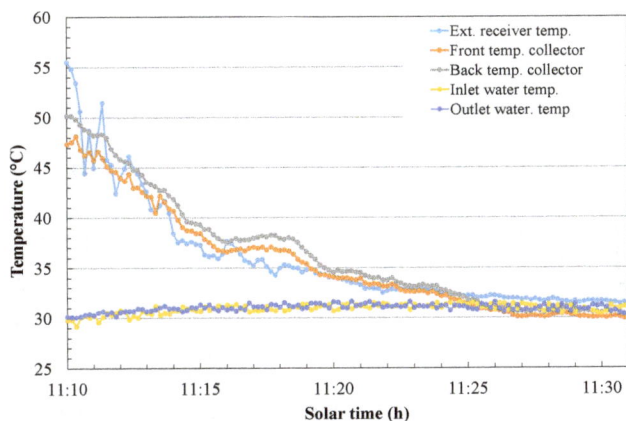

Figure 18. Variation of the temperature as a function of time without activation of the tracking system.

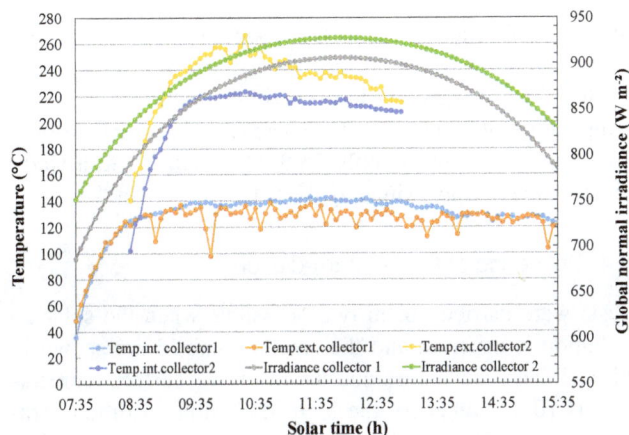

Figure 19. Variation of the internal and external receiver temperature (without water circulation).

and outlet of the receiver is approximately the same, 31.2 °C. With reference to the temperature of the absorber surface of the receiver, there has been a sharp decline in the early stages, varying around 15 °C during the first 5 min, by the fact that the concentration area of solar rays moves away from the center to the edges of the receiver, so the fins started to cool. Since the concentration area moves away from the center with time, the trend is that the receiver temperature will become in equilibrium with the ambient temperature. This is why after the first 5 min it is observed a decrease of about 0.5 °C min^{-1} and after 15 min the temperature decreases 0.1 °C min^{-1}. This experimental test demonstrates that 1° in azimuth angle variation causes significant losses in the system, considering that on this day and time the sun "moves" at a speed of approximately 0.29 °C min^{-1}.

5.2 Influence of the reflective surface

In Fig. 19 it is clearly observed that with a solar concentrator with a higher reflectivity value, the receiver can reach about twice the external temperature and that the temperature difference between the internal and external receiver temperature is also higher. The higher the temperature of the absorber surface, the greater the convection losses, and thus the higher the temperature difference between the internal and external receiver surface. Looking at the external temperature during the day, it is noted that, after the maximum temperature is reached on the receiver surface, this remains approximately constant varying only with the increase of the wind speed. The decrease in temperature occurs mainly from 15:00 LT, since the irradiance will sharply reduce from that time of day. For these reasons, it is concluded that the system allows a good exploitation of solar radiation, keeping the maximum temperature for, approximately, 4 h.

Comparing the experimental results obtained with both collectors (Figs. 20, 21), it is observed that the thermal performance of the collector with a high reflectivity is higher when compared with the other collector. In fact, the maximum temperature recorded at the receiver absorber surface was only 57 °C against 142 °C in the test with a collector with a higher reflectivity. In addition, in the first test the maximum water temperature was 42.5 °C, reached after 6 h and 15 min of operation and in the second one this temperature was 51 °C reached after 3 h. This confirms that the application of a collector with a high reflectivity leads to a significant increase in system efficiency. On the other hand, comparing the theoretical maximum temperature presented in Fig. 3, it is clearly observed that independently of the reflective surface, the maximum temperature reached by the receiver in the experimental tests is lower. This fact is essentially due to the thermal inertia of the receiver's material which was not considered in the theoretical study, but also due the external factors, such as wind and clouds that interferes during the test.

5.3 Influence of the receiver position

The orientation of the receiver to the collector's surface was changed in order to assess the influence of the slump error of the receiver in the system performance. For that four receiver's slump angles were tested, the ideal position (aligned with the optical axis of the collector), and with offsets 10, 15 and 30° in relation to the ideal position. The results, presented in Fig. 22, show that the maximum water temperature reached by the system, in clear sky conditions, is 46 °C. This value is obtained first in the ideal position and decreases with the increase in the offset angle. In all cases the temperature difference is higher in the early stages, due to the high temperature reached by the fins inside the receiver immediately before the water circulation, stabilizing after an hour and a half. It is also observed that the worst case (30° offset) leads

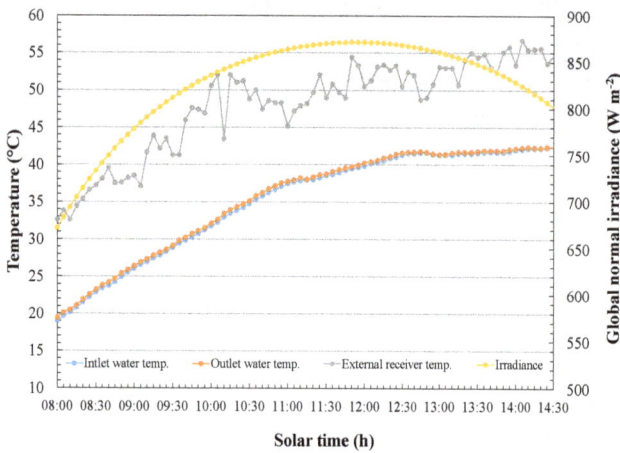

Figure 20. Variation of the system temperature as a function of time with a collector with a lower reflectivity.

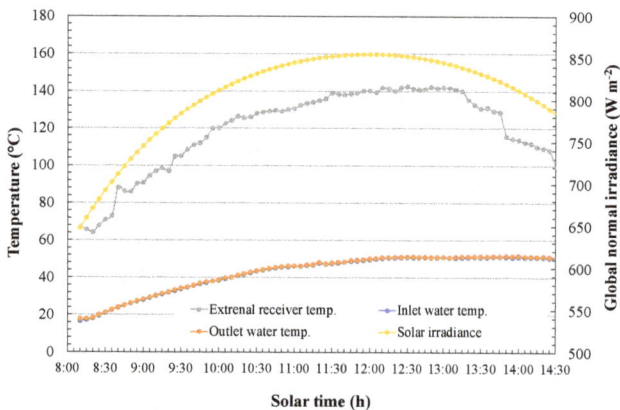

Figure 21. Variation of the system temperature as a function of time with a collector with a higher reflectivity.

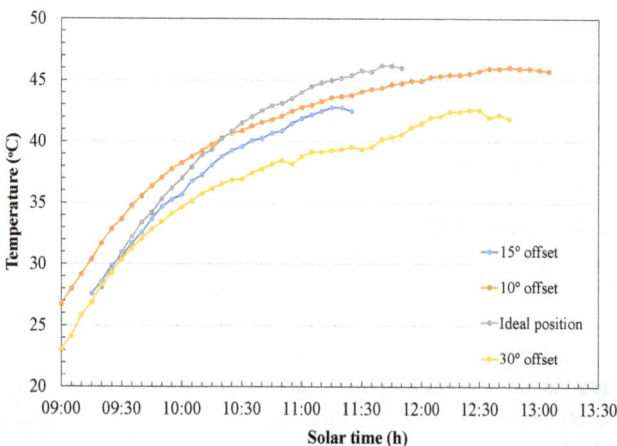

Figure 22. Variation of water temperature at external receiver surface over time.

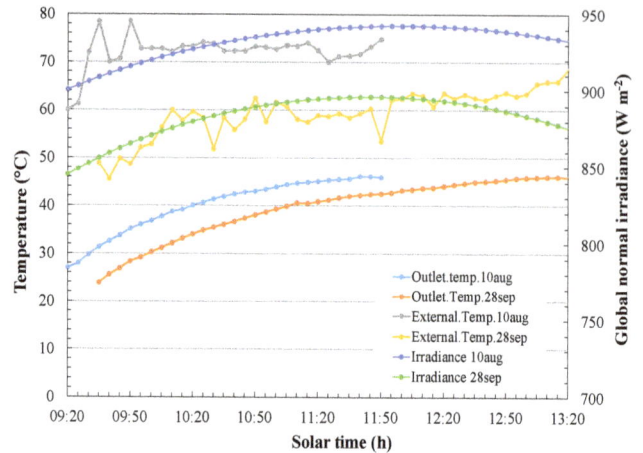

Figure 23. Temperatures recorded on 10 August and 28 September.

to a thermal efficiency losses of about 20%. The ideal position maximizes the receiver's surface temperature approximately 78.5 °C, leading to a maximum value of the water temperature of about 46.2 °C, whereas in the worst case the receiver temperature is only 75.8 °C, which led to a water temperature of approximately 42.5 °C.

5.4   Influence of the solar irradiation

Tests were carried out in two occasions when the solar irradiation is different and the results obtained are shown in Fig. 23. The data shows that the receiver's surface temperature is 10 °C higher for the high irradiation condition (irradiance is 46.3 W m^{-2} higher). The temperature reached by the water on 10 August is 46 °C, after 2 h and 30 min of system operation, while on 28 September, the same water temperature was reached after 4 h of operation. Regarding the receiver's efficiency recorded on 10 August this is, on average, 20% higher than that obtained on September (low irradiation). This efficiency difference is less in the collector, just 10% higher than August, as it can be observed in Fig. 24. Regarding the efficiency, it is important to mention that the maximum peaks observed are mainly due to the clouds, which block the direct solar radiation, and the wind speed variation. The little fluctuations are essentially due to errors associated to the thermocouples considering that the variation of temperature at each time is low. From the results, it can be concluded that a difference in solar irradiance of approximately 46 W m^{-2} leads to a significant loss of system efficiency.

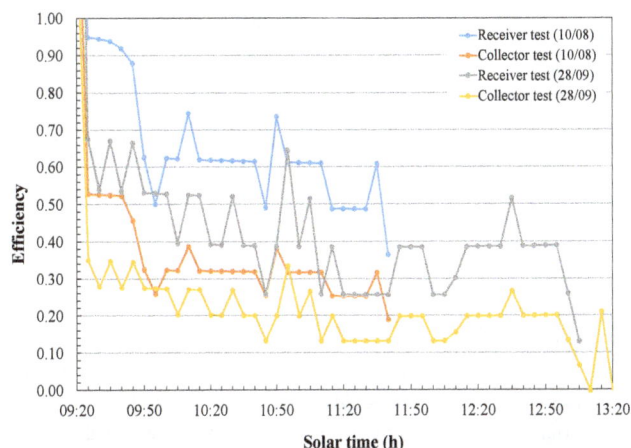

Figure 24. Efficiencies recorded on 10 August and 28 September.

6 Conclusions

In this paper, a solar dish concentrator and its tracking system was designed and constructed. During the study of this technology, three elements are highlighted, the collector, the receiver and the tracking system. Furthermore, it was mentioned that costs play an important role in this solar application, so, the development of a low cost system is crucial.

A relevant factor on performance analysis is the receiver position. This study concludes that, the higher the mismatch in relation to the optimal position, the lower the system performance, because, the solar radiation reflected by the concentrator is reduced. It shows that a small change of the optimal position can lead to an efficiency loss of up to 20 %. These results also highlight to the need of implementing an efficient tracking system which ensures the total track of the sun's path during the day. As the experimental test of the variation of the receiver temperature without tracking system demonstrates, 1° of sun motion, through the azimuth angle, is enough to create significant losses in the system. Comparing the results obtained in Sect. 5.1 with a full sun tracking, it is observed that if the tracking system is not activated, the thermal inertia of the absorber surface prevents the water temperature to rise to acceptable values. This is not the case in the others experiments, since the absorber surface reaches temperatures above 55 °C and maintains this temperature throughout the experiment. Another important factor is the atmospheric conditions. As it was observed from the results, the direct losses to the environment, clouds and wind lead to great losses of efficiency. Clouds minimize the radiation concentrated in the receiver and the wind increases the heat losses by convection. Since this factor cannot be controlled, it is important to ensure minimal errors during the design of all solar concentrator elements. However, it is important to refer that the structure of the solar receiver need to be improved, to minimize losses to the environment and increase the solar energy density that can be absorbed. In rela-

tion to the collector reflectivity, the experimental results confirm that an increase from 0.8 to 0.96 leads to a significant increase of the system efficiency. Comparing the two collector types, on the same conditions of clear sky and moderate wind speed, and considering that the incident irradiation is higher for the collector with a lower reflectivity, it appears that the efficiency can be 20 % higher on the collector with a higher reflectivity. Therefore the collector reflectivity is a fundamental characteristic in this type of projects.

This work shows that there is a good potential of use of solar dish concentrators. The collector may be selected depending on the fluid temperature to be achieved. The aim of this application can be the production of steam/vapour for the production of electrical energy, and for that it can be chosen a collector with a bigger diameter and a reflectivity close to the unit. On the other hand, if the aim was the production of hot water for daily consumption, it can be chosen a collector with smaller dimensions. Nonetheless it is required, for both situations, a reflectivity close to the unit.

Edited by: D. Pisla

References

Allaby, M.: Fog, Smog and Poisoned Rain, 1st Edn., VB Hermitage, USA, 2003.

Cengel, Y.: Heat and Mass Transfer Fundamental & Applications, 5th Edn., McGraw-Hill Education, USA, 2015.

Cortez, R.: Sistema de Seguimento Solar em Produção de Energia Fotovoltaica, Master thesis, Faculty of Engineering of Porto, Portugal, 2013.

CSP WORLD: CSP World Map, Spain, available at: http://www.cspworld.org/cspworldmap (consulted: 28 October 2015), 2015.

Duffie, A. and Beckman, W.: Solar Engineering of Thermal Processes, 2nd Edn., John Wiley & Sons, USA, 1991.

Fernandes, E. O., Marques, V. S., Rodrigues, J., Pimenta, C., Aguiar, C., Nunes, J., Rodrigues, B., Tirone, L., Veríssimo, M., Correia, M., Lopes, J. A. P., Mendes, J. F., Joyce, A., Cruz, J., Bicudo, C., Patrão, G., Mendes, C., Matos, J., Soares, H., Correia, N., Almeida, B., Serranho, H., Sarmento, A., Gonçalves, L. A., Estanqueiro, A., Curado, A., Barreto, M., Sá da Costa, A., and Schmidt, L.: Energias Renováveis, 1st Edn., Atelier Nunes e Pã, Lisbon, 2009.

Ferreira, A. C., Nunes, M. L., Teixeira, J. C. F., Martins, L. A. S. B., Teixeira, S. F. C. F.: Thermodynamic and Economic Optimization of a Solar-powered Stirling engine for Micro-Cogeneration Purposes, University of Minho, Portugal, Energy, 111, 1–17, doi:10.1016/j.energy.2016.05.091, 2016.

Fisher, D. and Gould, P.: Open-Source Hardware Is a Low-Cost Alternative for Scientific Instrumentation and Research, Modern Instrumentation, 1, 8–20, doi:10.4236/mi.2012.12002, 2012.

Fraunhofer ISE: Current and Future Cost of Photovoltaics, Long-term Scenarios for Market Development, System Prices and LCOE of Utility-Scale PV Systems, Study on behalf of Agora Energiewende, 2015.

Gorjan, S., Tavakkoli Hashjin, T., Ghobadian, B., and Banakar, A.: Thermal Performance of a Point-focus Solar Steam Generating System, 21st Annual International Conference on Mechanical Engineering, ISME2013, 7–9 May 2013, School of Mechanical Eng., K. N. Toosi University, Tehran, Iran, 2013.

Hess, J.: Analytic solutions for potential flow over a class of semi-infinite two-dimensional bodies having circular-arc noses, J. Fluid Mech., 60, 225–239, 1973.

Holman, J.: Heat Transfer, 4th Edn., International Student Edition, McGraw-Hill Kogakusha, Ltd, 1976.

Kalogirou, A.: Solar Energy Engineering: Processes and Systems, 2nd Edn., Academic Press, USA, 2014.

Mohamed, F. M., Jassim, A. S., Mahmood, Y. H., and Ahmed, M. A. K.: Design and Study of Portable Solar Dish Concentrator, International Journal of Recent Research and Review, 3, 52–59, 2012.

Mousazadeh, H., Keyhania, A., Javadib, A., Moblia, H., Abriniac, K., and Sharifi, A.: A Review of Principle and Sun-Tracking Methods for Maximizing Solar Systems Output. Renew. Sust. Energ. Rev., 13, 1800–1818, doi:10.1016/j.rser.2009.01.022, 2009.

Nogueira, H.: Manual das Energias Renováveis – o Futuro do Planeta, 1st Edn., AECOPS, Lisbon, 2010.

Kurup, P. and Turchi, C.: Initial Investigation into the Potential of CSP Industrial Process Heat for the Southwest United States, National Renewable Energy Laboratory, USA, available at: http://www.nrel.gov/docs/fy16osti/64709.pdf (last access: 15 December 2015), 2015.

Rabl, A.: Comparison of Solar Concentrators, Sol. Energy, 18, 93–111, doi:10.1016/0038-092X(76)90043-8, 1976.

Ripasso: Ripasso Stirling Hybrid, Sweden, available at: http://www.ripassoenergy.com/ (last access: 10 February 2016), 2016.

Robert, Y. M.: Wind Effects on Convective Heat Loss from a Cavity Receiver for a Parabolic Concentrating Solar Collector, Sandia National Laboratories, California, 1993.

Simbolotti, G.: Concentrating Solar Power, IEA-ETSAP and IRENA, available at: http://www.irena.org/DocumentDownloads/Publications/IRENA-ETSAPTechBriefE10ConcentratingSolarPower.pdf, last access: 10 September 2015, January 2013.

SolarGis: Maps of Global Irradiation, GeoModel Solar, Nova Scotia Canada, available at: http://solargis.info/doc/free-solar-radiation-maps-GHI, last access: 4 April 2015, 2010.

Solartron: How does a Solar Concentrator Solar Dish Work, Solartron Energy System Inc., available at: http://www.solartronenergy.com/solar-concentrator/how-does-a-solar-concentrator-work/, last access: 20 August 2015, 2010.

Stine, W. B. and Geyer, M.: Power from the Sun Book, USA, copyright© 2001 by Stine, W. B. and Geyer, M., available at: http://www.powerfromthesun.net/book.html, last access: 5 February 2015, 2001.

Steinfeld, A. and Schubnell, M.: Optimum Aperture Size and Operating Temperature of a Solar Cavity-Receive, Sol. Energy, 50, 16–25, 1993.

Thakkar, V.: Performance Analysis Methodology for Parabolic Dish Solar Concentrators for Process Heating Using Thermic Fluid, IOSR Journal of Mechanical and Civil Engineering (IOSR-JMCE), 12, 101–114, doi:10.9790/1684-1212101114, 2015.

17

An alternative design method for the double-layer combined die using autofrettage theory

Chengliang Hu, Fengyu Yang, Zhen Zhao, and Fan Zeng

Institute of Forming Technology & Equipment, Shanghai Jiaotong University, Shanghai 200030, China

Correspondence to: Zhen Zhao (zzhao@sjtu.edu.cn)

Abstract. The double-layer combined die is used for its longer life in forging. Autofrettage is a well-known elastic–plastic technology that increases the durability of thick-walled cylinders. This study explores an alternative design method of the double-layer combined die using autofrettage theory. An analytical solution for the autofrettage process of the double-layer combined die is obtained based on Lamé's equation. The relationship between the autofrettage pressure and the yield radius of the die insert is obtained, and expressions of residual stresses and displacements, which are directly related to geometric parameters, material properties and internal pressure, are derived. The finite-element simulation of a specific case is performed, and good agreement between theoretical calculations and simulation results is found. Furthermore, the effects of important parameters, including the ratio of the plastic area and yield strength of the die insert and the outer diameters of the die insert and stress ring, on the autofrettage effect are investigated. Compared with the conventional combined die, the autofrettaged die can bear larger working pressure, as expected. The use of the autofrettaged die can reduce the amount of expensive material required for the die insert and the working space of the die set, which would benefit the practical forging process.

1 Introduction

In the process of cold forging, a die usually needs to be of high strength to resist a large forging load. The accumulation of plastic deformation near points of stress concentration resulting from the cyclic loading conditions leads to fatigue damage and eventually to the generation of a crack on the surface of the die (Pedersen, 2000). A die insert with one or more massive stress rings, which is called a combined die or prestressed die, is thus designed to reduce the stress level during forging.

High radial and circumferential stresses strongly affect the elastic deformation and failure of forging dies, and the time and economic losses due to unpredicted die failure during service in forging are costly. Against this background, many studies (Joun et al., 2002; Yeo et al., 2001; Kwan and Wang, 2011; Yang et al., 2012) have attempted to optimize the dimensions of the die structure employing different optimization methods.

Usually, high-strength tool steel is used for the die insert, while normal steel is used for the stress rings. It was found

that a backward extrusion die prestressed with sintered carbide could be strengthened (Hur et al., 2002) and reduce elastic deformation (Hur et al., 2003). The strip-wound container with a winding core of tungsten carbide made it possible to create new innovative die designs (Groenbaek and Birker, 2000). The cited works introduced the stress ring with high-stiffness material or structure that could strengthen the combined die effectively, but the cost of a single set of the combined die increased.

Autofrettage is a well-known elastic–plastic technology that increases the durability of thick-walled cylinders, and is commonly used in high-pressure vessels. A cylinder tube is subjected to uniform internal pressure so that its wall becomes partially plastic, resulting in internal compressive residual stresses that increase the pressure capacity for the next loading (Majzoobi et al., 2003). The optimum autofrettage pressure is not constant but depends on the working pressure (Hojjati and Hassani, 2007). To improve the performance of a type-III hydrogen pressure vessel, the most appropriate autofrettage pressure has been determined by finite-element (FE) simulation (Son et al., 2012). The pressure reached in available commercial autofrettage processes is between 370 and 1200 MPa (http://www.felss.com).

A method of calculating the stress intensity factor was used to estimate the fatigue life of an autofrettage tube, and results showed that the fatigue life increased with the autofrettage level (Jahed et al., 2006). The ultimate load bearing capacity of the autofrettage cylinder was estimated theoretically (Zhu, 2008a, b). The autofrettage process led to tangible increases in the strength-to-weight ratio and fatigue life of the cannon tube (Anantharam and Kumar, 2014). A novel concept of an autofrettage compounded tube was proposed to give a high safe pressure and good fatigue life (Bhatnagar, 2013). Therefore, using autofrettage technology, the load bearing capacity, strength-to-weight ratio and fatigue life of cylinder parts can be improved.

Considering the advantages of autofrettage technology, it is desirable to explore the possibility of the application of the technology to the cold forging die. The effect law of strain hardening and the Bauschinger effect on the autofrettage process have been discussed by comparing the results of a bilinear kinematic hardening model and ideal elastoplastic model, and the optimum autofrettage pressure was determined to meet the high dimensional accuracy and strength requirements of a single-layer extrusion die (Qian et al., 2011). However, the single-layer die is seldom used in actual forging.

The present study obtained an analytical solution for the autofrettage process of the double-layer combined die and established formulae of the residual stresses and displacements. The simulation results for a specific case of the double-layer combined die after autofrettage were used to verify the reliability and accuracy of the theoretical derivation. On this basis, other cases were considered in studying the effects of the ratio of the plastic area and yield strength of the die insert and the outer diameters of the die insert and stress ring on the autofrettaged die. The results clarify the autofrettage effect and contribute to the design of the double-layer combined die for forging.

Different from previous design strategies – such as using stronger material for the die insert directly, optimization of the die dimensions including the interference fit, and using a stress ring with a high-stiffness material or structure – a novel design method of the combined die is proposed, where compressive stresses on the die insert are introduced by the autofrettage process to improve the strength of the combined die.

2 Theory and formulation

The combined die, which consists of insert die and stress ring, is first strengthened by the classical shrink fit, and the autofrettage process is then applied as a pre-process before the combined die is used in forging. During the autofrettage

process, an enormous uniform internal pressure is loaded on the double-layer combined die, so that there is uniform plastic deformation of the die. After unloading, there are residual stresses and a plastic area with certain thickness emerges. A theoretical discussion of the autofrettage process is now presented.

Several assumptions are made. (1) The double-layer combined die can be regarded as an axially constrained cylinder. (2) Because of the small axial deformation, the autofrettage process can be simplified as a plane-strain problem. (3) The material of the die insert is a perfectly elastic–plastic material and obeys the von Mises criterion. (4) The material is supposed to be incompressible. (5) Plastic deformation only occurs in the die insert, while the stress ring remains in the elastic state.

2.1 Stress in the loading process

In the loading process, the die insert can be subdivided as inner-plastic and outer-elastic layers considering the elastic–plastic deformation behavior of the combined die. The uniform autofrettage pressure is denoted p_0. The radius at the boundary between plastic and elastic areas is referred to as the yield radius r_p, the normal stress between the two areas is denoted q, and the normal stress between die insert and stress ring is denoted p_k^p, as shown in Fig. 1.

According to the equilibrium equation and the yield criterion

$$\frac{d\sigma_r}{dr} - \frac{\sigma_s}{r} = 0. \tag{1}$$

Under the assumptions mentioned above, the radial stress in the inner-plastic layer can be obtained according to the equilibrium, Eq. (1):

$$\sigma_r = \frac{2}{\sqrt{3}}\sigma_s \ln r + C (r_1 \leq r \leq r_p). \tag{2}$$

The constant C can be calculated using the boundary condition $\sigma_r = -p_0$ when $r = r_1$:

$$C = -p_0 - \frac{2}{\sqrt{3}}\sigma_s \ln r_1. \tag{3}$$

Then, σ_r in the plastic area of the die insert can then be solved as

$$\sigma_r = \frac{2}{\sqrt{3}}\sigma_s \ln \frac{r}{r_1} - p_0 (r_1 \leq r \leq r_p). \tag{4}$$

and σ_θ in the plastic area of the die insert can be solved as

$$\sigma_\theta = \sigma_r + \sigma_s = \frac{2}{\sqrt{3}}\sigma_s (1 + \ln \frac{r}{r_1}) - p_0 (r_1 \leq r \leq r_p). \tag{5}$$

According to the Eqs. (4) and (5), the stress components of plastic area are stationary (Xu and Liu, 1995), and they are

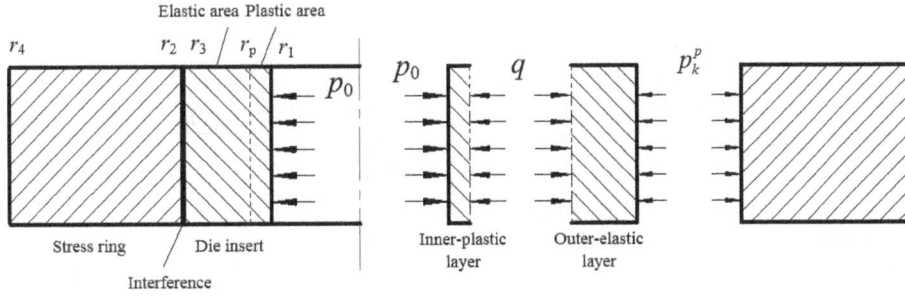

Figure 1. Double-layer combined die.

only related to internal pressure p_0, and has nothing to do with the stress of elastic area.

As shown in Fig. 1, the outer-elastic layer can be regarded as a thick cylinder under internal pressure q and external pressure. Using Lamé's formula, the external pressure is obtained according to the geometry boundary condition $r_2 + u_2 = r_3 + u_3$, where u_2 and u_3 are the displacements due to the interference fit between the die insert and stress ring. The external pressure is given by

$$p_k^p = \frac{r_3 - r_2 - q\Delta_1^p}{\Delta_2^p - \Delta_3}, \tag{6}$$

where $\Delta_1^p = 2r_2 \frac{(1-\mu_1^2)r_p^2}{E_1(r_2^2 - r_p^2)}$, $\Delta_2^p = \frac{(1+\mu_1)r_2^2}{E_1(r_2^2 - r_p^2)}[(2\mu_1 - 1)r_2 - \frac{r_p^2}{r_2}]$ and $\Delta_3 = \frac{(\mu_2 + 1)r_3^2}{E_2(r_4^2 - r_3^2)}[(1 - 2\mu_2)r_3 + \frac{r_4^2}{r_3}]$, E_1 and E_2 are respectively the elastic moduli of the materials of the die insert and stress ring, and μ_1 and μ_2 are respectively the Poisson ratios of the materials of the die insert and stress ring.

According to Lamé's equation, the stresses in the elastic area of the die insert are

$$\begin{cases} \sigma_r = \frac{r_p^2 r_2^2 (p_k^p - q)}{r_2^2 - r_p^2} \cdot \frac{1}{r^2} + \frac{r_p^2 q - r_2^2 p_k^p}{r_2^2 - r_p^2} \\ \sigma_\theta = -\frac{r_p^2 r_2^2 (p_k^p - q)}{r_2^2 - r_p^2} \cdot \frac{1}{r^2} + \frac{r_p^2 q - r_2^2 p_k^p}{r_2^2 - r_p^2} \end{cases} (r_p \leq r \leq r_2), \tag{7}$$

and the stresses of the stress ring are

$$\begin{cases} \sigma_r = -\frac{r_3^2 r_4^2 p_k^p}{r_4^2 - r_3^2} \cdot \frac{1}{r^2} + \frac{r_3^2 p_k^p}{r_4^2 - r_3^2} \\ \sigma_\theta = \frac{r_3^2 r_4^2 p_k^p}{r_4^2 - r_3^2} \cdot \frac{1}{r^2} + \frac{r_3^2 p_k^p}{r_4^2 - r_3^2} \end{cases} (r_3 \leq r \leq r_4). \tag{8}$$

In the outer-elastic layer, the die material just yields at the inner boundary. Therefore, the normal stress can be derived according to Lamé's equation:

$$\begin{aligned} q &= \frac{\sigma_s(r_2^2 - r_p^2)}{\sqrt{3}r_2^2} + p_k^p \\ &= (\frac{\sigma_s(r_2^2 - r_p^2)}{\sqrt{3}r_2^2} + \frac{r_3 - r_2}{\Delta_2^p - \Delta_3})/(1 + \frac{\Delta_1^p}{\Delta_2^p - \Delta_3}). \end{aligned} \tag{9}$$

In the inner-plastic layer, the normal stress at the outer boundary can be solved by substituting $\sigma_r = -q$ into Eq. (3) when $r = r_p$:

$$q = -\frac{2}{\sqrt{3}}\sigma_s \ln\frac{r_p}{r_1} + p_0. \tag{10}$$

The normal stress between the inner-plastic and outer-elastic layers should be the same. The relationship between the autofrettage pressure p_0 and the yield radius r_p can then be obtained as

$$\begin{aligned} p_0 &= (\frac{\sigma_s r_p^2(r_2^2 - r_p^2)}{\sqrt{3}r_p^2 r_2^2} + \frac{r_3 - r_2}{\Delta_2^p - \Delta_3})/(1 + \frac{\Delta_1^p}{\Delta_2^p - \Delta_3}) \\ &\quad + \frac{2}{\sqrt{3}}\sigma_s \ln\frac{r_p}{r_1}. \end{aligned} \tag{11}$$

2.2 Displacement in the loading process

According to Lamé's equation, the displacement of the stress ring is

$$\begin{aligned} u &= \frac{1+\mu_2}{E_2}[\frac{r_3^2 r_4^2 p_k^p}{r_4^2 - r_3^2}\frac{1}{r} \\ &\quad + (1 - 2\mu_2)\frac{r_3^2 p_k^p}{r_4^2 - r_3^2}r] \end{aligned} (r_3 \leq r \leq r_4). \tag{12}$$

and the displacement in the outer-elastic layer of the die insert is

$$\begin{aligned} u &= \frac{1+\mu_1}{E_1}[-\frac{r_p^2 r_2^2(p_k^p - q)}{r_2^2 - r_p^2}\frac{1}{r} \\ &\quad + (1 - 2\mu_1)\frac{r_p^2 q - r_2^2 p_k^p}{r_2^2 - r_p^2}r] \end{aligned} (r_p \leq r \leq r_2). \tag{13}$$

By considering the incompressible condition and geometric equation in the plane-strain problem, the displacement in the inner-plastic layer is obtained as

$$u = \frac{C_1}{r} (r_1 \leq r \leq r_p). \tag{14}$$

The displacement along the radius of the die insert should be continuous, and the displacements at the boundary between the inner-plastic and outer-elastic layers should thus be equal. When $r = r_p$, the displacement calculated using Eqs. (12) and (13) should be the same. The constant C_1 can then be determined as

$$C_1 = \frac{1+\mu_1}{E_1}\left[-\frac{r_p^2 r_2^2 (p_k^p - q)}{r_2^2 - r_p^2} + (1-2\mu_1)\frac{r_p^2 q - r_2^2 p_k^p}{r_2^2 - r_p^2} r_p^2\right]. \tag{15}$$

The displacement in the inner-plastic layer of the die insert is thus

$$u = \frac{1+\mu_1}{E_1 r}\left[-\frac{r_p^2 r_2^2 (p_k^p - q)}{r_2^2 - r_p^2} + (1-2\mu_1)\frac{r_p^2 q - r_2^2 p_k^p}{r_2^2 - r_p^2} r_p^2\right] \quad (r_p \le r \le r_2). \tag{16}$$

2.3 Residual stress after unloading

Residual stress is the difference between stresses of the loading and unloading processes:

$$\sigma_{ij}^r = \sigma_{ij} - \sigma_{ij}^e, \tag{17}$$

where σ_{ij}^r denotes the residual stresses, σ_{ij} denotes the stresses of the loading process and σ_{ij}^e denotes the stresses of the unloading process.

During the unloading process, the stresses in the die insert are

$$\begin{cases} \sigma_r^e = \dfrac{r_1^{e2} r_2^{e2} (p_k - p_0)}{r_2^{e2} - r_1^{e2}} \cdot \dfrac{1}{r^2} + \dfrac{r_1^{e2} p_0 - r_2^{e2} p_k}{r_2^{e2} - r_1^{e2}} \\[2ex] \sigma_\theta^e = -\dfrac{r_1^{e2} r_2^{e2} (p_k - p_0)}{r_2^{e2} - r_1^{e2}} \cdot \dfrac{1}{r^2} + \dfrac{r_1^{e2} p_0 - r_2^{e2} p_k}{r_2^{e2} - r_1^{e2}}, \end{cases} \tag{18}$$

where $\quad p_k = \frac{r_3^e - r_2^e - p_0 \Delta_1^e}{\Delta_2^e - \Delta_3^e}, \quad \Delta_1^e = 2r_2^e \frac{(1-\mu_1^2) r_1^e}{E_1 (r_2^{e2} - r_1^{e2})},$

$\Delta_2^e = \frac{(1+\mu_1) r_2^e}{E_1 (r_2^{e2} - r_1^{e2})}\left[(2\mu_1 - 1)r_2^e - \frac{r_1^{e2}}{r_2^e}\right], \Delta_3^e = \frac{(\mu_2 + 1) r_2^e}{E_2 (r_4^{e2} - r_3^{e2})}[(1 -$

$2\mu_2)r_3^e + \frac{r_4^{e2}}{r_3^e}]$ and $r_i^e = r_i + u_i$ ($i = 1, 2, 3, 4$). u_i is the displacement due to the interference fit between the die insert and stress ring.

The residual stresses in the die insert are then

$$\sigma_r^r = \begin{cases} (\frac{2}{\sqrt{3}}\sigma_s \ln\frac{r}{r_1^e} - p_0) \\[1ex] -[\frac{r_1^{e2} r_2^{e2}}{r_2^{e2} - r_1^{e2}}\frac{(p_k - p_0)}{} \cdot \frac{1}{r^2} + \frac{r_1^{e2} p_0 - r_2^{e2} p_k}{r_2^{e2} - r_1^{e2}}] & (r_1^e \le r \le r_p^e) \\[2ex] [\frac{r_p^{e2} r_2^{e2}}{r_2^{e2} - r_p^{e2}}\frac{(p_k - q)}{} \cdot \frac{1}{r^2} + \frac{r_p^{e2} q - r_2^{e2} p_k}{r_2^{e2} - r_p^{e2}}] \\[2ex] -[\frac{r_1^{e2} r_2^{e2}}{r_2^{e2} - r_1^{e2}}\frac{(p_k - p_0)}{} \cdot \frac{1}{r^2} + \frac{r_1^{e2} p_0 - r_2^{e2} p_k}{r_2^{e2} - r_1^{e2}}] & (r_p^e \le r \le r_2^e). \end{cases} \tag{19}$$

$$\sigma_\theta^r = \begin{cases} [\frac{2}{\sqrt{3}}\sigma_s(1 + \ln\frac{r}{r_1^e}) - p_0] \\[1ex] -[-\frac{r_1^{e2} r_2^{e2}}{r_2^{e2} - r_1^{e2}}\frac{(p_k - p_0)}{} \cdot \frac{1}{r^2} + \frac{r_1^{e2} p_0 - r_2^{e2} p_k}{r_2^{e2} - r_1^{e2}}] & (r_1^e \le r \le r_p^e) \\[2ex] [-\frac{r_p^{e2} r_2^{e2}}{r_2^{e2} - r_p^{e2}}\frac{(p_k - q)}{} \cdot \frac{1}{r^2} + \frac{r_p^{e2} q - r_2^{e2} p_k}{r_2^{e2} - r_p^{e2}}] \\[2ex] -[-\frac{r_1^{e2} r_2^{e2}}{r_2^{e2} - r_1^{e2}}\frac{(p_k - p_0)}{} \cdot \frac{1}{r^2} + \frac{r_1^{e2} p_0 - r_2^{e2} p_k}{r_2^{e2} - r_1^{e2}}] & (r_p^e \le r \le r_2^e). \end{cases} \tag{20}$$

The stresses in the stress ring during unloading can be expressed as

$$\begin{cases} \sigma_r^e = -\dfrac{r_3^{e2} r_4^{e2} p_k}{r_4^{e2} - r_3^{e2}} \cdot \dfrac{1}{r^2} + \dfrac{r_3^{e2} p_k}{r_4^{e2} - r_3^{e2}} \\[2ex] \sigma_\theta^e = \dfrac{r_3^{e2} r_4^{e2} p_k}{r_4^{e2} - r_3^{e2}} \cdot \dfrac{1}{r^2} + \dfrac{r_3^{e2} p_k}{r_4^{e2} - r_3^{e2}}. \end{cases} \tag{21}$$

The corresponding residual stresses are

$$\begin{cases} \sigma_r = (-\dfrac{r_3^{e2} r_4^{e2} p_k^p}{r_4^{e2} - r_3^{e2}} \cdot \dfrac{1}{r^2} + \dfrac{r_3^{e2} p_k^p}{r_4^{e2} - r_3^{e2}}) \\[2ex] \quad -(-\dfrac{r_3^{e2} r_4^{e2} p_k}{r_4^{e2} - r_3^{e2}} \cdot \dfrac{1}{r^2} + \dfrac{r_3^{e2} p_k}{r_4^{e2} - r_3^{e2}}) \\[2ex] \sigma_\theta = (\dfrac{r_3^{e2} r_4^{e2} p_k^p}{r_4^{e2} - r_3^{e2}} \cdot \dfrac{1}{r^2} + \dfrac{r_3^{e2} p_k^p}{r_4^{e2} - r_3^{e2}}) \\[2ex] \quad -(\dfrac{r_3^{e2} r_4^{e2} p_k}{r_4^{e2} - r_3^{e2}} \cdot \dfrac{1}{r^2} + \dfrac{r_3^{e2} p_k}{r_4^{e2} - r_3^{e2}}). \end{cases} \tag{22}$$

2.4 Residual displacement after unloading

The residual displacement is the difference between displacements of loading and unloading processes:

$$u^r = u - u^e, \tag{23}$$

where u^r is the residual displacement, u is the displacement of the loading process and u^e is the displacement of the unloading process.

During the unloading process, the displacement in the die insert is

$$u^e = \frac{1+\mu_1}{E_1}\left[-\frac{r_1^{e2} r_2^{e2} (p_k - p_0)}{r_2^{e2} - r_1^{e2}}\frac{1}{r} + (1-2\mu_1)\frac{r_1^{e2} p_0 - r_2^{e2} p_k}{r_2^{e2} - r_1^{e2}} r\right]. \tag{24}$$

The displacement in the stress ring during the unloading process is

$$u^e = \frac{1+\mu_2}{E_2}\left[\frac{r_3^{e2} r_4^{e2} p_k}{r_4^{e2} - r_3^{e2}}\frac{1}{r} + (1-2\mu_2)\frac{r_3^{e2} p_k}{r_4^{e2} - r_3^{e2}} r\right]. \tag{25}$$

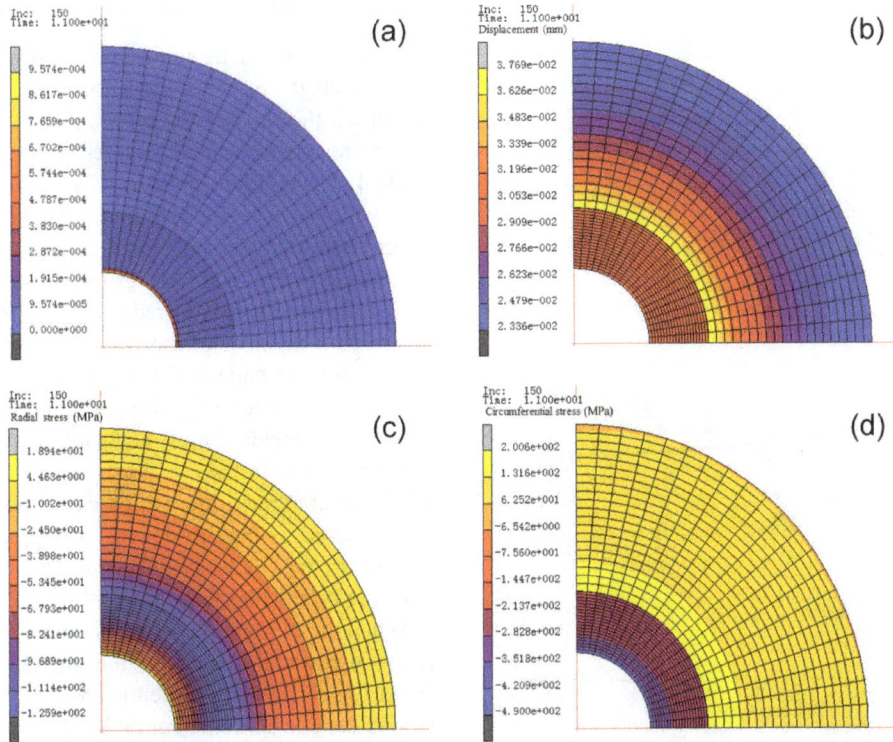

Figure 2. Distributions of (**a**) equivalent plastic strain, (**b**) residual displacement, (**c**) radial residual stress and (**d**) circumferential residual stress obtained from FE simulation results.

Table 1. Parameters of the combined die.

Parameters	Die insert	Stress ring
Elastic modulus (GPa)	212	212
Yield strength (MPa)	1300	900
Poisson's ratio	0.3	0.3
Inner diameter (mm)	20.00	36.00
External diameter (mm)	36.00	80.00
Interference fit (mm)	0.07	

3 FE simulation

To verify the reliability and accuracy of the theoretical derivation, a specific case of a double-layer combined die strengthened by the autofrettage process is analyzed in an FE simulation. The relative parameters of the combined die are listed in Table 1. In the simulated case, the autofrettage pressure is 940.0 MPa, 5.0 % of the die insert undergoes plastic deformation and the yield radius is 20.80 mm.

According to the parameters given above, a 1/4-scale geometric model was constructed using the commercial simulation software MSC MARC. The material model obeys elastic–plastic law perfectly. A total of 800 full integration quad elements were used and no remeshing was triggered during the simulation. During the autofrettage process,

the combined die was first subjected to inner pressure of 940.0 MPa until there was partial plasticity, and the pressure was then released to 0 MPa. As shown in Fig. 2a, 5.0 % of the die insert becomes plastic. The figure also presents residual stresses and residual displacement distribution on the combined die obtained from simulation results.

Meanwhile, the residual stresses and residual displacement in the autofrettaged double-layer combined die were calculated on a theoretical basis using the above formulas. As shown in Fig. 3, the radial residual stress is compressive stress, and it reduces with the radius of the die insert but increases with the radius of the stress ring. The circumferential residual stress on the die insert increases first rapidly then slowly with the radius, and the circumferential residual stress on the stress ring becomes tensile stress and decreases with the radius. The distributions of the residual stresses agree perfectly with the results obtained from FE simulation.

The residual stresses and displacements at the die boundaries are further extracted and listed in Table 2. The difference between theoretical and simulation values of the radial residual stress is 10.3 MPa at the inner diameter of the die insert, and the difference is only 0.8 MPa at the outer diameter of the stress ring. The absolute error of the values at the interface is less than 16.11 %. In the case of the circumferential residual stress, the absolute error between simulation and theoretical values is 5.62–9.49 %. The values of residual

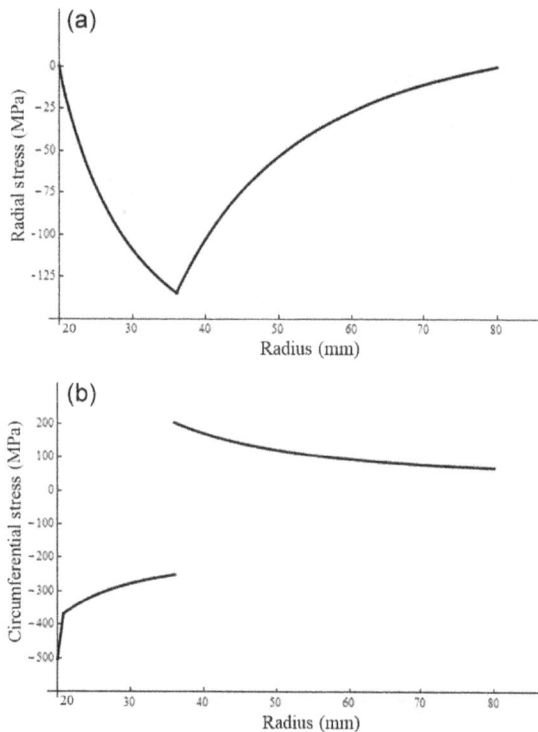

Figure 3. Distributions of (**a**) radial and (**b**) circumferential residual stresses obtained from theoretical calculation.

displacement at the inner and outer diameters of the insert die and stress ring are the same. There is good agreement between the theoretical calculations and FE simulation results.

4 Numerical results and discussion

For further analysis, nine cases of the combined die strengthened by the autofrettage process were numerically solved using the theory and formulas given in Sect. 2. The cases are listed in Table 3. The table gives important parameters, including the ratio of the plastic area and the yield strength of the die insert. The outer diameters of the die insert and stress ring were chosen as variables. Different ratios of the plastic area meant that there were different yield radii, requiring different autofrettage pressures. Here, the required autofrettage pressure was between 823.1 and 1144.4 MPa, which is a range of pressures that can be applied in practice (http://www.felss.com). The inner diameter of the stress ring should increase as the outer diameter of the die insert decreases.

For comparison, the case of the conventional combined die was chosen as case X, and the relative stresses were solved using the typical Lamé equation.

4.1 Effect of the ratio of the plastic area

The ratio of the plastic area on the die insert is the main index that reflects the effect of the autofrettage process. In cases I, II and III, the ratio of the plastic area varies from 3 to 8 %, and the results of residual stresses in these cases are given in Table 4. The residual stresses on the die insert are compressive, which improves the strength of the die insert. However, the circumferential residual stresses on the stress ring are tensile, which is not beneficial for the stress ring.

The absolute values of radial residual stresses slightly increase with the ratio of the plastic area. However, the value of the circumferential residual stress at the inner diameter of the die insert increases with the ratio of the plastic area. The compressive residual stress at the internal boundary of the die insert after autofrettage is much larger than that in the conventional combined die, which can better offset the tensile stress caused by the subsequent working process. Therefore, the autofrettage process can reduce the risk of longitudinal cracking on the internal surface of the die insert.

When the ratio of the plastic area reaches 8 %, the circumferential residual stress near the internal surface of the die insert is −572.0 MPa, which is around 1.5 times that generated in the conventional combined die.

At the same time, the circumferential residual stress near the internal surface of the stress ring increases to 204.3 MPa, and the increase in amplitude is 1.6 % relative to the conventional combined die. The greater circumferential tensile stress can increase the failure of the stress ring.

4.2 Effect of yield strength

In the forging process, the strength of the die steel is an important factor affecting the die life. In the case studies, the yield strength of the stress ring is 900 MPa and three different tool steels with yield strengths of 1100, 1300 and 1650 MPa are selected as the material of the die insert. Therefore, theoretical calculations were made for the different yield strengths of the die insert, and the results are presented in Table 5.

Table 5 shows that the radial residual stress changes little with a variation in yield strength. Meanwhile, the circumferential residual stress on the die insert increases with the yield strength. As the yield strength increases from 1100 to 1650 MPa, the circumferential compressive stress near the internal surface of the die insert increases by about 10.34 % and reaches 535.6 MPa, while the circumferential tensile stress near the internal surface of the stress ring remains almost the same.

In the case of the conventional combined die, prestress does not affect the yield strength of the die insert and stress ring because the prestressing behavior is a kind of elastic deformation effect originating from the interference fit. The residual stress is much smaller than that of the combined die after the autofrettage process. The circumferential residual

Table 2. Residual stresses and displacements at die boundaries.

	Calculation methods	Die insert		Stress ring	
		r_1	r_2	r_3	r_4
Radial residual stress (MPa)	theoretical	0	−134.7	−134.7	0
	simulation	10.3	−121.5	−113.0	−0.8
Circumferential residual stress (MPa)	theoretical	−503.6	−250.7	202.8	68.1
	simulation	−475.3	−226.9	187.6	61.9
Residual displacement (mm)	theoretical	−0.03	−0.03	0.04	0.02
	simulation	−0.03	−0.03	0.04	0.02

Table 3. Parameters of cases selected for further analysis.

Case	Die insert				Stress ring	Die insert		Stress ring	
	Ratio of plastic area	Yield radius (mm)	Autofrettage pressure (MPa)	Yield strength (MPa)	Yield strength (MPa)	r_1 (mm)	r_2 (mm)	r_3 (mm)	r_4 (mm)
I	3 %	20.48	918.1	1300	900	20	36.07	36	80
II	5 %	20.80	940.0	1300	900	20	36.07	36	80
III	8 %	21.29	972.0	1300	900	20	36.07	36	80
IV	5 %	20.80	823.1	1100	900	20	36.07	36	80
V	5 %	20.80	1144.4	1650	900	20	36.07	36	80
VI	5 %	20.75	944.4	1300	900	20	35.07	36	80
VII	5 %	20.70	949.2	1300	900	20	34.07	36	80
VIII	5 %	20.80	926.7	1300	900	20	36.07	36	75
VX	5 %	20.80	910.5	1300	900	20	36.07	36	70
X	–	–	–	1300	900	20	36.07	36	80

stress at the inner surface is only 79.30 % of that generated from the autofrettage process even when the yield strength of the die insert is the minimum 1100 MPa.

Taking the increase in circumferential residual stress as an index, it is seen that an increase in the ratio of the plastic area of the die insert is more effective than a change to a die material with higher yield strength. As shown in case V, the required autofrettage pressure reaches 1144.4 MPa when the yield strength of the die insert is 1650 MPa, and such pressure is difficult to realize in an actual autofrettage process. Therefore, a die insert made of lower-strength material could be used in the autofrettaged die to obtain the good effect of the prestress state.

4.3 Effect of die dimensions

In further study, the effect of die dimensions on the combined die after autofrettage was investigated. Both the residual stresses after autofrettage and the stresses during the working process were calculated, and the values at the inner diameters of the die insert and stress ring, which correspond to the zone most likely to fail during the working process, are listed in Table 6. The working pressure is assumed as 900 MPa, and the equivalent stress during the working process is computed to evaluate the die failure. Gener-

ally, the die will fail when the equivalent stress reaches the yield strength of the die material. In the cases considered, the working stresses are equal to the sum of residual stresses from autofrettage and the new stresses generated by the assumed working pressure. The equivalent stress is simply calculated according to the von Mises yield criterion as

$$\bar{\sigma} = \frac{\sqrt{3}}{2}(\sigma_\theta - \sigma_r). \tag{26}$$

In the case of the conventional combined die, the maximum working pressure is determined as 884.3 MPa when the equivalent stress of the die insert reaches a yield strength of 1300 MPa according to Lamé's formula and the von Mises criterion. Compared with the conventional combined die, the autofrettaged combined die can bear larger working pressure, as expected.

In cases VI and VII, the outer diameter of the die insert is 35.0 and 34.0 mm, respectively, and the variation in residual stress after autofrettage is within 10.2 MPa. During the working process, the circumferential stresses of the die insert become tensile stresses, and the stress of the stress ring increases with decreasing inner diameter of the ring. The equivalent stress on the internal surface of the die insert hardly changes and is far below the yield strength. However, when the outer diameter of the die insert is reduced to

Table 4. Residual stresses in cases with different ratios of the plastic area.

Stress	Case	Die insert			Stress ring	
		r_1	r_p	r_2	r_3	r_4
Radial residual stress (MPa)	I	0	−9.9	−134.3	−134.3	0
	II	0	−17.0	−134.7	−134.7	0
	III	0	−28.2	−135.7	−135.7	0
	X	0	–	−133.3	−133.3	0
Circumferential residual stress (MPa)	I	−457.1	−375.7	−251.3	202.2	67.9
	II	−503.6	−368.4	−250.7	202.8	68.1
	III	−572.0	−356.8	−249.3	204.3	68.6
	X	−384.9	–	−251.6	201.0	67.7

Table 5. Residual stresses in cases with different yield strengths.

Stress	Case	Die insert			Stress ring	
		r_1	r_p	r_2	r_3	r_4
Radial residual stress (MPa)	IV	0	−16.6	−134.5	−134.5	0
	II	0	−17.0	−134.7	−134.7	0
	V	0	−17.6	−135.0	−135.0	0
	X	0	–	−133.3	−133.3	0
Circumferential residual stress (MPa)	IV	−485.4	−368.7	−250.8	202.6	68.1
	II	−503.6	−368.4	−250.7	202.8	68.1
	V	−535.6	−367.9	−250.5	203.3	68.3
	X	−384.9	–	−251.6	201.0	67.7

34.0 mm, the equivalent stress of the stress ring increases to 864.7 MPa, which is close to the yield strength of 900 MPa. From a different view, a smaller die insert can be adopted after autofrettage to reduce the amount of expensive material that is usually used for the die insert of the conventional combined die. Taking case VII as an example, more than 15.0 % of the material used for the die insert can be saved.

In cases VIII and VX, the outer diameter of the stress ring is reduced to 75.0 and 70.0 mm, while the residual stresses after autofrettage increase. During the working process, the increase in amplitude of the equivalent stress near the inner surface of the die insert is larger than that of the stress ring. When the outer diameter of the stress ring is reduced to 70.0 mm, the equivalent stress of the die insert increases to 1280.3 MPa, and is very close to the yield strength of the die insert. Therefore, the autofrettaged die with a smaller outer-diameter stress ring can be used to substitute the conventional combined die, resulting in a material savings. More importantly, working space can be saved if the outer diameter of the stress ring can be reduced, especially for multi-stage forging dies.

5 Conclusions

1. To improve the strength of the combined die, an alternative method of designing the combined die was proposed to introduce compressive stresses on the die insert through the autofrettage process. The analytical solution for the autofrettage process of a double-layer combined die was obtained. The relationship between the autofrettage pressure and the yield radius of the die insert was determined, and expressions of residual stresses and displacements directly related to geometric parameters, material properties and internal pressure were derived. This theoretical derivation can guide the design of the autofrettaged die.

2. A specific case of a double-layer combined die strengthened in the autofrettage process was simulated employing the FE method. In the case of radial residual stress, the absolute error between simulation and theoretical values at the interface between the die insert and stress ring is less than 16.11 %. In the case of the circumferential residual stress, the absolute error is 5.62–9.49 %. In the case of the residual displacement, simulation and theoretical values are the same. The good agreement between theoretical calculations and FE simulation results

Table 6. Results for different external diameters of the insert die and stress ring.

Process	Stress	Position	VII	VI	II	VIII	VX	X
After autofrettage process	Radial residual stress (MPa)	r_1	0	0	0	0	0	0
		r_3	−138.7	−136.9	−134.7	−131.2	−126.8	−133.3
	Circumferential residual stress (MPa)	r_1	−523.2	−513.0	−503.6	−493.6	−481.0	−384.9
		r_3	199.5	201.4	202.8	209.4	217.7	201.0
Working process	Working pressure (MPa)				900			884.3
	Radial stress (MPa)	r_1	−900.0	−900.0	−900.0	−900.0	−900.0	−884.3
		r_3	−409.1	−388.7	−369.5	−359.9	−347.9	−364.0
	Circumferential stress (MPa)	r_1	496.3	506.5	515.9	543.7	578.4	616.8
		r_3	589.4	572.7	557.2	575.4	598.1	549.1
	Equivalent stress (MPa)	r_1	1209.2	1218.0	1226.2	1250.3	1280.3	1300.0
		r_3	864.7	832.7	802.5	809.9	819.2	790.8

verifies the reliability and accuracy of the theoretical derivation.

3. In further analysis, nine cases of the combined die strengthened in the autofrettage process were numerically solved using the theoretical derivation and four key parameters, namely the ratio of the plastic area and the yield strength of the die insert and the outer diameters of the die insert and stress ring, and the effects of the parameters on the autofrettage effect were investigated. For comparison, the relative stresses in the case of the conventional combined die were solved using the typical Lamé equation.

4. The circumferential residual stress at the inner diameter of the die insert increases with the ratio of the plastic area and the yield strength of the die insert. Taking the increase in the circumferential residual stress at the inner diameter of the die as an index, it was seen that an increase in the ratio of the plastic area of the die insert is more effective than a change to a die material with higher yield strength.

5. Compared with the conventional combined die, the autofrettaged die could bear larger working pressure, as expected. The autofrettaged die with a smaller-outer-diameter die insert could be used to substitute the conventional combined die, reducing the amount of expensive material required for the die insert. The autofrettaged die with a smaller-outer-diameter stress ring could be an alternative design to the conventional die, resulting in more working space, especially for multi-stage forging dies.

Competing interests. The authors declare that they have no conflict of interest.

Acknowledgements. This work was supported by the National Natural Science Foundation of China (no. 51475294).

Edited by: Chin-Hsing Kuo

References

Anantharam, K. and Kumar, B. S. K.: Design analysis of high pressure cylinders subjected to autofrettage process using MATLAB and ANSYS, Int. J. Eng. Res. Technol., ESRSA Publications, November, 955–961, 2014.

Bhatnagar, R. M.: Modelling, validation and design of autofrettage and compound cylinder, Eur. J. Mech. A-Solid, 39, 17–25, 2013.

Groenbaek, J. and Birker, T.: Innovations in cold forging die design, J. Mater. Process Tech., 98, 155–161, 2000.

Hojjati, M. H. and Hassani, A.: Theoretical and finite-element modeling of autofrettage process in strain-hardening thick-walled cylinders, Int. J. Pres. Ves. Pip., 84, 310–319, 2007.

Hur, K. D., Choi, Y., and Yeo, H. T.: Design for stiffness reinforcement in backward extrusion die, J. Mater. Process Tech., 130, 411–415, 2002.

Hur, K. D., Choi, Y., and Yeo, H. T.: A design method for cold backward extrusion using FE analysis, Finite Elem. Anal. Des., 40, 173–185, 2003.

Jahed, H., Farshi, B., and Hosseini, M.: Fatigue life prediction of autofrettage tubes using actual material behaviour, Int. J. Pres. Ves. Pip., 83, 749–755, 2006.

Joun, M. S., Lee, M. C., and Park, J. M.: Finite element analysis of prestressed die set in cold forging, Int. J. Mach. Tool. Manu., 42, 1213–1222, 2002.

Kwan, C. T. and Wang, C. C.: An optimal pre-stress die design of cold backward extrusion by RSM Method, Structural Longevity, 5, 25–32, 2011.

Majzoobi, G. H., Farrahi, G. H., and Mahmoudi, A. H.: A finite element simulation and an experimental study of autofrettage for strain hardened thick-walled cylinders, Mat. Sci. Eng. A-Struct., 359, 326–331, 2003.

Pedersen, T. Ø.: Numerical modelling of cyclic plasticity and fatigue damage in cold-forging tools, Int. J. Mech. Sci., 42, 799–818, 2000.

Qian, L. Y., Liu, Q. K., Han, Y., Li, H., and Wang, W. L.: Research on autofrettage technology for extrusion die based on bilinear hardening model[J], Chin. J. Mech. Eng., 47, 26–31, 2011 (in Chinese).

Son, D. S., Hong, J. H., and Chang, S. H.: Determination of the autofrettage pressure and estimation of material failures of a Type III hydrogen pressure vessel by using finite element analysis, Int. J. Hydrogen Energ., 37, 12771–12781, 2012.

Xu, B. Y. and Liu, X. S.: Applied Elasto-plastic Mechanics, Tsinghua University Press, Beijing, 195 pp., 1995 (in Chinese).

Yang, Q. H., Chen, X., Meng, B., and Pan, J.: Optimum design of combined cold extrusion die for bevel gear[J], IACSIT Int. J. Eng. Technol., 4, 348–351, 2012.

Yeo, H. T., Choi, Y., and Hur, K. D.: Analysis and design of the prestressed cold extrusion die, Int. J. Adv. Manuf. Tech., 18, 54–61, 2001.

Zhu, R. L.: Results resulting from autofrettage of cylinder, Chin. J. Mech. Eng.-En., 21, 105–110, 2008a.

Zhu, R. L.: Ultimate load-bearing capacity of cylinder derived from autofrettage under ideal condition, Chin. J. Mech. Eng.-En., 21, 80–87, 2008b.

Permissions

List of Contributors

Zirong Luo and Jianzhong Shang
School of Mechatronics Engineering and Automation, National University of Defence Technology, 410073 Changsha, China

Guowu Wei
School of Computing, Science and Engineering, University of Salford, Salford, M5 4WT, UK

Lei Ren
School of Mechanical, Aerospace and Civil Engineering, University of Manchester, Manchester, M13 9PL, UK

Bingxiao Ding, Xiao Xiao and Yirui Tang
Department of Electromechanical Engineering, University of Macau, Taipa, Macao SAR, China

Yangmin Li
Department of Electromechanical Engineering, University of Macau, Taipa, Macao SAR, China
Department of Industrial and Systems Engineering, The Hong Kong Polytechnic University, Hung Hom, Hong Kong SAR, China
Tianjin Key Laboratory for Advanced Mechatronic System Design and Intelligent Control, Tianjin University of Technology, Tianjin, China

Bin Li
Tianjin Key Laboratory for Advanced Mechatronic System Design and Intelligent Control, Tianjin University of Technology, Tianjin, China

Chin-Hsing Kuo, Jyun-Wei Su and Lin-Chi Wu
Department of Mechanical Engineering, National Taiwan University of Science and Technology, Taipei 106, Taiwan

Cihat Bora Yigit and Pinar Boyraz
Department of Mechanical Engineering, Istanbul Technical University, Inonu Cd. No:65, 34437, Beyoglu, Istanbul, Turkey

Arif Gok
Amasya University, Technology Faculty, Mechanical Engineering, 05000 Amasya, Turkey

Sermet Inal
Dumlupinar University, School of Medicine, Department of Orthopaedics and Traumatology, Campus of Evliya Celebi, 43100 Kutahya, Turkey

Ferruh Taspinar
Dumlupinar University, School of Health Science, Department of Physiotherapy and Rehabilitation, 43100 Kutahya, Turkey

Eyyup Gulbandilar
Eskisehir Osmangazi University, Faculty of Engineering&Architecture, Department of Computer Engineering, Meselik Campus, 26480 Eskisehir, Turkey

Kadir Gok
Manisa Celal Bayar University, Hasan Ferdi Turgutlu Teknoloji Fakültesi, Mechanical and Manufacturing Engineering, 45400 Manisa, Turkey

Tuanjie Li, Hangjia Dong and Lei Zhang
School of Electromechanical Engineering, Xidian University, Xi'an, 710071, China

Jie Jiang
Engineering College, Honghe University, Mengzi, 661100, China

Doina Pisla, Paul Tucan, Bogdan Gherman, Calin Vaida and Nicolae Plitea
CESTER – Research Center for Industrial Robots Simulation and Testing, Technical University of Cluj Napoca, Cluj-Napoca, Romania

Nicolae Crisan and Iulia Andras
University of Medicine and Pharmacy, Cluj-Napoca, Romania

Salvador Cardona Foix, Lluïsa Jordi Nebot, and Joan Puig-Ortiz
Mechanical Engineering Department, ETSEIB, Universitat Politècnica de Catalunya, Barcelona, 08028, Spain

Changcheng Hu, Hongbao Dong and Hua Yuan
School of Naval Architecture and Ocean Engineering, Huazhong University of Science and Technology, Wuhan, 430074, P. R. China

Yao Zhao
School of Naval Architecture and Ocean Engineering, Huazhong University of Science and Technology, Wuhan, 430074, P. R. China
Collaborative Innovation Center for Advanced Ship and Deep-Sea Exploration (CISSE), Shanghai, 200240, P. R. China

Shuang Xu, Alessandro Ferraris, Andrea Giancarlo Airale, and Massimiliana Carello
Mechanical and Aerospace Engineering Department, Politecnico di Torino, C.so Duca degli Abruzzi 24 Turin, 10129, Italy

Vieroslav Molnár, Gabriel Fedorko and Nikoleta Husáková
Technical University of Kosice, Letna 9, 042 00 Kosice, Slovak Republic

Ján Král' Jr.
Faculty of Mechanical Engineering, Technical University of Kosice, Letna 9, 042 00 Kosice, Slovak Republic

Mirosław Ferdynus
Faculty of Mechanical Engineering, Lublin University of Technology, Nadbystrzycka 36, 20-616 Lublin, Poland

Ali Tolga Bozdana
Associate Professor, Mechanical Engineering Department, University of Gaziantep, Gaziantep, Turkey

Nazar Kais Al-Karkhi
Lecturer, Automated Manufacturing Engineering, University of Baghdad, Baghdad, Iraq

Libo Meng
Intelligent Robotics Institute, School of Mechatronical Engineering, Beijing Institute of Technology, 5 Nandajie, Zhongguancun, Haidian, Beijing 100081, China
Key Laboratory of Biomimetic Robots and Systems, Ministry of Education, State Key Laboratory of Intelligent Control and Decision of Complex Systems, Beijing Institute of Technology, 5 Nandajie, Zhongguancun, Haidian, Beijing 100081, China

Zhangguo Yu, Xuechao Chen and Qiang Huang
Intelligent Robotics Institute, School of Mechatronical Engineering, Beijing Institute of Technology, 5 Nandajie, Zhongguancun, Haidian, Beijing 100081, China

Key Laboratory of Biomimetic Robots and Systems, Ministry of Education, State Key Laboratory of Intelligent Control and Decision of Complex Systems, Beijing Institute of Technology, 5 Nandajie, Zhongguancun, Haidian, Beijing 100081, China
Beijing Advanced Innovation Center for Intelligent Robots and Systems, Beijing Institute of Technology, China

Marco Ceccarelli
LARM: Laboratory of Robotics and Mechatronics, DICeM, University of Cassino and South Latium, Via Di Biasio 43, 03043 Cassino, Fr, Italy
Beijing Advanced Innovation Center for Intelligent Robots and Systems, Beijing Institute of Technology, China

Just L. Herder
Dept. Precision and Microsystems Engineering, Delft University of Technology, 2628CD Delft, the Netherlands

Giuseppe Radaelli
Dept. Precision and Microsystems Engineering, Delft University of Technology, 2628CD Delft, the Netherlands
Laevo BV, 2628CA Delft, the Netherlands

Yi Yang, Yaping Tian, Yan Peng and Huayan Pu
School of Mechatronic Engineering and Automation, Shanghai University, Shanghai, 200444, China

Flávia V. Barbosa and José C. F. Teixeira
Department of Mechanical Engineering, School of Engineering, University of Minho, Guimarães, 4800-058, Portugal

João L. Afonso and Filipe B. Rodrigues
Department of Industrial Electronics, School of Engineering, University of Minho, Guimarães, 4800-058, Portugal

Chengliang Hu, Fengyu Yang, Zhen Zhao, and Fan Zeng
Institute of Forming Technology & Equipment, Shanghai Jiaotong University, Shanghai 200030, China

Index

www.ingramcontent.com/pod-product-compliance
Lightning Source LLC
Chambersburg PA
CBHW080703200326
41458CB00013B/4941